Advance Praise for *A Good War*

"The COVID-19 crisis and Seth Klein's timely book demonstrate heroic measures that become possible in times of crisis. The coronavirus crisis is not yet assessable by historians. World War II presented the West with an existential crisis for which government institutions, corporations and civil society were utterly unprepared. Yet in astonishingly rapid order, society, government and industry were united in resolve and enabled unprecedented government action and leadership. Read this inspiring book to realize giving up is not an option and 'can't be done' is not an excuse." — DAVID SUZUKI

"One of the great privileges of my life is having Seth Klein as a brother. Here you will see why. This is the roadmap out of climate crisis that Canadians have been waiting for. Serious, specific and madly inspiring, *A Good War* will fill you with the confidence and courage required to fight like hell for the future we all deserve." — NAOMI KLEIN, activist, journalist and author of *This Changes Everything* and *The Shock Doctrine*

"Seth Klein's critically important book tells the climate truth: our climate emergency needs a WW2-style mobilization. In this strategic call to action we learn about Canada's impressive war production, and how we can join the war against climate devastation. Your children's futures are on every page. Read, organize, act!" — RAFFI CAVOUKIAN, singer, author and founder of Raffi Foundation for Child Honouring

"This is a truly great book. Few people have thought as deeply or with as much precision about the climate crisis as Seth Klein. Drawing on the lessons that history provides, he has provided a roadmap for how Canada should respond. We do need to fight the climate crisis as if it was a war, but how lucky we are: we need neither to kill or be killed, merely to do the hard but satisfying work of building a working planet." — BILL MCKIBBEN, author and founder of 350.org

"This is an important book, exactly what we need right now. Seth does more than set out the urgency of our climate challenge, he draws a careful study of Canada's extraordinary and rapid transformation during WW2 to give us a battle plan to address it. Big change is possible, he reminds us. We have done it before. Seth's book combines urgency and optimism, ambition and clear-eyed understanding of the obstacles to change. It should become an essential resource for policy makers but also for all of us who want to be part of the solution. It is also an excellent and inspiring read." — ALEX HIMELFARB, former clerk of the Privy Council and professor of sociology

"Pacifists are not known for evoking war-like rhetoric, so it is one of life's quirks that I first met an 18-year-old Seth Klein as he toured Canada calling for nuclear disarmament. But, like me, he sees that the only way for human civilization to survive the climate emergency is through an 'all hands on deck' approach that has only been seen in times of war — or recently in pandemic. Saving ourselves is a huge challenge. Marshaling the facts, the hope and the path forward is essential work for which writer and policy wonk Seth Klein is ideally suited." — ELIZABETH MAY, MP and former leader of the Green Party of Canada

"In this ambitious and informative book, Seth Klein intelligently connects past mobilization for war to the sweeping state measures now required to counter global warming." — PETER NEARY, Professor Emeritus, Department of History, University of Western Ontario

"In his examination of Canada's climate action landscape, Seth Klein recognizes the deeply rooted social inequality in Canada that continues to be an obstacle for mobilizing a unified front against climate change. Highlighting the power and authority of Indigenous-led climate activism, he makes the striking case for recognizing and building upon the intersections between the fight for climate action and the fight for Indigenous Title and Rights." — GRAND CHIEF STEWART PHILLIP, president of the Union of B.C. Indian Chiefs

"A call to action to protect our very existence on this planet — exactly the book we need right now. This book should be on the curriculum in every school across the country and required reading for every aspiring and current politician. Seth brilliantly challenges us to embrace the possible, while rejecting the current politics of ho-hum incrementalism. The threat posed by catastrophic climate change will eclipse everything we've faced to date, and our response will be the defining task of our era." — PAUL M. TAYLOR, anti-poverty activist and executive director of FoodShare Toronto

"This is the blueprint for rapid societal transformation that we've all been waiting for. Klein's book is not just about thinking big, it's about doing BIG. During the Second World War, Canada was nimble and consistently hit above its weight: mobilizing media, industry and the citizenry to face the biggest challenge of its time. Today, armed with science and far greater technological capacity, we can do it again. *A Good War* is a fascinating, timely and strategic guide to creating a better world." — ZIYA TONG, author of *The Reality Bubble* and science broadcaster (former co-host of Discovery Channel's *Daily Planet*)

"A magnificent job. A climate emergency manifesto that brings it all together: history, politics, economics, humanity, mobilization and science, with a wisdom of the actions we must undertake to win this good war." — LIBBY DAVIES, former NDP MP

A GOOD WAR

A GOOD WAR

MOBILIZING CANADA FOR THE CLIMATE EMERGENCY

SETH KLEIN

ECW

Published by ECW Press
665 Gerrard Street East
Toronto, Ontario, Canada M4M 1Y2
416-694-3348 / info@ecwpress.com

Editor for the Press: Susan Renouf
Cover design: David A. Gee

The opinions, policy recommendations and any errors are those of the author, and do not necessarily reflect the views of the book's funders or reviewers.

This book was written on unceded Coast Salish territory, including the lands belonging to the xʷməθkwəy̓əm (Musqueam), Sḵwx̱wú7mesh (Squamish) and səl̓ílwətaʔɬ/Selilwitulh (Tsleil-Waututh) Nations, and the lands of the Shíshálh (Sechelt) Nation.

LIBRARY AND ARCHIVES CANADA CATALOGUING IN PUBLICATION

Title: A good war : mobilizing Canada for the climate emergency / Seth Klein.

Names: Klein, Seth, 1968- author.

Description: Includes bibliographical references and index.

Identifiers: Canadiana (print) 20200248669
Canadiana (ebook) 20200248871

ISBN 978-1-77041-545-4 (softcover)
ISBN 978-1-77305-592-3 (PDF)
ISBN 978-1-77305-591-6 (EPUB)

Subjects: LCSH: Environmental policy—Canada. | LCSH: Climate change mitigation—Economic aspects—Canada. | LCSH: Sustainable development—Canada. | LCSH: Environmental economics—Canada. | LCSH: Economic policy—Environmental aspects—Canada. | LCSH: Climatic changes—Government policy—Canada. | LCSH: Environmental policy—Canada—Citizen participation. | LCSH: Climate change mitigation—Canada—Citizen participation.

Classification: LCC HC79.E5 K54 2020
DDC 338.9/270971—dc23

The publication of *A Good War* has been funded in part by the Government of Canada. *Ce livre est financé en partie par le gouvernement du Canada.* We acknowledge the contribution of the Government of Ontario through the Ontario Book Publishing Tax Credit, and through Ontario Creates for the marketing of this book.

The research and writing of this book was supported by funding from: the McConnell Foundation, the Metcalf Foundation, the Corporate Mapping Project (jointly led by the University of Victoria, the Canadian Centre for Policy Alternatives and the Parkland Institute, and funded primarily by the Social Science and Humanities Research Council of Canada), the David Suzuki Foundation, the British Columbia Teachers Federation, and Simon Fraser University's Urban Studies Program.

PRINTED AND BOUND IN CANADA

PRINTING: MARQUIS 5 4 3 2 1

MIX
Paper from responsible sources
FSC www.fsc.org FSC® C103567

For Christine,
who inspires me as she mobilizes others.

And for Zoe and Aaron,
my love and anxiety for whom motivated this project.

Joseph Sydney Hallam

Meital Smith

TABLE OF CONTENTS

PREFACE

The climate emergency is upon us.

Ever since my wife, Christine, and I started living together, for a few days every August we join her parents at a place they rent each summer in the southern end of British Columbia's Okanagan region. It's a lovely spot on a lake I enjoy swimming in, and a nice tradition my wife's parents have established of unplugging and spending time with friends and family.

The south Okanagan is always hot in the summer. But in the last few years, it's been different. During our visits in the summers of 2017 and 2018, wildfires throughout B.C. blanketed the area with smoke. With the full sun screened from view, the days were a little cooler. That came as some welcome relief. But it also felt a little apocalyptic as we sat outside only to have ash fall from the sky.

As our 2019 visit approached, the wildfire season in B.C. had proven less severe than the previous two years. We joked that we would finally see a return to normal skies and ash-free outdoor meals. Then, the night before our arrival, a fire broke out on the ridge behind where we visit. Not knowing what to expect, we made the drive nonetheless.

When we arrived, we came upon a scene unlike any our family had ever experienced. Residents in the area had been put on evacuation alert. All afternoon, a fleet of four water-bomber float planes had been scooping up water from the lake directly in front of our rental house, flying in low over the homes in formation, collecting water in their pontoons and then doubling back to the fire on the mountain ridge behind us to dump their loads. Right up until sundown, four helicopters likewise circled back to the lake in five-minute rotations, dropping massive buckets on long cords to collect water, and then racing back to the mountain to release their cargo. This continued all week. Swimmers and boaters had to stay close to shore to keep the path clear for the aircraft. And the noise was overwhelming. Our peaceful family time suddenly felt more like a scene from *Apocalypse Now*, with people trying to go about their routine but for the deafening sound of the helicopters. We had to huddle together and shout to communicate. As darkness fell, the hills behind us were dotted with flames, and the sky glowed ominously red.

This is what people now nervously joke about as "the new normal," although it is, in truth, not normal and but a taste of things to come. These weather events we are increasingly experiencing — and the war-like mood they create — represent attacks on our soil. They are a call to mobilize.

Compared to other places, my home province of British Columbia has been lucky. As I complete this book, Australia, which has been wrestling with record heat for years, has become the latest country to confront the terrifying reality of the climate emergency. Australia has just experienced a wildfire season unlike any before — over two dozen people killed, approximately six million hectares burned (an area larger than Switzerland), an estimated half billion wild animals perished, over 1,400 homes destroyed, tens of thousands evacuated, whole coastal communities in New South Wales cut off from road access as flames surrounded them. The impacts that climate scientists have warned of for years are now here.

While Australia's current government ranks among the world's leading climate policy foot-draggers, there is no question the 2019–2020 bushfires will have a major impact on the country's politics and the public discourse on climate. Support for the tone-deaf administration of Prime Minister Scott Morrison plummeted in the wake of the catastrophe. A poll commissioned by the Australia Institute in November 2019 (even before the worst of the crisis had occurred) found that two-thirds of Australians believe their country is facing a climate emergency, and 63% agree that "governments should mobilise all of society to tackle climate change, like they did during the World Wars."[1]

In the face of the wildfire emergency, the Australian government was forced to deploy the most military assets since the Second World War.[2] Sadly, the emergency response was purely defensive, a rearguard action. Our governments have not yet seen fit to adopt a wartime-scale response that pre-emptively tackles the climate crisis. We mobilize to put fires out, but not to prevent them.

I suspect the wartime approach employed in this book makes some of you reading it uncomfortable. Me too.

I am an unlikely person to be writing a war story.

I am the child of war resisters.

My parents came to Canada from the United States in 1967. The Vietnam War was in full swing. In September of that year, after trying unsuccessfully to gain formal "conscientious objector" status, my father received his military induction notice. Further complicating matters, about a month earlier, my mother discovered she was pregnant — with me — and my folks had decided to hurriedly get married. Then, along with tens of thousands of other Americans, rather than accept military service or continue to live and pay taxes in a country engaged in an immoral war, my parents chose to come to Canada.

I am Canadian because of my parents' refusal to participate in war. And I am forever grateful for the choice they made. Coming to Canada in those days, and in those circumstances, was very different than it is today. During the Vietnam War, a network of peace activists existed to help American draft resisters make their way to Canada — good folks who helped these young Americans cross the border, offered temporary shelter and assisted these immigrants in settling in a new country.

My parents had been living in New York City. My mother, Bonnie, was beginning her career as a documentary filmmaker, and my father,

Michael, was a pediatric resident at Albert Einstein College of Medicine in the Bronx. The Montreal Council to Aid War Resisters advised my parents to fly into what was then Dorval Airport (now Montreal-Trudeau International Airport) after midnight. They were told that the immigration officers on duty late at night were more likely to be French-Canadian, since in those days the Francophones generally got the crappy nighttime shifts, and the French-Canadians were much more likely than their English-Canadian counterparts to oppose the war.

So that's what they did. And sure enough, my parents were met by a Francophone immigration officer who, upon hearing their declaration (and with much more discretion than exists today), gave them landed immigrant status in 20 minutes and a kiss on both cheeks for good measure. Imagine that.

My family's war resistance goes even further back. My Jewish great-grandparents all escaped Tsarist Russia, fleeing the pogroms. My paternal great-grandfather, fearing conscription into the Russian monarch's military, set sail for America in the early 1900s and later brought over his family.

My own social activism started as a high-school student in Montreal in the 1980s, where I became engaged in the peace and disarmament movement near the end of the Cold War. That's where I cut my political teeth. It was an era when many felt the possibility of a cataclysmic nuclear war was very real. In fact, according to research at the time by the Children's Mental Health Research Group at McMaster University, 67% of Canadian teens believed a nuclear war was likely in their lifetimes, and the same percent thought there was little or nothing they could do about it.[3] That was the existential threat — a very real one — faced by an earlier generation.

And in that context, a group of Montreal teens including me created a youth disarmament group we called SAGE (in English, an acronym for Students Against Global Extermination; and in French, the far more elegant Solidarité Anti-Guerre Étudiante). As I made my way through grades 10, 11 and first-year CEGEP, I would frequently skip school to give presentations in other schools about the dangers of nuclear weapons and what young people could do to turn the tide.

Then in 1986–1987, when I was 18, four of us from SAGE (Maxime Faille, Désirée McGraw, Alison Carpenter and I), feeling the urgency of the issue, decided to take a year out of our studies and travel the country, speaking in schools and organizing youth peace groups.

We spent the summer of 1986 organizing the tour, followed by nine months on the road in an old red station wagon. Looking back, I find it unfathomable that we executed the SAGE Youth Nuclear Disarmament Tour in the days before email or cellphones. But we did. Like many tours before and since, we started out in St. John's, Newfoundland (where our stop was hosted by a youth peace group, including a 16-year-old kid named Rick Mercer), and ended the tour in Victoria, B.C. Along the way, we spoke to about 1 in every 20 high school students in the country and started well over 100 school and city-based youth peace groups. Occasionally, I still meet someone politically active today whose social activism started in the wake of the SAGE tour over 30 years ago.

For much of my life, I wasn't keen on military history. I've always had an ambivalent relationship with Remembrance Day, wanting to honour the sacrifices made, but sometimes uncomfortable with the glorification of war.

Like I said, I come to this analogy uneasily, and there is no small irony in me writing a book invoking a wartime call to action. Yet here I am, doing just that.

I am now convinced that to confront the climate emergency a wartime approach is needed, and moreover, that our wartime experience should be embraced as an instructive story. Climate breakdown requires a new mindset — to mobilize all of society, galvanize our politics and fundamentally remake our economy.

WHY THIS BOOK

William Rees, an ecological economist, professor emeritus at the University of British Columbia and creator of the ecological footprint concept, has said, "The ecologically necessary is politically infeasible, but the politically feasible is ecologically irrelevant."

That may indeed describe the current political reality, but is it necessarily true?

I've been wrestling with this paradox for many years. I served for 22 years as the founding British Columbia director of the Canadian Centre for Policy Alternatives (CCPA), Canada's foremost progressive public policy think tank. The CCPA is a non-governmental research institute committed to social, economic and environmental justice. Over my two decades with the centre, I gained deep and wide policy knowledge at the provincial and federal level. The CCPA has a long record of analyzing and critiquing government policy, but also of developing and proposing realizable policy alternatives — ones that would see us take better care of one another and the environment. Consequently, I have lived for a long time within the aggravating rift between what I'm convinced *should* happen and what our governments are willing to adopt.

My own research and writing at the CCPA focused on fiscal policy, progressive tax reform, welfare policy, poverty, inequality, economic security, job creation and climate policy — areas that all come together in this book. Of particular relevance to this endeavour, the CCPA's B.C. office has long had a focus on the interconnections between inequality and climate change. Indeed, we were among the first in Canada to start exploring these links. The CCPA-B.C. has co-hosted two major multi-year research alliances that have produced a huge body of knowledge on that broad subject: the Climate Justice Project, which since 2007 has produced dozens of reports that map out how our society can become carbon-zero in a manner that also reduces inequality, includes just transition for workers and enhances social justice; and the Corporate Mapping Project, launched in 2015, which seeks to investigate and document the power and influence of the fossil fuel industry in western Canada.

Many worry what climate action will mean for economic and job security. So do I. I come to the climate issue through the lens of social justice. I am a founder of the British Columbia Poverty Reduction Coalition and one of the architects of the methodology for calculating the Living Wage for Families, now used in dozens of communities across Canada. And I have worked in partnership with the labour movement for years.

The province and country where I reside are often held up as international leaders when it comes to the climate fight. And relatively speaking, they are. But that's not saying much.

All of us who heed the warnings of climate scientists are increasingly alarmed as we stare down the harrowing gap between what the science says is necessary and what our politics seems prepared to entertain, a frustrating phenomenon I call the "new climate denialism."[4] Traditional climate denialism simply refuses to accept the reality of human-induced climate change. In contrast, this new and insidious form of denialism manifests in the fossil fuel industry and our political leaders assuring us that they understand and accept the scientific warnings about climate change, but they are in denial about what this scientific reality means for policy or they continue to block progress in less visible ways.

The consequence: despite decades of calls to action, our greenhouse gas (GHG) emissions are not on a path to stave off a horrific future for our children and future generations. The accompanying chart tracks Canada's GHG emissions going back to the year 2000. What is evident is that, in the face of the defining challenge of our time, our politics are not rising to the task at hand.

Let this deeply disturbing chart sink in. And then let us all agree — political leaders, civil servants, environmental organizations, academics

CANADA'S GHG EMISSIONS

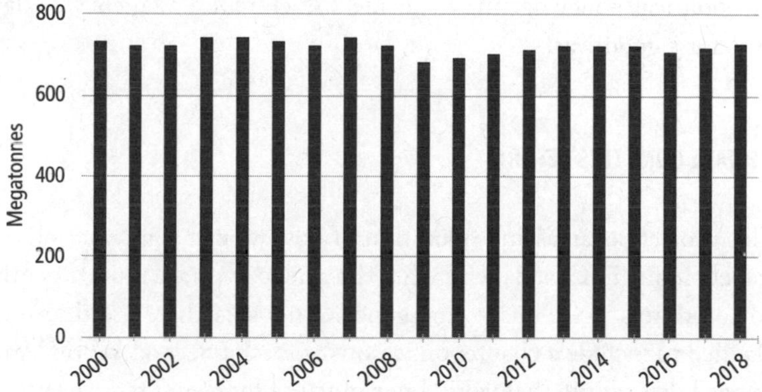

Source: Environment and Climate Change Canada: Tables-IPCC-Sector-Canada.

and policy wonks, labour leaders, socially responsible business leaders — that what we have been doing is simply not working. We have run out the clock with distracting debates about incremental changes. But where it matters most — actual GHG emissions — we have accomplished precious little. With the exception of a slight downturn in emissions during the 2008–2009 recession, we have made almost no progress, and frequently have slipped backwards.

Yet we have been at this for years. Canada initially committed to take action on climate change in 1992, when we signed the United Nations Framework Convention on Climate Change (UNFCCC) at the Rio Earth Summit. But our emissions continued to rise substantially right through the 1990s. We recommitted with the Kyoto Protocol (1997, ratified by Canada in 2002) and again at the international climate meetings in Copenhagen (2009) and Paris (2015). The chart, however, speaks for itself. We have failed to meaningfully bend the curve on carbon pollution.

At least things have more or less flatlined, you might say; our emissions are no longer rising. But as the great climate change warrior and founder of 350.org Bill McKibben has said, "Winning slowly on climate change is just another way of losing." Politics might be all about compromise and the art of the possible. But there is no bargaining with the laws of nature, and nature is now telling us something fierce.

And so a new approach is needed. We need a "wartime" mindset and political/policy agenda to tackle the climate emergency — it's time for a good war.

WE HAVE DONE THIS BEFORE

This project began as an exploration of how we can align our politics and economy in Canada with what the science says we must urgently do to address the climate emergency. And it is that. I had always planned to include a chapter on lessons from the Second World War. But as I delved into that work, I began to see more and more parallels between our wartime experience and the current crisis, and ultimately decided to structure this entire book around lessons from Canada's Second World War experience. Not because I get all weirdly animated about war. Nor is it because I think we need a metaphor about sacrifice, and certainly not because I think there is anything glorious or appealing about war. Rather, it is because I see in the history of our wartime experience a helpful — and indeed hopeful — reminder that *we have done this before*. We have mobilized in common cause across society to confront an existential threat. And in doing so, we have retooled our entire economy in the space of a few short years.

I'm far from the first person to say we need a wartime-scale mobilization to confront climate change.* But usually this comparison is

---

* The earliest Canadian example I have found exploring the connections between WWII mobilization and the climate challenge (and quite possibly the earliest piece to do so internationally) is a short journal article from 2001 by retired Memorial University academic Dennis Bartels, "Wartime Mobilization to Counter Severe Global Climate Change" (*Human Ecology* Special Issue 10, January 1, 2001). American researcher and founder of the Earth Policy Institute Lester R. Brown also draws upon WWII lessons in his Plan B books (the latest version being *Plan B 4.0: Mobilizing to Save Civilization*, New York: W.W. Norton and Company, 2009). Australian researchers David Spratt and Philip Sutton invoke the wartime frame in their 2009 book, *Climate Code Red* (Melbourne: Scribe Publications, 2009). Paul Gilding developed a "One Degree War Plan" in his 2011 book, *The Great Disruption* (New York: Bloomsbury, 2011). Academics Laurence L. Delina and Mark Diesendorf explored this parallel in a 2013 journal article ("Is Wartime Mobilisation a Suitable Policy Model for Rapid National Climate Mitigation?," *Energy Policy* 58 (July 2013): 371–380). Bill McKibben explored this theme in an American context in a 2016 essay for *The New Republic* entitled, "A World at War: We're Under Attack from Climate Change — and Our Only Hope Is to Mobilize Like We Did in WWII" (August 15, 2016). Also in 2016, Laurence Delina, drawing upon his earlier article and Ph.D. work, published a book entitled *Strategies for Rapid Climate Mitigation: Wartime Mobilisation as a Model for Action?* (New York: Routledge, 2016). It is an academic book and does not deal specifically with Canada (although it offers some Canadian examples). There is now a U.S.-based NGO, The Climate Mobilization, that is entirely structured around American WWII lessons for confronting the climate emergency. They have developed a "Climate Mobilization Victory Plan" (lead author Ezra Silk) that I cite in later chapters.

made as a passing reference. No one to date has, in the Canadian context, delved into the similarities and lessons in detail.

I invite you here to explore what wartime-scale mobilization could actually mean. In the chapters that follow, we will jump back and forth in time between stories of what Canada did during the war and what we now face. And in these comparisons, we will answer questions such as:

- How was public opinion rallied to support mobilization during the war, and how might it be galvanized again?
- Who or what was/is blocking needed mobilization?
- How was social solidarity secured across class, race and gender, and how can we do so again?
- How was national unity established across Canada's many provinces and regions with their varying views and interests, and can we successfully achieve this again as we move off fossil fuels?
- How did we collectively transform our economy and marshal all our resources to produce what was needed, and how might we do so again?
- What was the role of individual households and businesses during the war, and what must it be again?
- How did we collectively pay for that transformation and mobilization, and can we mobilize the necessary finances once again?
- What supports were offered for returning soldiers, and what can be learned for just transition for fossil fuel workers today?
- What was the role of Indigenous people in the war, and what is it in today's transformation?
- What was the role of youth and social movements then, and what is it today?
- Importantly, what are the war's cautionary tales, the warnings of things that brought us shame — the internments, the quashing of civil rights, the environmental pollution caused by wartime industry — that we do not wish to repeat?

- What must we remember about how Canada responded to refugees during the war, as we plan for the inevitable climate migration crises of the future?
- And critically, what sort of political leadership do we require to see us through this challenge?

This book is an invitation to our political leaders, to reflect on the leaders who saw us through the Second World War and to consider who they want to be, and how they wish to be remembered, as we undertake this defining task of our lives. My hope is that this book might embolden them to be more politically daring than we have seen to date, because that is what this moment demands.

And much like the trials that tested the character of past generations, this book is also an invitation to all of us to reflect on who *we* want to be as we together confront this crisis.

Confronting the climate emergency is not precisely the same as war and the battle against fascism. There are differences, of course. But I am arguing that our wartime experience provides very instructive lessons about how to confront an existential threat.

As you read this book, my hope is that you will marvel, as I have while researching and writing it, at the scale and scope and speed of what Canada did during the war years. And that you will find inspiration that we are capable of once again accomplishing something amazing — that we can do ourselves proud and, like then, that we can come out the other end of this transformation not only with a safer environment, but with a better and more just society than the one we are leaving behind.

This is not a book about climate science. It takes the urgent science and the impacts of climate breakdown as a given. Rather, it is a book about politics, history and policy innovation. It takes as inspiration Canada's Second World War experience and also finds encouragement from a few other countries that, unlike Canada, are starting to treat this crisis as the emergency that it is. I also draw heavily upon interviews conducted with politicians, academics, activists, Indigenous leaders, labour leaders and others.

I spend some time in the early chapters surveying the principal barriers to transformative climate action. But by and large, I choose to focus on what can be done to overcome these barriers. I believe you will find this an unusually hopeful book, given the subject matter.

Effectively tackling the climate crisis is not a technical or policy problem — we know what is needed to transition to a zero-carbon society, and the technology required is largely ready to go. Rather, the challenge we face is a political one. Climate solutions persistently encounter a political wall; the prevailing assumption within the leadership of our dominant political parties appears to be that if our political leaders were to articulate (let alone undertake) what the climate science tells us is necessary, it would be political suicide. And so they don't.

This book explores whether we can successfully align our politics with climate science, and the conditions under which it may be possible to practise such a bold politics that is well-received by the public. It outlines what a truly meaningful and hopeful climate program can look like in Canada and makes the case for why our political leaders should embrace this generational mission.

Our sense of what is possible is contained by what we know. Hopefully this exploration of what we did the last time we faced an existential threat can serve to blow open our sense of political and transformative possibility.

Like many of you, as I read the latest scientific warnings, I'm afraid. In particular, I feel deep anxiety for my children, and about the state of the world we are leaving to those who will live throughout most of this century and beyond. All of us who take seriously these scientific realities wrestle with despair. The truth is that we don't know if we will win this fight — if we will rise to this challenge in time. But it is worth appreciating that those who rallied in the face of fascism 80 years ago likewise didn't know if they would win. We often forget that there was a good chunk of the war's early years during which the outcome was far from certain. Yet that generation rallied regardless, and in the process surprised themselves by what they were capable of achieving. That's the spirit we need today.

Post-script, April 2020:

How quickly things can change. This book had already been completed and sent off for copy edit when the global coronavirus pandemic was declared. Overlapping lessons abound between our Second World War experience, the response to the COVID-19 crisis and the climate emergency. Given this unprecedented global health emergency, I have added an epilogue to this book that speaks to how policy-makers have quickly drawn from our wartime experience in combatting the pandemic, and how the pandemic response has reinforced the core point of this book, namely, that once emergencies are truly recognized, what seemed politically impossible and economically off-limits can be quickly embraced. (The pandemic story is still unfolding as I write, so for the latest version of the epilogue, visit sethklein.ca, where I will keep these lessons updated.)

PART ONE

AGAIN AT THE CROSSROADS OF HISTORY

Toronto Public Library

CHAPTER 1

INTRODUCTION: CONFRONTING EXISTENTIAL THREATS, THEN AND NOW

"Life can only be understood backwards, but it must be lived forward."

— Søren Kierkegaard, existentialist philosopher

"We are too late in the game for gradualism, to incrementally reduce emissions. Or for individualism, the idea that 'I'll take care of my emissions, you take care of yours.' What we envision is a rapid transition of our entire economy and society, with all hands on deck, as most recently happened in our history during World War Two."[1]

— Margaret Klein Salamon, founder and director of U.S.-based non-governmental organization The Climate Mobilization

THE SECOND WORLD WAR — THE LAST TIME WE MOBILIZED

The Canada led by Prime Minister William Lyon Mackenzie King in 1939 was a country at the crossroads. A lifetime later, we find ourselves there again. This time it is a climate crossroads that poses a generational challenge.

At the outset of the First World War in 1914, when Britain declared war, Canada, as a Dominion of the British Empire, was automatically at war as well. But in 1939, when Britain declared war on Germany, events unfolded somewhat differently. In the years between the wars, Canada had asserted its independence. While there was little doubt that Canada would follow Britain into war, a new protocol was important to the Mackenzie King government. And so Canada declared war one week later, on September 10, confirmed by a vote of the House of Commons. The Canadian cabinet and parliament would be the ones making key decisions about *how* Canada would engage in the war — the nature and extent of our contributions, and whether we would deploy troops beyond the defense of our own borders.

Despite Canada's war declaration, it is worth recalling that, even as the winds of war gathered in the late 1930s, our leaders were reluctant to recognize what would ultimately be necessary.

As the fascist powers began their ascendency and territorial acquisitions, neither Britain nor Canada made any serious move to oppose them. The Canadian government did not support the over 1,200 Canadians who made up the Mackenzie–Papineau Battalion — the volunteers who went to fight against Franco's fascist army in the Spanish Civil War in the late 1930s (even a battalion half named after the rebel grandfather of Prime Minister Mackenzie King himself).[2] Nor did Canada support sanctions against Italy at the League of Nations when Mussolini's forces invaded Ethiopia. No meaningful effort was made to oppose Imperial Japan's invasion of China. And when Mackenzie King met Adolf Hitler on a diplomatic mission in 1937, he privately dismissed him as "no serious danger."

The British prime minister, Neville Chamberlain, was similarly inclined to hope for the best. After the Nazi invasion of Austria and the threatened annexation of the Sudetenland region of Czechoslovakia in 1938, Chamberlain met with Hitler in Munich and sought appeasement. He infamously returned from Munich with an agreement promising "peace in our time."[3] Mackenzie King thought this to be a great deal. The Canadian prime minister was supportive of Chamberlain's policy of appeasement and expressed little desire to defend self-determination in Eastern European countries, Asia or Africa. News of what was happening internally within Germany — the establishment of the concentration camps and attacks on the Jewish community — was largely brushed aside.

Canada was on the cusp of being completely transformed by its Second World War experience, yet right up to the eleventh hour Mackenzie King still hoped to avoid being dragged into another war.

Consider it an early form of threat denial. And should we not draw some solace from the fact that, right up to that late hour, those in positions of power, like now, resisted the truths before them?

But let us cut Mackenzie King some slack. The prospect of leading a nation into another war in 1939 couldn't have been easy or welcome. Despite his record as Canada's longest-serving prime minister, Mackenzie King was not particularly popular. Unlike then U.S. president, Franklin Delano Roosevelt, or the soon-to-be British prime minister, Winston Churchill, Mackenzie King was hardly charismatic, nor was he a particularly memorable public speaker.

But Mackenzie King's challenges were more than just personal. The memory of the First World War — the one that promised to be "the war to end all wars" — was still fresh. Politically, Mackenzie King had lived through the conscription crisis of that war — the battle over mandatory military service that had pitted French and English Canada against one another — and knew full well how that crisis had nearly torn apart both the Liberal Party and the country.

The First World War had bred resentments about not everyone pulling their weight, not just between those who enlisted to fight and those who did not, but also because some corporations and many of the wealthy had engaged in outrageous profiteering, making a financial killing while others paid the ultimate sacrifice.

Then, in the wake of the financial crash of 1929, the country's people had been through a decade of the Great Depression, and were still wrestling with economic despair, anxiety, and high unemployment (just as many Canadian residents, particularly in certain regions of the country, wrestle with economic insecurity and precariousness in the present).

In short, Canadian families had already lost a lot within recent memory. It was surely tough to ask such a weary population to make yet more sacrifices, particularly when the threat seemed so distant. Convincing the public to once again go to war represented a leadership challenge of the highest order.

The type of leadership challenge we face again today.

Despite the early reticence of Canada's leaders, Canada did declare war in 1939, and then set about galvanizing public support, marshaling the armed services, and retooling the economy to meet the unprecedented production needs of the war effort.

Among its first actions, the government invoked the War Measures Act, allowing it to bypass Parliament and the lengthy process of law-making and to use executive cabinet orders (what are called orders in council) to make things deemed necessary for the war to happen quickly, such as establishing price controls and organizing the military. Troublingly from a civil liberties perspective, the act permitted the government to censor news and information, to ban political parties and cultural/religious organizations deemed harmful to the war effort and to imprison dissenters and those considered "enemy aliens" without due process.

Lest there be any doubt among younger generations today, the Second World War did indeed represent an existential threat. While history is replete with wars that should never have happened, and foreign interventions by more powerful and imperialist nations that were insidious and unjust, the Second World War is rightly seen as a "necessary war." The regimes of Hitler and Imperial Japan sought domination. They were totalitarian antidemocratic states. They not only believed in racial superiority, but were intent on operationalizing that "supremacy" in all aspects of social life and state power, and, in the case of the Nazis, were fixed on the genocidal extermination of entire populations. Had Canada and the Allies not won the war and defeated the Axis fascist powers, our lives today would look very different. As Mackenzie King said in a CBC radio address to the nation on September 3, 1939 (the day Britain declared war): "There is no home in Canada, no family and no individual, whose fortune and freedom are not bound up in the current struggle. I appeal to my fellow Canadians to unite in a national effort, to save from destruction all that makes life worth living, and to preserve for future generations those liberties and institutions that others have bequeathed to us."

Over the centuries, there have been few *just* wars. But this was one.

One thing we forget — and this is surely a key lesson for us today — is that, for a good part of the war's initial years (1939–1941), it was entirely an open question as to which side would prevail. During the first half of 1940, Hitler's army moved with remarkable speed, quickly occupying Denmark, Norway, Belgium, Holland and Luxemburg. Then in June, the Nazis rolled into Paris and France surrendered. People following events back in Canada were stunned by the rapidity of these losses. Throughout 1940 and 1941, Britain was subjected to brutal air assault — the infamous nightly Blitz bombings that left London in shambles, forced people into underground shelters and led to the relocation of children to the countryside and overseas. It was a distinct possibility that Britain would fall, and that North America and what was left of the British Commonwealth would have to face the Axis powers alone.

As 1941 unfolded, Hitler's forces pushed south and east into the Soviet Union. By 1942, fascist governments — Hitler's Nazis, Italy's Mussolini dictatorship and the Franco regime in Spain — controlled virtually all of Europe and North Africa, save the U.K., the Soviet Union east of Moscow, and neutral Switzerland and Sweden, while the Imperial Japanese army occupied most of China's coastal regions and southeast Asia. In both the battles of the Atlantic and of the Pacific, Allied ships were being routinely sunk. The U.S. was still ramping up its wartime production, and the Soviet people were experiencing devastating losses. It was not at all obvious to people at the time that the Allies would eventually defeat the Axis powers. We know in hindsight how this story ended; they did not.

Today, we too live in an ambiguous time. We don't know if we will win the climate fight, or if we will adequately act in time. But like those who sought to defeat fascism in the Second World War, we have to rally to the cause regardless.

When urged to take climate action, there is a tendency for Canadians to say, "But we're only a small population, and our efforts don't matter." This feels particularly true when the U.S. (under President Donald Trump) seems to be moving in the opposite direction. But here too the analogy is apt; after all, when it came time to finally confront Nazism, Canada did not wait until the U.S. finally joined the war effort in late 1941 — we threw ourselves into the fight two years earlier. And unlike the U.S., which joined the war only after the Japanese assault on Pearl Harbor, Canada's war declaration did not require an attack on our own soil.

The wartime example is also informative because the climate crisis, like the Second World War, can be confronted only by all of us together. Both are inherently collective enterprises. Yet so far, a major shortcoming of both our governments and traditional environmental organizations has been that solutions have been pitched in either individualistic or market terms. We are told to fix this ourselves as consumers or to embrace the capacity of the market to develop and adopt new technologies if given the right price signals and incentives. Neither of these approaches can or will work on their own. They would not have won victory in the Second Would War, and they cannot achieve what we must quickly accomplish now. Rather, a cross-society, mandated, all-hands-on-deck strategy will be needed to prevent the worst effects of climate change in the short time we have. Every person, every business and every level of government will be needed to succeed.

So, it is worth asking: What can be learned from the last time we collectively mobilized and completely retooled our economy in short order? How might these lessons be put into service once again? And can the Second Would War example — an ambitious mobilization that produced full employment — provide a renewed sense of national purpose in undertaking a common project?

There are of course limitations to the wartime comparison. But reminding ourselves of the scope and speed of what Canada did in the Second World War expands our sense of what is possible. It invites us to break free of the narrow thinking and assumptions that guide so much of climate policy planning today, and to think creatively and boldly about new solutions with a "we can do it" attitude.

TODAY'S CLIMATE CROSSROADS MOMENT: IT'S AN EMERGENCY!

While the threat today may move in slower motion than war, the climate crisis we face isn't really all that different. Only now we need governments that can lead us not into battle against other nations, but rather, into the fight for our collective future. The value of seeing the crisis we face as an emergency is that it forces a new mindset — it jolts us out of a business-as-usual mode and demands that we steadfastly focus our attention on addressing the urgent task at hand.

Some of you may be unsure this comparison really works. During the Second World War, after all, people understood the threat to be a clear and present danger. The climate crisis is more abstract. Most of our lives still look and feel unchanged and untouched, at least for now. As former Ontario Environmental Commissioner Dianne Saxe told me, "Carbon pollution is an invisible foe. The Second World War had a clear enemy and the prospect of a clear victory after which it would stop. People can't see climate pollution, so they don't think it's real."

Yet the threat before us today is indeed an emergency. The imperative for urgent action mounts daily. As Bill McKibben has written, "The question is not, are we in a world war? The question is, will we fight back? And if we do, can we actually defeat an enemy as powerful and inexorable as the laws of physics?"[4]

This book does not go into detail on the threat before us. If you wish to delve deeply into what the best science predicts, there are plenty of great resources readily available. Read the latest reports of the UN's Intergovernmental Panel on Climate Change (IPCC). Read the works of Bill McKibben, particularly his 2019 book, *Falter*. Read the work of climate scientist Michael E. Mann. Read the terrifying warnings of Australian climate researchers David Spratt and Philip Sutton, such as *Climate Code Red*. Read David Wallace-Wells's deeply disturbing *The Uninhabitable Earth*. And, of course, read *This Changes Everything* and *On Fire*, the insightful work of Naomi Klein (yes, my sister).

Suffice it to say, the science is clear. The October 2018 report of the IPCC, jointly authored by the world's top climate scientists, gave us 12 years to at least halve our emissions (a 50% reduction by 2030), and until mid-century to become net carbon-zero, in order to have a decent chance of keeping global temperature rise under 1.5°C above preindustrial levels. (We are already at 1° of temperature rise.) Failure to do so, the IPCC warns, will have catastrophic and terrifying results:

- Massive disruptions to food systems, both on land and in the sea, with related losses of biodiversity, species and ecosystems, the breakdown of whole food chains and mass die-offs of coral reefs due to ocean acidification, with resulting threats of starvation for some, and food shortages and huge run-ups in food prices for others;
- More extreme heat events, with deadly consequences for the most vulnerable among us at home and abroad;
- Many more major weather events — from forest fires to floods and hurricanes to droughts — resulting in billions of dollars in damage to property, months of smoke-filled skies (particularly harmful for the very young and old, and to those who suffer from asthma or other respiratory illnesses), massive disruptions to agriculture, devastating environmental impacts and countless lost lives;
- Greater and faster sea-level rise, threatening coastal communities, flooding major cities, forcing billions of dollars to be spent in defensive infrastructure and, in some cases, the loss of entire island nations;
- Lost water sources, again with incalculable impacts on the food supply and human health, and causing new geopolitical tensions and wars;
- Spikes in illnesses carried by insects that thrive with global warming, such as Lyme disease, West Nile virus and the Zika virus (borne by mosquitos, this terrifying virus, if contracted by pregnant women, can shrink the brains of developing fetuses);
- And related to all this — mass human climate displacement and migration.

The human, ecological and economic costs will be profound and devastating. And the higher the temperature rise we permit, the more destructive and deadly the impacts. Were global temperature rise to reach 4°C, hundreds of millions of lives could be lost globally.[5]

Still unknown is the degree to which feedback loops will compound these effects and produce runaway impacts. For example, as warmer winters fail to kill pests and diseases — such as the mountain pine beetle infestation that has devastated so much of British Columbia's interior forests — our forests die and become a carbon source rather than a carbon sink. Or when melting permafrost in the north releases ancient carbon and methane stores (methane being a much more potent GHG than CO_2). Or when acidification and dying coral reefs undermine our oceans' ability to absorb atmospheric CO_2. Or when melting ice no longer reflects the sun's light away from the earth. All these dynamics are already underway and have the potential to dramatically speed up global temperature rise and the rate at which we accumulate GHGs in the atmosphere.

If we fail to act quickly, then over the course of the rest of this century, things start to get horrific — a world that is unlivable and catastrophic for many, deeply disruptive for all others and quite possibly ungovernable.

ATTACKS ON OUR SOIL

During the war, people could track the unfolding of the conflict on a map — they could see the fascist conquest of territory from 1938 to 1942 and, conversely, as the tide started to turn, they could track the slow advance of the Allies. Today we need a new map to track the climate attacks on our territory — the extreme weather events and the proposed new fossil fuel infrastructure projects.

The challenge is to make clear that climate change isn't an abstract threat — somewhere else or sometime in the future. It is here, it is now, and it has profound consequences for people we love, up to and including death. It is a threat to our security and well-being equal to or greater than any war or terrorist attack, for which the public is generally quite prepared to mobilize.

The more general and global list of climate impacts above isn't simply a theoretical set of threats. These impacts — these attacks — are already playing out on Canadian soil. The climate crisis finds expression in increased flooding across the country, such as the 2013 Alberta floods that left five dead, forced the evacuation of 100,000 people and caused infrastructure and property damage of over $6 billion. It manifests in ever-more frequent and devastating forest fires, such as those that ravaged B.C. in the summers of 2017 and 2018, and of course the infamous and terrifying 2016 Alberta wildfire — "The Beast" — that forced the evacuation of Fort McMurray in the heart of the oil sands and resulted in $3.6 billion in property damage — including the destruction of 2,400 homes and businesses and over 18,000 vehicles — and total economic costs of almost $10 billion, making it the most costly catastrophe of modern Canadian history. The crisis is evident in deadly heat waves like we saw in eastern Canada in the summer of 2018, which hit Quebec particularly hard and was responsible for as many as 70 deaths. The attacks on our soil include increasingly common water shortages, storm surges, hurricanes, coastal erosion, melting permafrost and ice in Canada's north, disrupted animal patterns and fish returns, more numerous asthma cases and, increasingly, impacts on mental health. In all these cases, those who are poor, elderly, suffer from chronic diseases or are members of marginalized communities feel these impacts more acutely. (For a more complete listing of Climate Attacks on Canadian Soil see Appendix I online at sethklein.ca.)

Yes, there is the usual caveat that no one event can be specifically attributed to climate change. But what is now beyond dispute is that both the frequency and severity of these events is a product of climate change. Rising global temperatures are resulting in drier periods (making various locales more susceptible to drought and fire), in ocean warming and in the atmosphere capturing more water and then releasing it in more frequent fierce events. As climate scientist Katharine Hayhoe explains, "'Was it caused by climate change?' is the most common question when we hear about an extreme event. But when it comes to hurricanes, that's the wrong question. The right one is 'How much worse did climate change make it?'. . . As the ocean warms, hurricanes are intensifying faster, getting stronger, have a lot more rainfall associated with them, getting bigger, moving more slowly, and then there's sea level rise, which makes storm surges worse."[6]

For us in Canada, the spring of 2019 brought additional bad news from leading federal scientists and academic experts warning that temperature rise in Canada's north has been and will continue to be about three times the global average.[7]

What the above impacts drive home is that the climate crisis isn't only an environmental issue — it's a health issue and, fundamentally, a social justice and human rights issue. Which is to say, it's about people.

And of course, for our fellow human beings in more vulnerable parts of the world, these climate and weather-related events have been and will be far more deadly and devastating.

Fighting the climate crisis isn't some elite-driven concern, as it is sometimes portrayed. It is our collective moral responsibility. And blocking effective and meaningful climate action is the work of scoundrels — irresponsible and cynical political and economic interests who would do harm to the well-being of our children and grandchildren. Those who sow the seeds of climate denial and doubt and who block progress on climate action must be opposed or ignored. Just like we did in the Second World War. Taking action doesn't require everyone's agreement. Nor can we wait for universal agreement when faced with an emergency.

The public intellectual and MIT professor Noam Chomsky recently commented: "I start my classes these last couple of years by simply pointing out to the students that they have to make a choice that no one in human history has ever made. They have to decide whether organized human society is going to survive. Even when the Nazis were on the rampage, you didn't have to face that question. Now you do."[8]

At a global level, the current emergency has been particularly powerfully articulated by Greta Thunberg, the extraordinary young Swedish climate activist who ignited the worldwide student climate strike movement in the fall of 2018. In January 2019 she was invited to address the World Economic Forum in Davos, Switzerland — the annual gathering of the richest and most powerful people on Earth. Then age 16, she looked at those assembled and said in her quiet, calm and deliberate manner: "Adults keep saying we owe it to the young people to give them hope. But I don't want your hope. I don't want you to be hopeful. I want you to panic. I want you to feel the fear I feel every day. And then I want you to act. I want you to act as you would in a crisis. I want you to act as if the house was on fire. Because it is."

THE BATTLE PLAN: KEY LESSONS FROM THE SECOND WORLD WAR

To execute a successful battle, we need a plan — a roadmap to guide us through the stages of climate mobilization. From my study of Canada's Second World War experience, and in particular how we successfully mobilized on the home front, the following key strategic lessons emerge:

Adopt an emergency wartime mindset, prepared to do what it takes to win. Something powerful happens when we approach a crisis by naming the emergency and the need for wartime-scale action. It creates a new sense of shared purpose, a renewed unity across Canada's confederation, and liberates a level of political action that seemed previously impossible. Economic ideas deemed off-limits become newly considered, and we open ourselves up to fresh ways of thinking. We see the attacks on our soil for what they are. And we become collectively willing to see our governments adopt *mandatory* policies, replacing voluntary measures that merely incentivize and encourage change with clear timelines and regulatory fiat in order to drive change and meet ambitious targets.

Rally the public at every turn. Many assume that at the outbreak of the Second World War everyone understood the threat and was ready to rally to Mackenzie King's call. But that was not so. It took leadership to mobilize the public. In frequency and tone, in words and in action, the climate mobilization needs to look and sound and feel like an emergency. If our governments are not behaving as if the situation is an emergency, then they are effectively communicating to the public that it is not. As occurred in the war, our governments need to develop and execute multifaceted advertising programs that boost the level of public "climate literacy" and outline and explain their policy responses. The news media and educational institutions need to reimagine their approach to this crisis, and we must demand that they do so. We need to marshal the cultural and entertainment sectors, which requires major public funding for arts and culture initiatives that seek to rally the public. And we need to better include the public in decision-making as we refine our climate policies, through the use of citizen assemblies and other means of democratic engagement.

Inequality is toxic to social solidarity and mass mobilization. A successful mobilization requires that people make common cause across class, race and gender, and that the public have confidence that sacrifices are being made by the rich as well as middle- and modest-income people. During the First World War, inequality undermined such efforts. Consequently, at the outset of the Second World War, the government took bold steps to lessen inequality and limit excess profits. Such measures are needed again today.

Embrace economic planning and create the economic institutions needed to get the job done. During the Second World War, starting from a base of virtually nothing, the Canadian economy and its labour force pumped out planes, military vehicles, ships and armaments at a speed and scale that is simply mind-blowing. Remarkably, the Canadian government (under the leadership of C.D. Howe) established 28 crown corporations to meet the supply and munitions requirements of the war effort. That is just one example of what the government was prepared to do to transform the Canadian economy to meet wartime production needs. The private sector had a key role to play in that economic transition, but vitally, it was not allowed to determine the allocation of scarce resources. In a time of emergency, we don't leave such decisions to the market. Throughout most of the war years, the production and sale of the private automobile, in both Canada and the U.S., was effectively banned; instead those auto factories were operating full tilt to churn out wartime vehicles.

Howe's department undertook detailed economic planning to ensure wartime production was prioritized, conducting a national inventory of wartime supply needs and production capacity and coordinating the supply chains of all core war production inputs (machine tools, rubber, metals, timber, coal, oil and more). The climate emergency demands a similar approach to economic planning. We must again conduct an inventory of conversion needs, determining how many heat pumps, solar arrays, wind farms, electric buses, etc., we will need to electrify virtually everything and end our reliance on fossil fuels. We will need a new generation of crown corporations to then ensure those items are manufactured and deployed at the requisite scale. We will require huge public investments in green and social infrastructure to expedite the transformation of our economy and communities. And as we did in the war, we will need to mobilize labour to get this job done, banishing unemployment in the years to come.

Spend what it takes to win. A benefit of an emergency or wartime mentality is that it forces governments out of an austerity mindset and liberates the public purse. The Second World War saw an explosion in government spending. In order to finance the war effort, the government issued new public Victory Bonds and new forms of progressive taxation were instituted. Yet these new taxes and what remains to this day historic levels of public debt did not produce economic disaster, as is so often claimed. On the contrary, they heralded an era of record economic performance. As we confront the climate emergency, financing the transformation before us requires that we employ similar tools.

Leave no one behind. The Second World War saw over one million Canadians enlist in military service and a similar number employed in munitions production (far more than are employed in fossil fuel industries today). After the war, all those people had to be reintegrated into a peacetime economy. That too required careful economic planning, and the development of new programs for returning soldiers, from income support to housing to post-secondary training. Those post-war programs weren't simply the result of government largesse and goodwill; they stemmed from the demands of labour and social movements, who after the ravages of the Depression and war insisted on a new deal. The ambition of these initiatives provides a model for what a just transition can look like today; they should inspire us to develop robust programs for all workers whose economic and employment security is currently tied to the fossil fuel economy, with a special focus on those provinces and regions most reliant on oil and gas production.

Reject the straightjacket of neoliberal economic thinking. The previous lessons all share a common thread — the casting off of free-market economic ideas and assumptions that have kept us from doing what we need to do in the face of the climate emergency. During the war, given the urgency and scale of the task, both the general public and private-sector leaders understood that the economic transformation had to be state-led. Canada's Second World War government was by and large a free-market oriented administration (indeed, that orientation had severely constrained government action during the Depression of the 1930s, at the price of great hardship). But in the face of the urgent need to confront fascism, its leaders were no longer ideologically rigid. They were prepared to embrace a level of economic planning, public investment and public enterprise that seemed previously unimaginable.

Transform government. Once an extended emergency is truly recognized, all the institutions and machinery of government are focused on the task of confronting it. During the Second World War, Mackenzie King appointed a powerful war subcommittee of cabinet to oversee the government's efforts. We need a Climate Emergency War Cabinet Committee today, and a Climate Emergency Secretariat in the Prime Minister's Office and each premier's office, coordinating our emergency response as a whole-of-government approach. Just as we have created a governance architecture for fiscal planning, budgeting, budget consultations and accountability in the present, so too we need to build similar systems for carbon budgeting. We need new federal-provincial-municipal cost-shared programs focused on the climate crisis, including a new federal Climate Emergency Just Transition Transfer to collaboratively fund new green infrastructure and job training initiatives, with funding going disproportionately to the provinces with the most heavy-lifting to do in this transition. We need to breathe a new, ambitious spirit into the civil service. During the war, C.D. Howe created end runs around the existing civil service to expedite wartime production. That was effective but also produced its own problems. The challenge now is to transform the public service — to recruit and promote the people willing and able to make bold things happen quickly. We need visionary and creative people in key leadership positions in the civil service and to bring in outside experts, civil society leaders and entrepreneurs as needed to drive change and oversee the necessary scale-up. And we need all political parties to advance policy agendas that are truly consistent with what the science demands of us.

Indigenous leadership, culture and title and rights are central to winning. Indigenous people played an important role in the Second World War. Today, their role in successfully confronting the climate crisis is pivotal. As our mainstream politics dithers and dodges meaningful and coherent climate action, the assertion of Indigenous title and rights is buying us time, slowing and blocking new fossil fuel projects until our larger politics come into compliance with the climate science. Some of Canada's most inspiring renewable energy projects are also happening under First Nations' leadership. It is imperative to both honour and support such efforts, first by embedding the UN Declaration on the Rights of Indigenous Peoples into law at all levels of government, and second by ensuring that Indigenous communities and nations are full partners in the development of our climate emergency plans.

Everyone has to do their bit. The Second World War was a total war effort. It was not merely prosecuted by government, the military and war manufacturing firms. All households played their part. Every company in the country made adjustments. All institutions were engaged. The same is true today. Households will need to shift their consumption, their transportation and how they heat their homes. All companies and institutions, public and private, need transition plans. Thousands of young people want a role to play, and many could find meaning in a new national Youth Climate Corps. And social movements will need to keep governments' feet to the fire at every stage.

This time, human rights must not be sacrificed. The government's invocation of the War Measures Act in 1939 came at too high a price. People were imprisoned and interned without due process. Communities were forcibly relocated. Civil liberties were forsaken. Canada's wartime experience offers cautionary tales of what *not* to do. The current crisis gives us an historic opportunity to avoid the sins of the past, and to engage in a form of emergency mobilization that is collaborative rather than coercive.

Canada is not an island. We don't win wars by ourselves, and neither can we opt out when justice demands our engagement. Canada's population is relatively small, yet we have punched above our weight before — we certainly did in the Second World War — and we can again. This lesson applies at multiple levels. First, while Canada's domestic GHG emissions may be small at a global level, we are also a major international exporter of fossil fuels. Second, in addition to taking climate action at home, Canada must embrace our responsibilities to the rest of the world. During the Second World War, Canada was extremely generous with our financial transfers to various Allies, despite unprecedented demands at home. Our historic per capita GHG emissions have been disproportionately high, carbon pollution does not stop at our borders, and we are one of the world's wealthiest countries. Given all this, it is incumbent on Canada to substantially boost our financial transfers to poorer countries, particularly in those regions hardest hit by the climate crisis and extreme weather. This is not a matter of charity, but of necessity and justice. Third, we must make right one of the most shameful chapters of Canada's Second World War legacy — the response to refugees. Before, during and after the war, Canada refused to open its doors to people fleeing persecution, particularly Jews seeking to escape Nazi-occupied Europe. In the coming decades, the crises of people displaced by climate impacts will surely be a defining issue. This time, we need to act with honour.

When necessary, real leaders throw out the rule book, and they are the heroes. Stay alert throughout this book to people who, in the face of a humanitarian crisis, defy orders and the norms of their time and circumstance — they are the ones who change the course of events. These are some of the people we remember from the Second World War, and they will be the people history again recalls as climate emergency champions.

Know thine enemy. Before engaging in battle, we need to know what we are up against. The enemy was clear in the Second World War — today, less so. We face numerous barriers to change, particularly a fossil fuel industry that has done much to block climate action. One of the most insidious barriers is a dynamic I call the "new climate denialism," along with its various manifestations, peddlers and enablers. The new climate denialism currently dominates our politics, and it is the new modus operandi of the fossil fuel industry. We will start our mobilization roadmap here.

In the pages that follow, we will "put meat on the bones" of each of these lessons. This book is an historical excavation — an unearthing of what we are capable of when we collectively approach an emergency with a new mindset, not only with respect to economic change, but with a new spirit of collaboration and purpose.

WAR AS A SENSE OF PURPOSE

Allow me to speak directly to a core doubt that is likely occurring to you at this point — a "yeah, but" voice in the back of your mind. You share a growing sense of alarm about the climate crisis. You want to believe that we can rally in time. But you question whether our governments, acting on our collective behalf, can really do what needs to be done. Years of disappointment, downsizing, deregulating, privatizing and globalizing, of doing the bidding of various corporate interests, and sometimes of downright corruption and cronyism, have left you skeptical that we can collectively accomplish what was done in the Second World War. We meet the contemporary threat in an era of cynicism and distrust, in which the very idea of acting together has been significantly eroded along with trust in each other and our governments. Our sense of collective ambition has been sapped, replaced by a prevailing culture of impossibility.

I get all that. But come on this journey with me regardless.

Why should you suspend your doubts about our collective capacity to rise to this task? Because wars transform us.

Emergencies and disasters can bring out the worst in us. But they can also bring out our best. The worthiest kind of leadership — whether at a grassroots or community level or a big-P political level — seeks to animate our best selves. And in the face of crisis, such leadership does not shy away but rather invites us to embrace a sense of shared purpose as we collectively tackle the emergency.

American writer Rebecca Solnit, in her 2009 book, *A Paradise Built in Hell: The Extraordinary Communities That Arise in Disaster*, details the beautiful expressions of human caring and solidarity that have marked terrible events — the San Francisco earthquakes of 1906 and 1989, the Halifax explosion of 1917 and the hurricane that hit that city in 2003, Hurricane Katrina and the catastrophe it inflicted upon New Orleans in 2005, and many others.[9] Our mainstream media takeaways of most of these events are of destruction and chaos. But as Solnit notes:

> In the wake of an earthquake, a bombing, or a major storm, most people are altruistic, urgently engaged in caring for themselves and those around them, strangers and neighbours as well as friends and loved ones. The image of the selfish, panicky, or regressively savage human being in times of disaster has little truth to it. Decades of meticulous sociological research on behaviour in disasters, from the bombings of World War II to floods, tornadoes, earthquakes, and storms across the continent and around the world, have demonstrated this . . . Horrible in itself, disaster is sometimes the door back into paradise, the paradise at least in which we are who we hope to be.[10]

Solnit describes visiting Halifax shortly after the disastrous hurricane that hit the city in 2003 and talking with a man who recalled the post-hurricane experience thus: "He spoke of the few days when everything was disrupted, and he lit up with happiness as he did so. In his neighbourhood all the people had come out of their houses to speak with each other, aid each other, improvise a community kitchen, make sure the elders were okay, and spend time together, no longer strangers."[11]

Today's climate crisis isn't just a demand that we remake our homes, infrastructure and economy. It's an opening — a chance to rebuild trust, community and democracy.

American philosopher William James published his famous essay "The Moral Equivalent of War" in 1910. James had just lived through the 1906 San Francisco earthquake (he was teaching at Stanford University) and was active in antiwar circles opposing the Spanish–American War and other imperialist wars. But while opposed to such wars, James understood that people need the sense of meaning, purpose and common struggle that comes with war. "He proposed something akin to the Peace Corps or the War on Poverty," writes Solnit, a form of youth conscription but directed towards something more worthy than war.[12] Watching how people responded so well to the San Francisco earthquake provided James with evidence that, in the face of disaster, people often behave magnificently, not only overcoming fear but also caring for one another — the emergency creates a sense of purpose and brings out our best selves. James sought examples of events or struggles that could "inflame the civic temper as past history has inflamed the military temper."

To which this book replies, look no further. Welcome to *A Good War*.

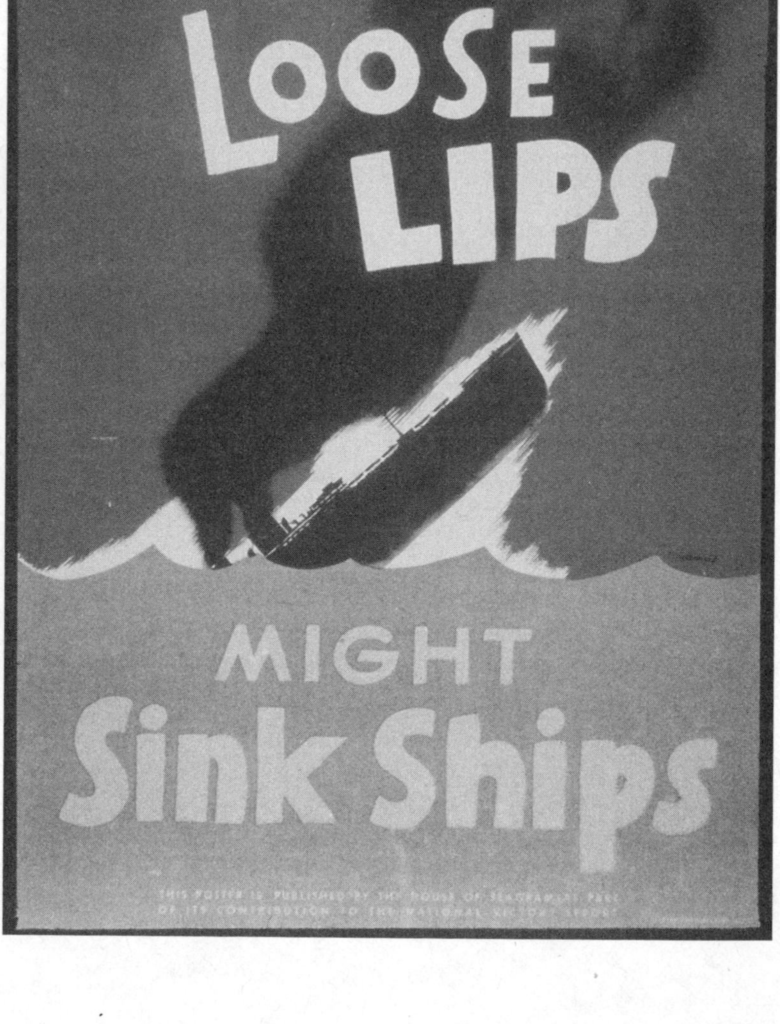

McGill University Archives

CHAPTER 2

What We're Up Against: The New Climate Denialism in Canada

"No country would find 173 billion barrels of oil in the ground and just leave them there. The resource will be developed."

— Prime Minister Justin Trudeau, speaking at a Global Energy and Environment Leadership Award Dinner in Houston, Texas, on March 9, 2017

"As leaders we have a responsibility to fully articulate the risks our people face. If the politics are not favorable to speaking truthfully, then clearly we must devote more energy to changing the politics."

— Marlene Moses, then ambassador to the United Nations for the South Pacific island nation of Nauru, 2012

"The crisis consists precisely in the fact that the old is dying and the new cannot be born; in this interregnum a great variety of morbid symptoms appear."

— Italian Marxist theorist Antonio Gramsci, imprisoned by Mussolini's fascist regime

OUR PHONY WAR

As fascist forces grew in power and territory over the 1930s, many knew that war was coming.

When the Nazis invaded Poland on September 1, 1939, a line was finally crossed. Britain and France declared war on Germany in response. A week later, so did Canada. And then . . . not a lot really happened.

Historians have dubbed much of the first year of the Second World War "The Phony War." Neither the U.K. nor Canada, nor any other nation actually came to Poland's rescue. No immediate battlefront emerged. For the next nine months, there was not much fighting. Everyone knew it was coming, but there remained much preparing and training to do. No one knew quite how and when to get things started. As the Nazis took control of more and more countries, the Allies observed a level of German capacity that caught them off guard. Leadership in Britain seemed rudderless until the Conservative party forced out Neville Chamberlain and replaced him with Winston Churchill (leading a coalition "unity" government) in May 1940.

Canada too got off to a slow start, and Mackenzie King was frequently criticized for failing to adequately prosecute the war.[1] Uniforms were not yet available. Supplies were minimal. Training was slow to come together. George Chow, a Chinese Canadian artillery man who participated in the Normandy campaign and one of the few living Second World War combat veterans, recalls signing up in the first year of the war at the age of 18 in Victoria "for the adventure" (and despite the fact that Chinese Canadians had not yet been granted the vote in Canada). He told me he remembers being transferred a week later to the Seaforth Armoury in Vancouver for basic training, where they had to learn gun skills using broom handles, given the shortage of actual weapons. "I didn't see the Bofors gun [the anti-aircraft gun Chow would operate] until 1942."

The government was busy negotiating financial issues with Britain and keen to point out that Britain had dragged Canada into this fight. From a cost perspective, the Mackenzie King government was hoping, it seemed, to engage in what historian Jack Granatstein termed a "limited liability" war. While there had never really been any doubt that Canada would follow Britain in *declaring* war, and would if needed defend our own territory, in those early months it remained an open question as to whether Canada would send troops overseas and, if so, to what extent.

The defeat of France in June 1940 changed things. It was a stunning wake-up call and shifted the national psyche. Suddenly, things got much more serious.

And so we find ourselves in a similar awkward interregnum today. Much like Mackenzie King's government declaring war in 1939 but then not initially doing much, in July 2019, the House of Commons passed a Climate Emergency motion introduced by the Justin Trudeau government, and then proceeded with business as usual, reapproving a large fossil fuel pipeline the very next day. It was political symbolism, by and large.

Which prompts the question: what will our "fall of France" moment be — the event or series of events that fundamentally shift the popular zeitgeist and makes possible that which feels politically impossible today? Or might that already have occurred?

Like racers kicking the dirt at the start line, we await clarity on when the race will begin and who will line up where. This is the time of our Phony War. Most of us know the battle for our lives must soon get underway, and most of our leaders are now talking tough on climate. But we're not quite sure how to begin in earnest. Just like we abandoned Poland to her fate, so too we look away today with uncomfortable shame from parts of the world already experiencing the worst ravages of climate change and extreme weather. Everyone's waiting for some sort of starting gun to go off. Or for someone to lead.

But as with the Second World War, the Phony War will not last. Ours is about to end.

TO THE BARRICADES!

Before delving into how we can mobilize for the climate emergency, it is necessary to understand the present reality — to appreciate the barricades we must now scale or cast aside.

The impediments we face to truly bold climate action operate in three domains:[2]

Political barriers to change: These include a failure on the part of our dominant political parties and governments — even those that purport to appreciate the scientific imperative — to entertain transformative policies. It encompasses the systematic refusal of our governments to recognize and respect Indigenous title and rights, especially when Indigenous communities are saying no to fossil fuel activity and infrastructure on their territories. But it also refers to more structural political barriers, such as how our electoral system and Canadian federalism — the division of powers between our federal and provincial governments — all conspire to block forward momentum on climate.

Economic barriers to change: These include factors such as how inequality undermines social solidarity and thus blocks progress, how economic and job insecurity fosters a reluctance to embrace transition, and the ways in which regional economic differences complicate agreement on climate action. It includes the ways in which we are all currently reliant on fossil fuels. Most significantly, this dimension captures the insidious manner in which the economic and political power of the oil and gas industry and its corporate partners have profoundly stalled meaningful climate action.

Cultural barriers to change: These include dynamics we traditionally consider when we think of popular culture, such as mainstream media bias. But it is also the status quo advantage of dominant political-cultural ideas, such as neoliberalism, individualism or a general loss of faith in our collective capacity to accomplish great things. These pervasive ideas block our ambition and lead to a failure of imagination just when we most need creative and bold solutions to the climate emergency.

The chapters to come will address all of these barriers to transformative change. But first we must name and understand an overarching force we are up against.

Whenever I've told someone about the book I am writing and its wartime framework, almost invariably the point is made that every war has an enemy. "Who is the enemy in this battle?" I'm frequently asked. "Or does this fight even have one?" A reasonable question. And in truth, one of the questions I find hardest to answer. As I previously shared, my political roots are in the peace movement — I get uneasy talking about "enemies" and all that implies.

It feels most comfortable to say this modern-day story doesn't need an enemy — it's a battle we are all waging together, not against another nation, but rather in defence of our shared future and against climate disruption.

At the most basic level, the enemy, as Bill McKibben proposes, is simply carbon and methane — the GHGs that bear us no ill will but neither do they have an ethic to which we can appeal. They merely do what they do once we release their ancient stores, trapping heat in the Earth's atmosphere. A deadly and invisible foe.

Or maybe the enemy is all of us — our individual and collective dependence on fossil fuels and our reluctance to give up some of what these fuels allow. Or maybe it is our complacency — the status quo always has an unfair advantage; we are by nature anxious of change, and prefer the world we know, with all its inadequacies, to a future we can't imagine. Or perhaps it's a pervasive feeling of hopelessness, that loss of faith in our collective capacity to rise to this challenge.

Iconic environmentalist and scientist David Suzuki adheres to a variant of the "all of us" view. In a conversation we had, Suzuki proposed that the enemy is a human mindset of domination, freed of constraints by new technological capabilities, operating in tandem with an economic system where the logic of corporations is hardwired to support the human desire to conquer and exploit nature. Consequently, Suzuki sees the enemy as "the collective impact of all of us."

There are, I think, elements of truth to all these outlooks.

If this story does have a villain, though, or at least "collaborators" (if that language feels more palatable), it is the fossil fuel corporations. The industry itself — the corporations and their leaders, not its workers — has outright lied for decades about the truth of climate change and has done everything in its power to systematically deny, delay and divert the need for climate action at every turn. The fossil fuel industry, in pursuit of its financial self-interest, has become expert at preying on our fears, misgivings and desires.

While we are all currently reliant on fossil fuels — all complicit in some way given the economy in which we live — we are not all the enemy in this fight. Naomi Oreskes, Harvard professor of the history of science, notes that during the U.S. Civil War and the battle against slavery, "people in the North wore clothes made of cotton picked by slaves. But that did not make them hypocrites when they joined the abolition movement. It just meant that they were also part of the slave economy, and they knew it. That is why they acted to change the *system*, not just their clothes."[3]

But here's my main take on this thorny question. The primary "enemy" or force we confront at this juncture is an insidious way of thinking and practice that plagues our politics and economic policy-making — a dynamic I call the new climate denialism.

While conventional climate denialism simply refuses to accept the reality of human-induced climate change (think of Donald Trump or Maxime Bernier), this new form of denialism sees our political leaders, the fossil fuel industry, as well as leading media outlets and pundits assure us that they understand and accept the scientific warnings about climate change, but then they promote and practice a politics and policy agenda that fails to align with what the scientific consensus says we must do.[4] Others, perhaps more kindly, calls this "all-the-above policy-making," the inclination of political leaders to tell the public that we need not make hard choices. Whatever term one uses, the result is the continuing sorry state of Canadian political non-leadership on the defining challenge of our time.

In the balance of this chapter, I focus on how the new climate denialism plays out in our politics and how it is practised by the oil and gas industry itself.

THE POLITICS OF OUR DISCONNECT

We are living in a time of deep disconnection — a gaping void between what is scientifically/ecologically necessary and what is considered politically possible, one that breeds cynicism and despair.

There is no shortage of examples:

- The Trudeau government's failure to present a plan commensurate with its much-vaunted Paris climate agreement commitments, while doubling down and anteing-up on the Trans Mountain pipeline expansion and fossil fuel production.
- The former Rachel Notley NDP government in Alberta released a climate plan with great fanfare in November 2015, including the introduction of a carbon tax and a commitment to phase out coal-fired electricity by 2030, while simultaneously permitting a 40% increase in oil sands emissions and relentlessly pursuing new bitumen pipeline capacity. The Notley plan rightly gained kudos for its accelerated phaseout of coal. But of the 18 coal-fired electricity generators all but three are being converted to natural gas, another fossil fuel.
- The B.C. NDP government of John Horgan seeks to reclaim climate leadership (and sustain its governing partnership with the B.C. Greens) with a welcome *Clean BC* climate plan, while simultaneously pursuing a new liquified natural gas (LNG) industry and overseeing a dramatic ramp-up in natural gas fracking — a huge new source of GHG emissions. The math on these two competing paths with respect to GHG reductions cannot be made to work.
- Newfoundland and Labrador started producing offshore oil in 1997, ironically the same year the Kyoto Protocol was signed. Oil and gas now represent approximately 25% of the province's gross domestic product (GDP) and about 3% of its employment. Transition ought to be a major concern. Yet in Newfoundland and Labrador's May 2019 provincial

election, the issue barely received a mention. "I assure you the future in our offshore has tremendous potential, and we are working hard to make sure that we can realize that potential," Liberal premier Dwight Ball told a conference of the Newfoundland and Labrador Oil and Gas Industries Association a month after his 2019 re-election.[5] The province's current goal is to double oil production.

- A fixation with carbon pricing, a good but incremental policy that will not achieve what is needed at the speed required, has consumed the political and media oxygen at the expense of other more systemic and bold changes. Politicians, academic economists and many environmental organizations have placed great stock in carbon pricing, expending huge political capital and even (in the case of the Trudeau government) trading oil sands and pipeline expansion for agreement on a carbon tax.[6] All for a mere $50/tonne carbon tax (equivalent to only about 11 cents a litre at the pump). I'm not opposed to carbon pricing; it's one useful tool in the policy toolbox. But it will not get us where we need to go — and I believe much of the public suspects as much.

These are all manifestations of the new climate denialism. We elect governments that promise climate action, they deliver underwhelming and contradictory policies, and then get replaced by right-wing governments that undo what little progress we've seen. This has been the Canadian story of the last three lost decades. What is clear from this record is that politics as usual won't cut it, and the status quo is a recipe for disaster. The incrementalism of the last 30 years will see us burn.

The list above does not bother to include the non-climate plans of the federal Conservatives under Stephen Harper and Andrew Scheer, Ontario's Doug Ford government, Alberta's Jason Kenney government or Saskatchewan's Scott Moe government — the conservative leaders who graced the cover of *Maclean's* magazine in November 2018, posing as a posse of tough guys ready to fight the carbon tax.[7] *Maclean's* dubbed these men "The Resistance," but Second World War–era resistance fighters they are not, as they give aid and comfort to the most powerful vested interests in our country. These people are not serious about climate leadership. And they stand in stark contrast to past Conservative leaders with an admirable record of rising to confront emergencies and of being early movers in the battle against climate change. My focus, rather, is on those leaders, parties and governments who purport to be climate leaders. And my interest is in how their climate programs can become truly coherent and convincing.

We need to crack this puzzle. If a major change in government, such as the elections of Trudeau, Horgan or Notley, fails to bring about needed transformative change in climate policy, then something isn't working.

In researching this book, I interviewed numerous politicians and political insiders — specifically those who I believe understand the climate science — seeking to gain insight into how and why the new climate denialism prevails in our politics, even among progressive governments. I took a special interest in the federal and B.C. governments' climate plans because they are widely considered the highwater mark with respect to Canadian climate action. Those interviews shed some light on what needs to change, and on why a wartime approach may provide a path out of this frustrating conundrum.

The Federal Liberal Government

After many months of negotiations with the provinces and territories, the federal Liberal government released its Pan-Canadian Framework on Clean Growth and Climate Change in November 2016. It was co-published with the governments of all three territories and all provinces except Manitoba and Saskatchewan (whose conservative governments at the time refused to sign on).

The national plan contains a number of welcome elements. Among them:

- Its centrepiece policy is a price on carbon, starting at $20 per tonne of CO_2 in 2019 and rising annually until reaching $50 per tonne in 2022. The provinces were all encouraged to establish their own carbon price system to comply with this policy (if they didn't have one already). But the "stick" in the plan, euphemistically referred to as the "federal backstop," is the federal government's assurance/threat that if a province fails to bring in a suitable carbon price system, the feds would do it for them.

- A commitment to phase out the use of coal for electricity production across Canada by 2030.
- New methane emission regulations seeking to tighten up on leaks from natural gas production, although questions remain about the industry's self-monitoring, underreporting and thus enforcement of these regulations.
- Commitments to invest in electric vehicle (EV) infrastructure and to encourage the purchase of EVs.
- Help for small off-grid communities to get off diesel power.
- Promotion of new GHG-reduction technologies for industry, backed up by applying the carbon price to any industrial emissions that exceed a certain standard (which is to strengthen over time) set for industries like steel, cement, aluminum, chemicals, oil and gas, metals, and pulp and paper.

Sounds good at first blush. The problem with the plan, at least when it was released, is that it still maintains the same inadequate target set by the previous Harper government — a 30% reduction in GHG emissions from 2005 levels by 2030, a target that falls well below what the IPCC now says is necessary. In the wake of the 2019 federal election, the re-elected Liberal government announced plans to enact a Just Transition Act and to legislate a commitment to hit net-zero emissions by 2050, along with new legally binding five-year milestones to reach that target. This is a welcome development, but the details remain unknown at the time of writing. The Pan-Canadian Plan itself acknowledges that it is merely a "foundation" for meeting the 2030 goal, and that additional policies and regulations will be needed to meet even that deficient target.

During the Liberals' first mandate, numerous arm's-length federal agencies produced critical assessments of the plan's strength. In 2018, the federal Ministry of Environment and Climate Change issued a report estimating that, considering all government climate plans in Canada combined, Canada was on track to lower its annual emissions to 592 megatonnes CO2 equivalent by 2030 (well short of the target of 513 megatonnes).[8] The following year, similar warnings were issued by Julie Gelfand, Canada's former independent environment commissioner,[9] and by the office of the Parliamentary Budget Officer.[10] So far, Canada's GHG emissions have increased in both years since the pan-Canadian plan was signed — 2017 and 2018, the last year for which we have data. The incremental approach is not off to a great start.

Crucially, none of this speaks to Canada's role as an *exporter* of fossil fuels. All the above is to occur even while oil sands production, LNG and their associated fracking and pipelines continue to expand. That's no small omission. The oil sands produced 81 megatonnes of carbon pollution in 2017 alone — more than all the GHG emissions of either Quebec or British Columbia. Were it not for the expansion of oil sands production in Alberta and Saskatchewan over the last two decades Canada's GHG emissions would be on a solid downward track.[11]

What are we to make of this?

I met with federal Liberal Member of Parliament Joyce Murray in the spring of 2019, shortly after she was promoted to Trudeau's cabinet. A veteran politician, Murray represents the riding of Vancouver Quadra.

Unlike many politicians, Murray's bona fides as an environmentalist go way back. Her academic master's thesis tackled climate change in 1992. She and her husband founded a tree-planting company. "I consider climate change to be the biggest threat we have as a human civilization," she shared, and she contends the issue is what motivated her to go into politics. Murray gets the climate crisis, which is precisely why I was keen to interview her.

Yet now, speaking as a member of the federal cabinet, she insists "the current federal government is on track" to meet its Paris commitments. "We are going to continue to record the predicted outcomes of reducing greenhouse gases emissions from the measures we are taking, the measures that we have announced, and we will put in place more measures to close that gap."

Murray argues that real plans are more important than targets. "The more useful conversation is really about what are you preparing to do about it, and are you prepared to use political capital to do difficult things." Fair point. But is the federal government doing that? Beyond the fight over carbon pricing, is the federal government really rallying the public to embrace the transition before us?

No one is saying we have to shut down the oil and gas industry and turn off the oil sands taps tomorrow. Rather, in order to meet the IPCC's targets, we have 20–30 years to fully wind down Canada's extraction of fossil fuels. But green-lighting *new* fossil fuel infrastructure, like major new pipelines or new LNG plants, is another matter. Yet that is precisely what the federal Liberal government has done. New LNG plants will give life to an entirely new fossil fuel industry, and would indisputably drive up the extraction of fracked gas. Murray and her government insist that the Trans Mountain pipeline expansion will have a neutral effect on oil sands production and emissions. But as UBC political science professor Kathryn Harrison argues, "The federal government's own estimate is that the pipeline's annual upstream emissions — i.e., emissions resulting from extraction, processing and transportation of crude within Canada — will be 13 to 15 million tonnes, equivalent to two million cars."

Harrison notes further, "The tar sands have accounted for three quarters of Canada's emissions growth since 1990. It's also the sector that accounts for almost all projected growth going forward." Most importantly, Harrison warns, "The longer-term challenge looms even larger. A pipeline is an investment in long-lasting infrastructure. Yet Canada's 2030 target is just the first step. It will be ever-harder to make the deeper cuts needed after 2030 (if not before) if we chain ourselves to new pipeline infrastructure and associated heavy oil production expected to operate for decades to come."[12]

Murray also asserts the pipeline is needed to diversify our global market for oils sands bitumen and secure a better price. But this too is part of a false-hope narrative propagated by the federal and Alberta governments. As veteran energy analyst and earth scientist David Hughes has demonstrated, the price for heavy crude in Asia is not higher than in the U.S., even before accounting for the higher transportation costs to Asia.[13]

Murray contends, "The approach that our government is taking is not that there will be no more development of any fossil fuel. It is that we need to count those emissions and we will need to make [GHG] reductions [elsewhere] that account for that." The problem with such an approach is that, if one accepts the IPCC goal of reaching carbon-zero by 2050, the math of trading-off one sector of the economy against another starts to break down; there's not much horse trading to be done over nothing.

Ultimately, Murray sees a logic in not moving too fast. "Whether it is a pipeline or LNG, I think if you go too radical on this, I do not think that is the best choice in the long run. I think that risks dividing people, and I really think that the way you go forward on something that is difficult and as complex as this is you have to continue trying to find a way to bring people together on it. That may be frustrating to those who would like to see bigger and more radical moves, but those will put people's backs up, people who are paying the cost or not seeing the opportunities for themselves . . . Is it going to work to be alienating a big chunk of the public? I think that to try to move too fast without bringing people along is to lose."

And so the government has steadfastly not moved too fast.

Tick tock.

But in time of war, we did move fast — very fast. And in the wake of the 2018 IPPC report and extreme weather events, maybe the public is further ahead than many politicians give them credit for, a point to which we will return.

British Columbia's NDP Minority Government

A year and a half into the term of B.C.'s NDP government, in December 2018, the province released its new climate plan — *Clean BC*.[14]

As with the federal plan, it contains many welcome elements, and it is a great improvement over what the province had seen to that point. The plan was enthusiastically endorsed by the B.C. Green Party and widely praised by key B.C. environmental groups. Core elements of *Clean BC* include:

- Maintaining annual increases to B.C.'s flagship carbon tax, along with enhancements to the offsetting carbon tax credit for low-income households. But, refreshingly, the plan de-emphasizes the role and importance of the carbon tax and focuses instead on regulatory policy measures.
- Specific dates by which certain things will be banned or required. For example, the sale of new fossil fuel vehicles will be banned as of 2040, and all new buildings must be net carbon-zero by 2032. Firm dates are good, although these dates are set too far into the future.
- Promises to continue expanding the electric vehicle charging network and to subsidize EV purchases.
- Increased spending for building retrofits, and new rebates for those switching to electric heat pumps.
- The requirement that all natural gas used in buildings must be 15% "renewable" gas by 2030, meaning, the gas must be captured from waste or agriculture rather than extracted from the earth, although it is frustrating that this 15% target is so modest.

Clean BC is, quite likely, the most aggressive and comprehensive provincial or federal climate plan in Canada (hence my focus on it here). And yet, like the federal government, the B.C. government's targets are no longer aligned with what the latest IPCC report says we must do. B.C.'s legislated targets are to reduce GHG emissions by 40% by 2030 (from 2007 levels), and by 80% by 2050. The IPCC now says we need to hit 50% by 2030 and carbon-zero by 2050.

The difference between a 2050 target of carbon-zero versus reducing GHGs by 80% may feel largely academic, given the extended time frame. But the difference matters greatly. The problem with a target of 80% reductions by 2050 is that so many of us — both individuals and businesses — falsely presume that what we do or plan to do can be made to fit in the remaining 20% of emissions room. A carbon-zero target disabuses us of this notion.

As with the federal plan, the B.C. plan only lays out steps to get three-quarters of the way to the province's 2030 GHG reduction target, with a commitment to outline how to close the remaining 25% gap over the following two years (still to be determined at the time of writing). But those next steps are going to be hard.

The lofty commitments of *Clean BC* are not yet reflected in the B.C. Budget, where one must always see if fine words are backed up with real dollars. CCPA senior economist Marc Lee, a long-time analyst of B.C. climate and fiscal policies, calculates the three-year B.C. Budget plan will see the province spend only about 0.1% of provincial GDP on climate-related expenditures.[15]

The framing of the plan is very positive — *Clean BC*! It rightly says our future can look nice, with plentiful employment opportunities, if we do this. It does not, however, communicate that we face a climate emergency. Indeed, the plan never once uses the terms "climate emergency" or "climate crisis" or "climate breakdown."

Most significantly, B.C.'s climate plan is fundamentally at odds with the province's LNG plans. Earlier in the same year that the plan was unveiled, the provincial and federal governments had happily joined industry leaders and some Indigenous leaders in celebrating the final investment decision of LNG Canada, an international consortium led by PetroChina and Shell that is building a massive new LNG plant in Kitimat, B.C. "The largest private-sector investment in Canadian history," both the federal and B.C. governments repeated ad nauseam.

The problem is that the project represents another huge "carbon bomb" — a massive new source of domestic GHG emissions. Just phase one of the LNG Canada project, along with its "upstream" impacts from extracting and transporting fracked gas, will add between 4 and 6 megatonnes of GHGs to B.C.'s annual emissions.[16] All this when the government has committed to reduce *total* provincial GHGs to 12 megatonnes by 2050. These folks aren't making their job any easier.

Our governments have sought to portray LNG as a climate-friendly alternative, helping to reduce coal use in Asia. But there is no reason to believe exported LNG will displace coal; it is just as likely to delay the use of renewables or displace nuclear power in Asia (which would result in more GHGs). And while natural gas is indeed cleaner-burning than coal, this "benefit" does not account for how that gas is extracted, produced and transported. In B.C., about 90% of the natural gas produced is now extracted by fracking. The liquification process itself is extraordinarily energy-intensive, as the gas must be cooled to -162°C. When the entire production process is considered, including methane leaks, today's unconventional natural gas now has a GHG profile very similar to coal.[17]

While the work of landing the LNG Canada investment was largely led by the Horgan government, which sweetened the pot with numerous subsidies and tax reductions well beyond what the previous Liberal government of Christy Clark had offered,[18] the federal Trudeau government also anted up a major subsidy to help secure the deal. About half the much-vaunted LNG Canada investment will actually occur outside Canada, mainly in China, where the steel modules for the plant will be built. The federal government kindly offered to exempt those steel imports from tariffs, a subsidy worth approximately $1 billion.[19]

How are we to make sense of the pace and contradictions of B.C.'s climate policies?

B.C.'s minister of environment and climate change strategy, George Heyman, has a rare Second World War story. His parents were Polish Jews who escaped the Holocaust with the aid of a nearby Japanese vice-counsel in Lithuania, Chiune Sugihara. Sugihara is not as well-known as the German industrialist Oskar Schindler, whose extraordinary efforts to save over one thousand Jews during the Holocaust were immortalized in the Steven Spielberg film *Schindler's List*. But Sugihara's efforts were of a similar nature and of a much larger scale. He is credited with saving the lives of some six thousand Jews by granting them visas to travel to Japan, contrary to the orders of his supervisors. At the height of the crisis, he handwrote as many transit visas in a day as would normally be produced in a month

— a remarkable act of disobedience and risk-taking in the face of an emergency. Heyman's parents were among those saved. They travelled initially through the Soviet Union to Japan, before landing in Vancouver, where Heyman was born.

Take note of Chiune Sugihara. What made him extraordinary was not that he was a good person amidst the chaos — we are surrounded by such people. What set him apart as a hero of the Second World War was a rare willingness to break the rules.[20]

Like Minister Murray, one would be hard pressed to question Heyman's environmental bona fides. While working for the provincial forest service in northern B.C., Heyman became active in his union, the B.C. Government and Services Employees' Union (BCGEU), eventually rising to president in 1999, a role he held for nine years. Under Heyman's tenure, the BCGEU adopted some of the most progressive environmental positions in the province's labour movement. Then, in 2008, he walked away from his comfortable and well-resourced union position and took on a job as executive director of the Sierra Club B.C., a terrific albeit much smaller organization with an uncompromising track record of climate campaigning. In 2013, Heyman made the move to electoral politics, becoming MLA for the riding of Vancouver-Fairview. When the NDP formed government in the summer of 2017, Heyman became the cabinet minister responsible for driving forward the climate file. For the record, I consider him a long-time friend.

And yet, when we met for a formal interview for this book in summer 2019, Heyman's answers to my questions were remarkably similar to Murray's.

Heyman carefully tracks the scientific reports and takes them seriously. But he does not think a fear-based approach to climate is effective, and he is not inclined to use wartime language. "I think if one is truly scared, one gets a little frozen. I do not think that is a productive response. So, I try to think about what is my responsibility, what is in my power to do, and how can I best do it in a way that builds support and understanding."

I asked Heyman about my premise that a gap exists between what the science says we must do and what our politics seems prepared to entertain. He answered, "It is important to understand the politics . . . I am not going to pretend it is not skewed by populism, by weird stuff on social media, by big money and such that can control a lot of opinion shaping. But at its basic form, politics is about what people want governments or politicians to do . . . We are busy finding the way that we can demonstrably model the measures that will get us to our 2030 targets, and then we need to do 2040 and 2050. And obviously they are not separate. It is a continuum. But let us start with 2030 first. And at the same time, we are trying to communicate to people about why that is important to do, that we can do it in a way that is fair and equitable and supports individuals to make changes, and supports businesses to makes changes."

I also wanted clarity from the minister on how and why certain sector-specific targets were chosen within *Clean BC*. Why, for example, did the province legislate that all new buildings would be net carbon-zero by 2032? Why was that target not sooner?

"We used credible companies to model what was possible and that is what we came out with," replied Heyman. "That is what seemed to be feasible."

If that is indeed the case (and I'm not prepared to accept that it is), then we have a problem, because every building built between now and 2032 that is not carbon-zero will have to be subsequently retrofitted to become carbon-zero later in its life — to delay the inevitable is simply to punt the problem and cost to a future building owner. Notably, the City of Vancouver, which has a much more ambitious climate plan than the province, is requiring that all new buildings cease using natural gas or any other fossil fuel for space and water heating by January 2022.

And why is the province, under *Clean BC*, requiring that only 15% of domestic natural gas use be "renewable gas" by 2030?

"Industry told us that they thought they could comfortably do a lower number, and we worked with them to bump the number up," said Heyman. "And if we are able to adjust these times in the future, I think we will. But we did not want to pick a number that industry and others would say, 'We just cannot do that.' Or that it would drive the cost of gas to a very high level."

When I probed regarding the degree of lobbying and engagement with the fossil fuel industry, the minister replied, "I met with them periodically. Obviously staff consulted with them in more depth. But we decided that we could have more success if we found ways to work in cooperation, and in many ways we have . . . I think as a package we found one that they could work with us on."

I asked how important it was to the government that industry would give *Clean BC* the thumbs up. "We wanted [industry generally] to say 'This [plan] is credible; we want to work with government on this.' . . . People react badly if they get worried, and we want this to be a province-wide plan that has broad buy-in. We do not want any one sector saying, 'We cannot do this.' That does not mean that we did not want to have a strong, ambitious, credible plan. I think we have, arguably, the most ambitious plan in North America. Which is not to say that we may not be more ambitious in the future, as we calibrate how successful we are as well as what we are seeing. But we are also trying to model for the rest of Canada how you can have credible, concrete action on climate that is actually good for the economy, that helps diversify the economy, that people will support, and that creates a better today as well as a better tomorrow."

There is no doubt that Minister Heyman, like federal Minister Murray, cares passionately about climate and wants to make a difference. Both could have comfortably retired some time ago. But they remain in the political game, subject to all the frustrations and sometimes the targets of great anger, because they want to help bring about change and advance their respective climate plans.

What is so frustrating, however, with both these federal and provincial plans, is that the level of ambition is nowhere near where we need them to be. Even putting aside the reluctance to speak some hard truths on the future of fossil fuels, nothing is stopping these governments from substantially staffing up their climate action teams,[21] from undertaking much higher levels of climate infrastructure investment and, vitally, from using the regulatory power of the state to drive faster change. The net impact on job numbers would undoubtedly be positive. Yet they have not. These plans — which to date represent the most determined climate programs in Canada — are painfully slow. They do not reflect or communicate a sense of urgency.

Canadian federal and provincial governments have been willing to entertain modest and mainly voluntary demand-side solutions (policies that seek to encourage lower demand for fossil fuels), such as carbon pricing and consumer incentives to switch to electric vehicles and heating. But few governments are prepared to honestly address the supply side — our role as an extractor and exporter of some of the dirtiest fossil fuels on the planet, namely oil sands bitumen and fracked natural gas.

A perceived need to not push the public too hard or too fast serves as a basis for rationalizing a host of contradictory policies and messages. Over and over again, when I asked politicians and political insiders how we can speed up progress, I heard back variations of the rejoinder: "You have to meet the public where they are at, and then bring them along." When I asked Minister Murray what was blocking progress, rather than pointing to the oil and gas industry, or political caution, or some other interest setting up barriers, she replied that the majority of the public is simply not yet there. "Pressure from the public needs to be broader," Murray insisted.

But the notable thing about the leaders we remember from the Second World War is that they didn't seek to meet the public where they were. Rather, *they took the public where they needed to go*.

If our current leaders believe we face a climate emergency, then they need to act and speak like it's a damn emergency. We need them to name it, speak continually about it, and rally us at every turn. Because that's what you do in a crisis.

In a similar vein, the argument made by so many of these "progressive" governments is that only compromise and a middle path can succeed politically. Yet quite aside from the fact that one cannot compromise or bargain with the immutable laws of nature, this justification for incremental and incoherent climate policy is simply not working.

In a speech in Edmonton the night after Trudeau reapproved the Trans Mountain pipeline expansion, Avi Lewis, a co-author of *The*

Leap Manifesto (and, full disclosure, my brother-in-law), summed up the untenable nature of the federal Liberals' strategy thus:

> Trudeau's grand bargain is now essentially dead. And it's not, as the punditocracy insists, because of right-wing provincial governments. This traditional Liberal strategy — to have it all ways, tell everyone what they want to hear, take the reasonable center path between two extremes — is manifestly failing in real time, both in its offer to the public and its ability to get anything done. This is a moment when the sheer contradictions of liberal centrism are on display like never before . . .
>
> We, all of us, have been locked into four years of battling this impossible, absurd argument. The one that says we could only have a national climate strategy if we accepted a new pipeline to export bitumen and increase Canada's oil production and emissions. And we can now say — what should have been painfully obvious at the time — that this was a stupendous strategic miscalculation. Because of course it didn't work. It may have taken a few precious years that we absolutely could not afford, and likely upwards of $10 billion dollars of public money that should have gone to many more worthy things, but I think it's fair to say at this point in history that buying a pipeline and ramming it through unceded territory so that we can have an utterly insufficient market half-measure like a carbon tax . . . it's just not an argument that anyone can make with a straight face anymore. But it cost us four precious years of decade zero. This is what it looks like when politics-as-usual slams into this unprecedented historical turning point for the climate.

Others too are losing patience. Veteran environmental campaigner Tzeporah Berman has spent years leading some of this country's most visible environmental protests, but also working behind the scenes advising on official climate plans in B.C. and Alberta. Over the years she's lauded various government climate plans put forward by governments of varying political stripes, seeing them as necessary steps in the transition process.

But in early 2019, as the National Energy Board once again recommended approving the Trans Mountain pipeline expansion, her patience ran out. "Something deep inside of me snapped," Berman wrote on her Facebook page, in a post she called "Turning the corner on denial." And then she laid out her new commitment:

> I am done with smiling in the face of incrementalism. I am done with massaging communications to be "positive" to deny the horror we are facing. I am done with not calling out small measures framed as "climate leadership" when the house is, literally on fire.
>
> I get that people like Minister McKenna, Prime Minister Trudeau and many of my old friends from the climate movement who work for them think they are "doing what's possible" and that "demand-side policies" are all we can do. They justify themselves by saying the world still needs oil and gas despite the clear IPCC findings that we need steep reductions from all industries and all countries now. Enough. If you truly understand the science you need to — as Greta says — "panic!"...
>
> I will strive to open my heart to create common purpose but I will no longer give an inch of room to incrementalism in the face of disaster. It is not just wrong or too slow or unjust. It is immoral. It is about hundreds of thousands, if not millions, of lives.[22]

Sadly, the political status quo in Canada today also includes a backlash in which, to quote Catherine Abreu of Climate Action Network Canada, "climate action is weaponized by the political right," used as a wedge issue for electoral gain. And we've seen a toxic Yellow Vests counter-campaign that mixes aggressive support for the oil sands with anti-immigrant extremism.

While the progressive politicians cited above may be reticent to invoke warlike language, those campaigning *against* climate action have no such qualms. Alberta premier Jason Kenney has established an energy "war room" to defend his province's fossil fuel extraction. If someone is declaring war on "enemies of the oil sands," will the battle be enjoined? Our federal government desires to appease the fossil fuel industry and resistant provinces — seeking to strike compromises to gain their support. But surely those days are over. In the U.K., once war was declared, the Brits felt compelled to replace Neville Chamberlain with Winston Churchill — a political leader prepared to meet the fight.

"PEACE IN OUR TIME"? APPEASING THE OIL AND GAS SECTOR

In a battle, you need to know who and what you're up against and how they operate.

The new climate denialism at play with respect to the fossil fuel industry operates at two levels. First, there are the industry leaders themselves, who claim to accept the climate science, but then advance merely incremental and market-oriented climate policies that are unaligned with that science. Second, and I would contend more problematic at this late stage in the crisis, are the political leaders and policy-makers who still continually seek to *appease* the oil and gas industry. They want to compromise and win the industry's favour and goodwill, in hopes that the industry does not pull up stakes and take their jobs away with them.

When I asked former NDP MP Libby Davies what she thought blocks meaningful action on climate, she reflected that a key barrier is an assumption in Canada about what constitutes good governance, and in particular that it presumes we have to "appease" (her language, interestingly) the corporate sector and private interests.

Our governments fear the oil and gas companies. They worry that if they fail to grant them adequate concessions, we will suffer the consequences. This climate emergency moment sees too many leaders keen to pre-emptively declare that "peace in our time" has been found with the fossil fuel industry. And just as the lead-up to war saw American corporate interests who advocated neutrality so that they could continue to do business with the Nazi regime, so too does the climate emergency face obstruction from corporations reluctant to give up their profits.

More and more people are seeing these parallels.

On August 3, 2019, in response to news that a federal court judge in the U.S. had found against oil companies for "beclouding" the scientific evidence to prevent climate action, none other than former Conservative Prime Minister Kim Campbell tweeted, "This is precisely why I have said that the oil companies have committed CRIMES AGAINST HUMANITY! All the factors are there: KNOWLEDGE of the truth and DELIBERATE action to CONCEAL ("becloud") the truth to save their profits while preparing to protect themselves! Nuremberg worthy!" The all-caps are Campbell's. It was a remarkable indictment from such a high-profile Conservative, invoking the famous Nuremberg war crimes trials of Nazi leaders at the end of the war. And as minister of justice and attorney general in the Brian Mulroney government (1990–1993), Campbell knows the law.

While others may not be ready to take their criticisms quite as far as Campbell, much of the public is deeply suspicious of the oil and gas sector's motives. In a national poll I commissioned from Abacus Data in July 2019, 2,000 Canadians were asked their opinions about the role of various industries in addressing climate change. By far and away the industry with the poorest public perception was oil and gas companies: only 12% of poll respondents believed the sector "sincerely wants to tackle climate change"; another 33% believe oil and gas companies "say they want to tackle climate change, but don't really want to"; and a notable 35% (a much larger share than for any other industry) said oil and gas companies "are actively trying to block climate action in Canada." (The balance had no opinion.) It seems Canadians are increasingly attuned to the new denialism.

Thankfully, the ways in which the fossil fuel industry has blocked progress on climate action have been well-documented elsewhere. I don't wish to go into detail on the subject, just enough to understand what has to change if we are to successfully align our politics with the climate emergency. (For more on how the oil and gas industry in

Canada practises the new climate denialism, see Appendix II, online at sethklein.ca.)

At a time when we urgently need to plan a managed wind-down of the fossil fuel industry, the oil and gas corporations have poured hundreds of millions of dollars into disinformation campaigns and institutes that sow doubt about the reality of human-induced climate change. And they have sought to focus "solutions" on modest market-based options like carbon pricing, seeking to keep stronger regulatory options off the table.

It is now well established that key oil and gas companies have known and understood how fossil fuel usage was putting the planet at risk since at least the 1970s. As the Exxon Knew campaign has shown, the U.S.-based multinational oil and gas giant — and one of the most profitable corporations in the world — was conducting research into fossil-fuel induced climate change 40 years ago, yet chose to spend millions of dollars spreading disinformation.[23] In the ensuing years, the global accumulation of GHGs has gone from a safe level below 350 parts per million (PPM), to dangerous levels over 400 PPM. In May 2019, a report from the Influence Map initiative calculated that, since the signing of the Paris climate accord in 2015, the five largest oil and gas companies — BP, Shell, ExxonMobil, Chevron and Total — collectively spent $200 million *each year* on global lobbying efforts to delay and block policies to tackle climate change.[24]

In the Canadian context, how the fossil fuel sector exerts its political and economic influence has been detailed in two outstanding books — Donald Gutstein's *The Big Stall: How Big Oil and Think Tanks Are Blocking Action on Climate Change in Canada*[25] and *Oil's Deep State: How the Petroleum Industry Undermines Democracy and Stops Action on Global Warming — in Alberta, and in Ottawa* by Kevin Taft.[26] And its malign influence continues to be daylighted by the Corporate Mapping Project (CMP).

Shannon Daub replaced me as the B.C. director of the Canadian Centre for Policy Alternatives. She is the initiator of the CMP and has served as its co-director, along with University of Victoria sociology professor William Carroll, since the project was launched in 2015. The CMP is a research and public engagement alliance that seeks to investigate and document the power and influence of the fossil fuel industry in western Canada. It is hosted by the University of Victoria and jointly led with the Canadian Centre for Policy Alternatives (B.C. and Saskatchewan offices) and the Alberta-based Parkland Institute. It brings together dozens of academics, researchers and non-governmental organizations (NGOs), with most of its funding coming from the Social Science and Humanities Research Council of Canada.[27]

I asked Daub how the industry seeks to block effective and meaningful climate policy. The CMP's research has led to the identification of different "modes" of corporate power by which policy progress is scuttled:

First, there is *political power*, "which is exercised through influence on the state by lobbying, political donations, and a revolving-door of personnel between government and industry."

Second, there is *economic power*, "the industry's control of capital itself and the commodity chains." The industry's ability to decide where it will locate production means it wields great influence with government, and is able to either promise jobs, or threaten to eliminate jobs if its wishes are not adhered to.

Third, there is *cultural power*, "which is about influencing opinion and ideas and how we think about things." This is manifested in the way that corporations flow money to universities, institutes, think tanks, right-wing bloggers and "astro-turf" groups — fake grassroots groups that are actually fronts for the industry. We see the results of these efforts play out in both the mainstream and social media. "This is much of the realm of the denial industry," explains Daub. "Cultural power is the one that I actually think is the most problematic and the hardest to always see, and the one that we really have the hardest time pushing back against."

And Daub adds a fourth mode, *colonial power*, "by which you have the fossil fuel industry and the state, together, over many, many decades, engaged in the process of dispossession of land and resources," and now encouraging Indigenous communities to "partner" in extraction activities. "The dispossession piece continues in the sense of denying traditional territory rights, like we see with the Wet'suwet'en and the Coastal GasLink pipeline. Dispossession takes more than one form.

There is dispossession of formal territory, but there is also the destruction of territory. So, for example, the First Nations with treaties in Alberta and northeast B.C. have territory rights that have been established through the treaty process, but they are not absolute and they are overwhelmed by the impact of development and development-permitting requests on their lands and the surrounding territory, and by the cumulative effects of industry development poisoning lands, disrupting wildlife migration patterns and impacting water. All of this fundamentally undermines their ability to have actual sovereignty."

"These modes work together and overlap," explains Daub. As a book compiling the CMP's research explains, these modes together constitute a "regime of obstruction" on climate and energy transformation.[28]

The CMP framework helps to explain how, even if a "progressive" government is elected, and even if it bans corporations and unions from making political donations (as we have thankfully seen in recent years), and even if it purports to "get" climate, it remains very challenging to resist the power of the fossil fuel sector. The industry still has many cards to play.

Stunningly, the CMP's research has revealed that much of the Climate "Leadership" Plan of the previous B.C. Liberal government was, quite literally, written in the boardroom of the Canadian Association of Petroleum Producers (CAPP) — in Calgary. The collaboration involved many meetings between senior representatives from over a dozen leading fossil fuel companies, along with senior government officials from B.C.'s Ministry of Natural Gas Development, the B.C. Climate Action Secretariat and BC Hydro.[29]

As for the current B.C. NDP government, Daub reports that "we took a look at the volume of lobbying in the year after the new government was elected, and we looked at the top ten most active lobbyists, and it continues apace."

A November 2019 CMP report dug into fossil fuel industry lobbying at the federal level, under both the Harper and Trudeau governments, and found an extraordinary level of interaction. Over a seven-year period between early 2011 and early 2018, the study discovered the fossil fuel industry recorded 11,452 lobbying contacts with federal government officials — more than six contacts per working day! The fossil fuel sector was found to "dominate the lobbying agenda," and there was minimal difference between the Harper and Trudeau years.[30]

If the Canadian Association of Petroleum Producers — the country's primary oil and gas industry lobby group — is not an outright enemy of meaningful climate action, they are certainly collaborators in obstructing progress, and our governments should cease trying to appease them. There is no "peace in our time" to be had with such people. If our climate policies and actions are not making CAPP and its members deeply anxious and upset, at this late hour, then they are not policies worth having.

LOOSE LIPS SINK SHIPS: ENDING THE NEW CLIMATE DENIALISM IN OUR MIDST

Let us put aside the small and diminishing rump of outright climate deniers that remain in Canada. True, they poison our social media feeds and open-line radio shows and seek to intimidate and silence those who advocate for climate action. But these people are no longer our main obstacle to change. I'm grateful that a dedicated contingent of science defenders takes the time to debunk the myths and claims of these climate deniers and to expose their financial backers. But the good news is that we have largely won this battle.

More troubling and problematic, I believe, is how elements of the new climate denialism find quite wide expression within the public at large.

Even the sizeable majority of us who accept the reality of human-induced climate change are, perhaps for understandable psychological reasons, inclined to seek explanations for why we can't or needn't take more significant action. We are all influenced by a prevailing culture of impossibility — that feeling that we are collectively unable to rise to this task. We mutter misgivings and excuses to ourselves and each other. I have been guilty of this myself. But we should be mindful of how what we say can give comfort to those who would undermine the task of rising to the climate emergency. As the ubiquitous Second World War slogan warned those who might inadvertently pass valuable intelligence to the Nazis, "Loose lips sink ships."

We say that we will switch away from fossil fuel use if and when we see all our neighbours doing likewise. We commit that we will change our transportation choices once the necessary alternative infrastructure is in place (better transit or more EV charging stations). We assure ourselves we would do more if the government cracked down on large industrial emitters. We all practise denial each time we take a flight.[31]

These are all understandable sentiments. After all, none of us wants to be a chump, endeavouring to lower our GHG emissions only to have others undo our best efforts. But these views speak to the value of a wartime approach. People will be more willing to take and support significant actions if and when they see that their fellow Canadians are also being compelled to do so, both our neighbours and industry. After all, a core reality of confronting both the war and the climate emergency is that we cannot do so on our own — they are inherently collective enterprises.

Sometimes we are simply resigned to hope for the best. True, we tell ourselves, climate change will have dreadful impacts around the world, but not so much where I live in this northern country. Maybe if we wait just a little longer, technological innovation will save us, a roll of the dice few of us would be prepared to entertain if nine out of ten doctors told us a particular course of action was needed to save the life of our own child or loved one.

And among the most common excuses we tell ourselves — but Canada has such a small population, even if we stopped emitting, what difference would it make? Let us dispense with this particularly pervasive justification for inaction.

True, Canada's emissions are small relative to China, the U.S., the European Union and India, all of whom have much larger populations than us. But on a per capita basis, Canadians are the highest GHG emitters in the world, we drive the most inefficient vehicles in the world, and Canada consistently ranks as one of the world's top ten overall emitters.[32]

Moreover, these statistics fail to capture our role as *exporters* of fossil fuels, as global peddlers of these addictive substances. And that role should deeply trouble us all. Canada is the world's sixth largest oil producer and the fourth largest gas producer, and production of both is still slated to increase through to the year 2040.[33]

In a 2017 report, Marc Lee, the long-time director of the CCPA's Climate Justice Project, documented the extent of the GHG emissions associated with Canada's fossil fuel exports.[34] His research finds that the GHG emissions associated with Canada's net fossil fuel exports — mainly oil sands exports, but also natural gas and to a lesser extent coal and other petroleum products — are nearly as high as Canada's total domestic emissions. And unlike domestic emissions, which have more or less flatlined, the GHG emissions associated with our exports continue to climb, rising 26% since the year 2000. While Canada may be home to only 0.5% of the world's population, we are the source of 3% of the world's fossil fuel supply. We are punching well beyond our weight, but not in a way that should make us proud.

Then there is all the oil and gas we know to be or is likely underground. Were Canada to extract and export all the fossil fuels in

CANADA'S EXTRACTED CARBON

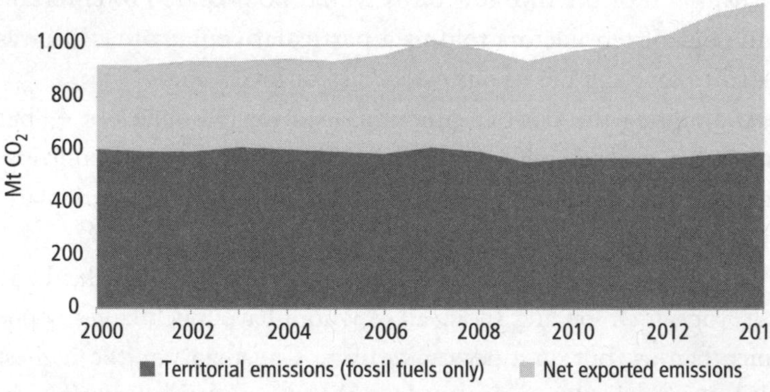

Source: Calculated by Marc Lee, based on data from Environment and Climate Change Canada: Tables-IPCC-Sector-Canada; Statistics Canada: Table 131-0001, Table 126-0001, Table 134-0004, National Inventory Report, Table A6–8, Statistics Canada Merchandise Trade Database.

our proven and probable reserves, the global consequences would be catastrophic. That is why many now understand that Canada's oil sands and a potential LNG industry represent global carbon bombs. That is why we must wind down the oil sands and other fossil fuel extraction, and why the only option for much of our reserves is to leave it in the ground. Lee calculates that if global temperature rise is to stand a 50% chance of staying under 1.5°, and Canada were to stay within its fair share of what remains of the global carbon budget to achieve this goal, then 86% of Canada's proven fossil fuel reserves need to remain unextracted.[35]

By international convention, however, only the GHG emissions associated with the domestic extraction and production of our oil and gas counts towards Canada's GHG emissions. The GHG emissions of everything we *export* is conveniently counted in the emissions of the country where the product is ultimately burned, even though it all ends up in the same shared atmosphere. True, we may benefit economically from those exports, but the "solution," we are told, lies in lowering demand elsewhere. "Blame them."

So said the nations that partook in the slave trade, even after banning the practice in their own countries. So says our own government today, as it sells weapons to Saudi Arabia. And so said the American multinational corporations — such as Ford, General Motors, Dow Chemical, IBM and various banks — that continued to sell equipment to and finance the Nazis, until forbidden to by law.

On the other hand, to extend some sympathy to Canada's leaders, they are caught in a sort of "prisoner's dilemma." Everyone knows that collectively we need to leave most of the world's fossil fuel reserves in the ground, but everyone is afraid that if they stand down, someone else will merely extract more. It's the chump dynamic at an international scale. For this reason, Lee proposes that what we ultimately need is "a framework of global fossil fuel supply management, within a carbon budget," in order to meet our Paris goals. In a similar vein, an innovative new international initiative, co-developed by Canada's Tzeporah Berman, has called for a global fossil fuel non-proliferation treaty.[36] In the absence of an international agreement such as this, countries like Canada simply find themselves in a race to extract as much as they can while they can still get away with it.

But we should not and cannot wait for global agreements such as these to materialize. There is every reason why Canada and Canadians can and must act now. Even better — we can lead.

Canada was an even smaller country at the time of the Second World War, with less economic and human capacity. Canada entered the war with a population of just 11.5 million people, a tiny military, and virtually no wartime production. Yet we did what we had to do, even though victory was far from certain. We did not wait for the United States to join the war but threw ourselves in two years earlier.

And by war's end, no one questioned the role and value of Canada's contribution, to either the battles or the necessary war production.

PART TWO

GALVANIZING PUBLIC SUPPORT AND SOCIAL SOLIDARITY

Canada War Museum

CHAPTER 3

Ready to Rally: Marshalling Public Opinion, Then and Now

"What if we covered the climate emergency like we did World War II?"

— journalist and broadcaster Bill Moyers

"Addressing climate change means fixing the way we produce energy. But maybe it also means addressing the problems with the way we produce stories."

— writer Rebecca Solnit[1]

THE PUBLIC MOOD IN 1939

In the rearview mirror, many of us assume that, at the outbreak of the Second World War, the public was all on board — that everyone understood the threat as a clear and present danger and were ready to rally to our government's call. But that was not so. First, the threat was not immediately obvious to Canadians. It was on the other side of two oceans. Second, the horrors and the burden of the First World War were still a fresh memory for many. Indeed, public monuments to the many lost in "The Great War" had only recently been unveiled[2] and awareness of the terrible cost to families and entire communities remained close to the surface. And, still contending with the long shadow of the Great Depression, people were understandably preoccupied with widespread economic insecurity and unemployment.

There was no public opinion polling data in 1939 from which to draw (the Gallup organization only came to Canada and started undertaking "modern" polling midway through the Second World War). But we know enough from the historic record and media reports at the time to know that public opinion was far from unanimous.

According to historian Jack Granatstein, Canada entered the war in September 1939 "in a state of half-hearted unity,"[3] with public opinion divided on its necessity. For most in English Canada it was simply accepted as a matter of course, though with little enthusiasm, that when Britain was at war, so too was Canada. But many others flatly rejected this view.

When the Mackenzie King government declared war that September, the official opposition Conservatives supported it, as did almost all MPs from the Cooperative Commonwealth Federation (CCF) party, the predecessor to today's NDP. Only the CCF leader at the time, J.S. Woodsworth, a lifelong pacifist, opposed Canada's entry into the war, along with three Liberal MPs from Quebec. And so, a virtually united Parliament would face this threat (a unity very unlike what we see in today's Parliament with respect to climate action).

Outside Parliament, however, it was a different story. Much like today's mixed and ambivalent public opinion on the climate emergency, the public was not yet all on board. Many Canadians were content to be isolationist, and right through the summer of 1939, the Canadian government had signaled its hope to avoid any European entanglements.

In Quebec there was widespread opposition to Canada joining the war, and numerous Quebec newspapers took editorial positions opposing Canada's involvement. "On the Prairies, too, many European immigrants remembered 1917 [and the conscription crisis] and were unhappy," notes Granatstein.[4] Many wondered why

Canada should be the only country in the Western hemisphere to join the war.

Politically, the general membership of the CCF (outside Parliament) was split on joining the war. In an attempt to keep unity within the party, the CCF arrived at an awkward compromise: they would support Canada making economic contributions to the war effort but oppose sending troops overseas. This half-hearted arrangement satisfied pretty much no one, a reality that likely feels all too familiar to NDP supporters today as the party seeks to sustain contradictory positions on climate and fossil fuels. Within a couple of years (after the death of Woodsworth) the CCF shifted and became strongly supportive of the war effort.

Communist Party supporters also initially opposed the war, given that Hitler and the Soviet leader Joseph Stalin had signed a non-aggression pact. Only when Hitler tore up that agreement and invaded the Soviet Union in 1941 did the position of hardline Communists in Canada make a 180-degree shift, and they went from pacifists to war proponents almost overnight. Canadian media reporting about the Soviet Union underwent a similarly abrupt about-face that year. While Hitler and Stalin had their pact in the early years of the war, wartime communications warned of communist infiltrators discouraging mobilization. But as soon as the USSR became part of the Allied cause, public information sources began singing the praises of Russian resistance, organization and fortitude.

Major church denominations were also divided on the matter. The Catholic Church leadership in Quebec kept largely quiet, given opposition within its membership. Within the Protestant churches, "too many clergymen remembered the hysteria of the 1914 war and in particular the role of the church in the 1917 election not to worry about being used again."[5] Some United Church ministers, albeit a minority, spoke out against war.

Many business leaders also opposed entry into war. The First World War experience was fresh in their memory too, and for many, it was an unpleasant recollection of state intervention into their affairs. In the U.S., many business leaders "feared that American entry into World War II might revitalize the left . . . They anticipated that [military] work might not be especially profitable, given that the defense sector was often the target of intense regulation,"[6] a sentiment no doubt shared by some business leaders north of the border.

During the 1930s, in most major Canadian cities, particularly in the province of Quebec, there was even an organized movement of Nazi sympathizers and fascists — small but not inconsequential groups that admired Hitler and were virulently anti-Semitic. In Quebec this movement had, according to RCMP estimates, approximately 6,000 members. There was a Canadian branch of the Nazi Party with a few thousand members (mostly German-Canadians), the largest group of which, with about 2,000 members, resided in Saskatchewan. A fascist Canadian Nationalist Party had a few thousand members, mainly in Ontario. And the Canadian Union of Fascists held a convention in Kingston in 1938 and a march in Toronto.[7]

The understandable reality is that the war had many people who simply wanted to look away, who denied the threat that was unfolding. People are, of course, entitled to hold whatever views they wish, but that doesn't mean our politics and public policy should accommodate them. At today's crossroads moment, our reality is that we've run out of time.

"LET'S GO CANADA!" RALLYING PUBLIC OPINION IN THE SECOND WORLD WAR

Given the state of public opinion in 1939, what is clear is that it took both leadership and circumstances, not least the staggeringly quick defeat of France, to bring the Canadian public on board, for enlistments into the armed services to take off and for the bulk of the Canadian public to truly view our involvement as "Canada's War" (as Granatstein titled his major history of the period). In the U.S., in contrast, it took two additional years and ultimately an attack on U.S. soil before U.S. leaders felt they could declare war with sufficient public support.

The importance of rallying public support for the war in Canada was made all the more critical because Mackenzie King, having seen how the conscription crisis of the First World War had torn apart the country and the Liberal Party, was deeply intent on avoiding mandatory conscription for overseas service. Canada adopted conscription in June 1940 only for "home defence," and so Canada needed to marshal wartime-level enlistment for overseas service on a *voluntary* basis. A huge challenge. (Only after the Normandy campaign of 1944 did pressure finally mount for Canada to send its domestic conscripts for overseas service, and by the time this came into effect, the war was virtually over.)

Yet by war's end, 1.1 million Canadians had enlisted in the armed services (including Newfoundlanders, who would join confederation in 1949). This is astonishing, considering the *total* Canadian population at the time was only about 11.5 million. Thus, nearly one in 10 Canadians volunteered for military service.[8] Of the total enlistment, about half served overseas. These were all people prepared to make the ultimate sacrifice during the war, and more than 44,000 of them did just that. The fact that so many volunteered is extraordinary, and clearly speaks to the successful mobilization of public opinion.

There is no doubt that wartime propaganda played an important role in building that public support, and the Mackenzie King government drew upon numerous agencies to rally the population. Wartime posters, ubiquitous at the time, called upon Canadians to enlist, buy Victory Bonds, save gas, recycle materials that could be used for the war, plant Victory Gardens and more. As in the First World War, wartime artists were sent overseas, including Alex Colville, Jack Shadbolt and Canada's first female wartime artist Molly Lamb Bobak, and their work was sent home to Canada to stir the souls of the public.

In a fortuitous bit of timing, both the Canadian Broadcasting Corporation (CBC) and the National Film Board (NFB) had been established in the years just before the war, and both institutions were now called upon to play their part. The CBC, then just a radio service, offered daily coverage of the war, with reporting from Europe led by senior foreign correspondent Matthew Halton.[9] And in an era when radio and entertainment options were limited, pretty much everyone shared in the nightly experience of hearing those reports.

The NFB produced dozens of wartime films in this period, mostly 15- to 20-minute shorts that could play in movie theatres before a feature film or at other community gatherings, but a few full-length movies as well.

Depending on one's age, the name of Canadian-born actor Lorne Greene conjures up different memories. If you are of my parents' generation, you likely think of the old TV western *Bonanza*, in which Greene played Pa Cartwright. If you are a child of the 1970s like me, you may well think of the original (and very cheesy) *Battlestar Galactica* series, in which Greene played Commander Adama. But before Greene went off to become a Hollywood star, he was a CBC news reader and served in that role right through the war. He was widely known then as "The Voice of Doom," as the Canadian evening news began each night with the latest update from the war, delivered by Greene's fabulous deep and dramatic voice. Greene also narrated most of the wartime documentaries produced by the NFB.

In 1939, upon the declaration of war, the government established the Bureau of Public Information (BPI), with the goal of building a national consensus within English Canada in support of wartime mobilization. (Building support in Quebec was a more challenging exercise.) The bureau's output was certainly propagandistic, seeking to animate feelings of patriotic fervour and to portray Hitler and the Nazis as evil incarnate, lusting for conquest. No one would accuse its advertising or the films it commissioned from the NFB of nuance. As historian and former federal civil servant William Young explains, "The Bureau began by painting the war as a fight to the death against an implacable aggressor. In the second instance, official propaganda played up Canada's material and military contribution to the allies. The final thrust of the Bureau's activities from 1939 to 1942 aimed at creating a sense of 'Canadianism' that would encompass ethnic groups in the English Canadian community. All these efforts, hoped the propagandists, would cement English Canadian attachment to the nation, eliminate domestic conflicts and mobilize the population in support of the war policies."[10]

Midway through the war, in 1942, information officials recognized that the all-too-unsubtle approach of the BPI was proving inadequate — the public's sophistication was greater than the public information bureaucrats had originally assumed, and the BPI's propaganda was increasingly the subject of satire and critique. Consequently, the mobilization and public information function was shifted to a new entity, the Wartime Information Board (WIB). This new agency adopted a more sophisticated and nuanced style. The WIB understood that different groups within Canadian society were differently motivated, and the board developed a new approach "devised by adult educators and social scientists who believed that support for the war effort would grow from a sense of national social goals. Convincing Canadians that the government respected the differences between workers, soldiers and businessmen, for example, became the basis of this effort to create a stronger attachment to the country."[11]

The limited success of early propaganda efforts led some influential Canadians to press for "a different propaganda approach that expressed Canadian nationhood less traditionally," writes Young:

> As early as mid-1940, a group of public servants, members of Parliament and academics, all participants in the Canadian Institute of International Affairs, lobbied for recognition that "a democratic spirit must infuse the wartime instruments of regimentation or the war will be lost on the home front. . . ." For Canada, they concluded, "the dynamic can be found in a common national purpose to create a genuinely democratic society." . . . Across Canadian society, they began to see the idea of a "people's war" catching hold, the belief in the futility of war unless it resulted in a better post-war life for all. Another group with influential connections, the Canadian Association for Adult Education, urged recognition of this public mood and wanted to build "a more dynamic popular conception of the war effort . . . in terms of the new world which can emerge from the war if there is an enlightened and effective national will to that end." By abandoning its preoccupation with patriotism, propaganda could educate the population about "the process through which a better society might evolve."[12]

These ideas, supported by opinion polls, did indeed get incorporated into government information efforts during the latter half of the war, particularly after John Grierson became general manager of the WIB in early 1943. Grierson was a pioneer of documentary film who had been recruited to Canada from Britain just before the war to establish and lead the National Film Board. Many of the NFB's propaganda films during the war were directed by him, including *Churchill's Island*, which won an Academy Award in 1942. In the face of rising popularity for the CCF, these more progressive ideas about how to motivate and mobilize the public — inviting people to think about the more just society that could emerge from the war — were absorbed into the government's post-war plans and the Liberal Party's platform.

As Young explains, "Prime Minister King gradually authorized Grierson and the Wartime Information Board to design a new set of propaganda programmes that would form a national consensus around a new set of issues. Grierson's officers believed that citizens would not support policies unless they could link the war to the fulfillment of basic needs, such as working conditions, health services and housing. This course, they felt, would promote common purpose and an individual appreciation of a relationship to the total national effort."[13] The WIB's output became more educational and informative, and less propagandistic, seeking to provide the public with reasoned explanations for policy choices, such as why wage and price controls were necessary. And the WIB sang the praises of labour cooperation and collaborative management, and of the vital role of women in the war effort. A distinct information program was targeted to those in military service, to engage them in dialogue and debate about core issues facing the country, such as health insurance plans and housing.

In 1941, the Gallup organization set up shop in Canada and began conducting opinion polls under the banner of the Canadian Institute of Public Opinion (CIPO). The WIB soon began contracting with CIPO for their own internal research purposes and would occasionally encourage CIPO to publicly release results that might aid the mobilization effort. CIPO began producing a Canada Speaks column, which they would sell to various newspapers.[14]

All these government and media efforts, combined with the unfolding of the war itself, did indeed have an effect. In 1942, the government went to the public with a plebiscite, seeking permission to be freed from Mackenzie King's 1939 promise that there would be no conscription for overseas service. A slogan of the time captured the government's request: "Conscription if necessary, but not necessarily conscription." 65% of Canadians voted to release the Mackenzie King government from its 1939 pledge — a sizable majority in Quebec was opposed, but a very large majority in English Canada voted yes — a clear indication that by this stage most people were ready to see Canada throw all of its resources into the war effort.

WHERE IS CANADIAN PUBLIC OPINION TODAY ON THE CLIMATE EMERGENCY?

For years, far too much of the political oxygen and polling on climate change has been consumed by the carbon tax/pricing debate. Carbon pricing, while sensible and useful, will not on its own get us where we need to go, and the focus on the topic has distracted us from the scale of action needed. Additionally, too often, polling questions individualize the challenge and solutions, rather than focusing on collective and governmental actions. Past polling has tended to over-test people's willingness to change their *personal* behaviour or to pay a carbon tax. But people increasingly understand that these "solutions" are insufficient. People rightly feel cynical when presented with voluntary solutions that don't match the scale of the challenge or that others around them are not undertaking, including industry. And on those occasions when climate polls have tackled policy questions, they have mostly tended to test incremental options, rather than bold, system-change solutions. The questions we ask, and the solutions we propose, matter.

As I conducted interviews with politicians and political insiders for this book, another question emerged. I kept encountering some variant of the line: "We have to meet the public where they're at, and they aren't there yet." But is that prevailing assumption about the public correct?

And what about the wartime-scale/emergency framework employed and recommended in this book? Environmental communicators and pollsters have historically recommended against using such crisis frames, contending that the public does not respond well to an alarmist or fear-based approach. That's why most official government climate plans never mention the climate crisis, but rather, focus on positive messaging and imagery. Yet I find the wartime frame to be a hopeful one, a source of inspiration and a reminder that we have done this before. And I believe the choice between fear and hope to be a false dichotomy; we are frequently motivated by a combination of both. Certainly, that was the case in the Second World War. Moreover, the reality is that we *do* face an emergency, and we do indeed need a wartime-scale response. All of which led me to wonder whether such framing resonates among the Canadian public and whether there is an appetite for solutions that require systems-level change.

Looking to test these questions and presumptions, I commissioned an extensive national public opinion poll from Abacus Data in July 2019. The results were fascinating. The full results of the online poll of 2,000 representative Canadians can be found on the Abacus website.*

Overall, the results are very hopeful, revealing that *the public is ahead of our politics*. A large share of Canadians is already deeply worried about the climate crisis and increasingly ready for bold and ambitious climate actions. It turns out Canadians support systemic solutions that go well beyond what our governments have so far been willing to undertake.

---

* Abacus conducted this national survey of 2,000 people between July 16 and 19, 2019. A random sample of panelists were invited to complete the survey online from a set of partner panels based on the Lucid exchange platform. The margin of error for a comparable probability-based random sample of the same size is +/− 2.19%, 19 times out of 20. The data were weighted according to census data to ensure that the sample matched Canada's population according to age, gender, educational attainment and region. The full public results from the poll can be found on the Abacus website at https://abacusdata.ca/wp-content/uploads/2019/08/Climate-Emergency-Polling-July-2019-RELEASE.pdf.

In the wake of the 2018 IPCC report (which provides a clear and urgent timeline) and recent extreme weather events, we are witnessing a shift in the public opinion terrain. Consequently, what may not have worked from a communications perspective even a few years ago may work now.

Here are some of the highlights from the Abacus poll:

- **The Canadian public is increasingly worried about climate change.** Three-quarters of respondents said they were worried, with 25% saying they "think about climate change often and are getting really anxious about it," and a further 49% saying they "think about it sometimes and are getting increasingly worried." In contrast, only 19% say they don't think about climate change often, and only 7% either don't believe climate change is real or something for us to worry about. Similarly, 82% rated climate change as an extremely serious or serious problem, second only to the rising cost of living.
- **42% believe climate change is now "an emergency," while a further 20% believe it will likely be an emergency within the next few years** (for a combined total of 62%). Even in Alberta, which registered the lowest level of support for this proposition, a combined total of 47% of people believe climate change is either an emergency or will likely be one in the next few years.
- **People are deeply anxious about what climate change means for the fate of our children and grandchildren.** When asked if climate change represents a "major threat to the future of our children and grandchildren," 81% responded that it does (49% strongly agree and a further 32% agree). Even 67% of Albertans agree with this statement.
- **For a majority of Canadians, climate change is no longer an abstract threat impacting people somewhere else or at some time in the future. They see it happening here and now.** When asked: "To what extent have you or someone close to you experienced the effects of climate change (such as living with the consequences of changing weather patterns or severe weather

events such as flooding, wildfires, droughts or intense heat waves)?" three-quarters of respondents said they or someone close to them had experienced the effects of climate change (13% of respondents said "in a major way," while 37% said "to some extent," and a further 23% said "in a minor way.") Only 21% said they had not experienced climate change at all, while 6% reported being unsure.

- **Canadians are ready for a major transition.** 44% of respondents agreed "in the future, we should produce energy and electricity using 100% clean and renewable sources, such as hydro, solar, wind, tidal and geothermal," while a further 37% support shifting in that direction but don't believe getting to 100% is possible. Even in Alberta these numbers clock in at 28% and 47% respectively.
- **The wartime frame resonates with many.** When asked about the statement "The climate emergency requires that our governments adopt a wartime-scale response, making major investments to retool our economy, and mobilizing everyone in society to transition off fossil fuels to renewable energy," 58% of respondents responded positively (21% strongly agreed while a further 37% agreed). Younger respondents (those between 18 and 44) were even more inclined to agree, with agreement levels closer to 65%. This wartime frame found particularly high resonance in Quebec, with 68% supporting this proposition.
- **People are ready for bold policies to move us off fossil fuels.** The poll listed a series of six major policy moves and asked people if they agreed or disagreed with these actions. The six policies, along with the results, are shown opposite.

These results are stunning. They show that when one combines "strongly support," "support" and "can accept," we find the public's willingness to go along with these bold actions to reduce greenhouse gases ranges from a low of 67% to a high of 84%. As yet, however, no federal or provincial government in Canada has been prepared to move this ambitiously.

SUPPORT FOR BOLD CLIMATE POLICIES

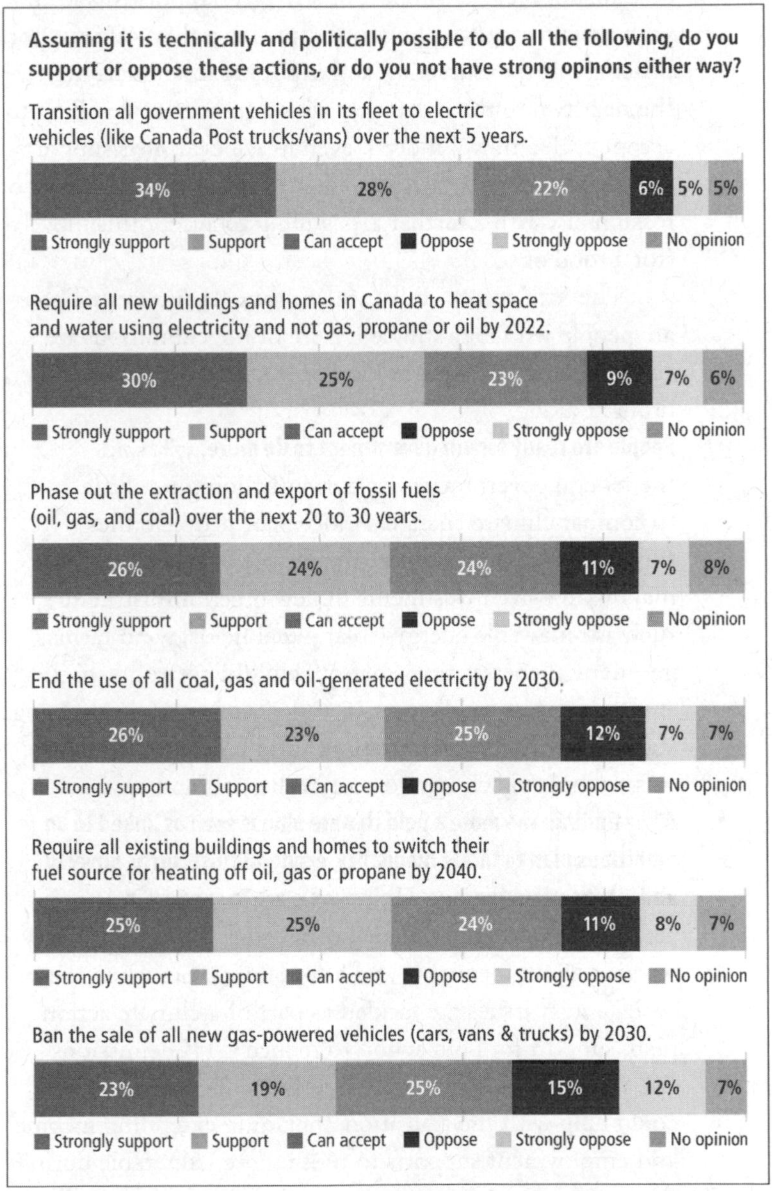

Source: Abacus Data, Summer 2019 Survey of 2000 residents of Canada.

Zeroing in on the policy of banning all new buildings and homes from using fossil fuels for heating by 2022, 55% of Canadians either support or strongly support this idea, with a further 23% willing to accept this policy. 50% support phasing out the extraction and export of fossil fuels over the next two to three decades, with a further 24% willing to accept such a move. Indeed, even in Alberta, 29% support or strongly support phasing out the extraction and export of fossil fuels, with a further 21% willing to accept this move (for a total of 50%).

The "can accept" folks are of particular interest. These are people who may still be unsure of how ambitious we can be, but with the right kind of leadership they could be brought along.

- **People are ready for our government to do more.** 57% said the federal government was currently doing too little to combat climate change. And 75% of people either support or strongly support the idea of "our governments making massive investments in new green infrastructure, such as renewable energy (solar panel fields, wind farms, geothermal energy, tidal energy), building retrofits, high-speed rail, mass public transit, and electric vehicle charging stations, as well as reforestation." (Note the word "massive" was deliberately chosen for use in the polling question.)
- **A key finding: the more a bold climate plan is seen as linked to an ambitious plan to tackle inequality, economic insecurity, poverty and job creation, the more likely people are to support it.** People are concerned about climate change, but they are also very worried about inequality and affordability. So, when these social equity issues are tackled as part of a climate action plan, support for bold action to reduce GHG emissions rises dramatically. The poll listed five policy actions that could help with the transition, including extending income and employment supports to those more vulnerable during the transition, and increasing taxes on the wealthy and corporations to help pay for the transition. It asked people

GOVERNMENT POLICIES IN SUPPORT FOR BOLD CLIMATE ACTION

Now we are going to show you some things governments could do to help the speedy transition of the economy and society away from fossil fuels. For each, tell us whether it is something that would make you more or less supportive of a bold and ambitious climate action plan?

If governments provided financial support to low and modest income households to help them transition away from fossil fuels.

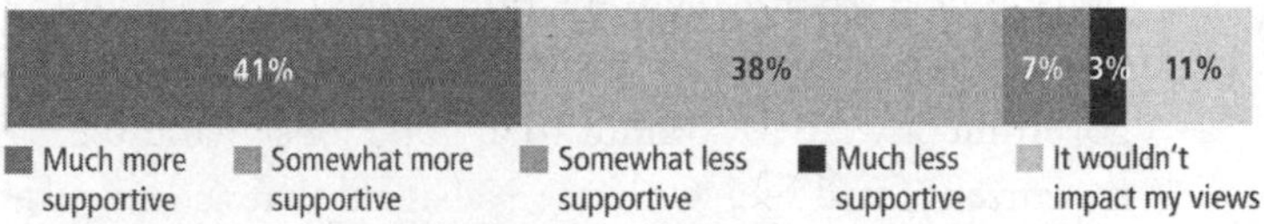

If your own income taxes didn't increase as a result of the plan.

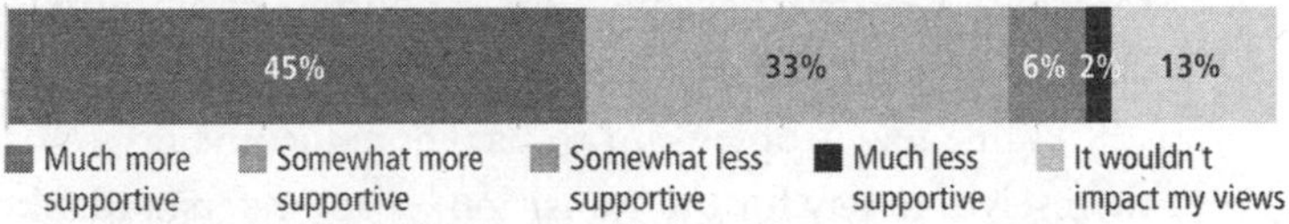

If the wealthy and large corporations were required to contribute more in taxes to help pay for this plan.

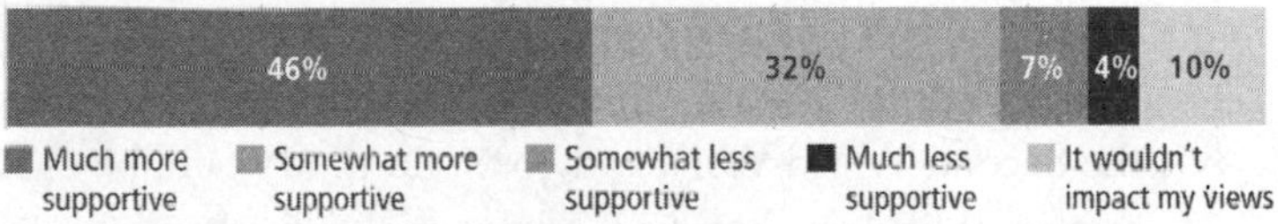

If the government committed to pay a sizeable "climate action dividend" (similar to the GST credit) to all low and modest income households to help offset rising energy costs.

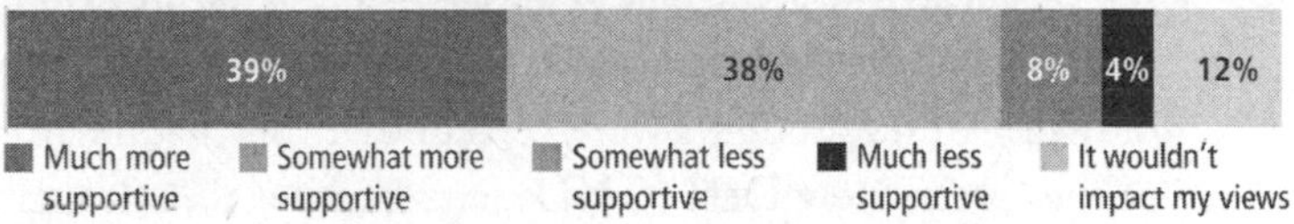

If the federal government committed to a "good jobs guarantee" for all people currently employed in the oil, gas and coal industry.

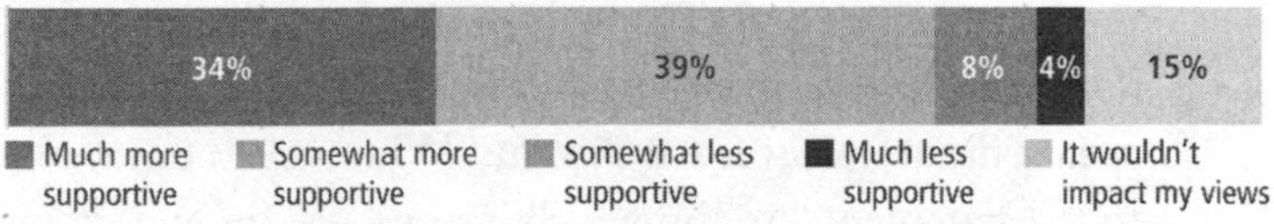

Source: Abacus Data, Summer 2019 Survey of 2000 residents of Canada.

if such policies would make them more or less supportive of bold and ambitious climate actions. Those five policy options and the responses are shown on the previous page.

If government provided financial support to low- and modest-income households to help them pay for the transition away from fossil fuels, 79% of people became more supportive of bold climate action. Similarly, if the government increased taxes on the wealthy and corporations to help pay for the transition, 78% of respondents became more supportive of a bold climate plan. And if the government were to commit to a "good jobs guarantee" for current fossil fuel workers — a signal that the government was ready to actively help with a just transition plan for workers — 73% became more supportive of ambitious climate action.

While few people want to pay more income taxes themselves to pay for the transition — an understandable response given the affordability challenges many are feeling — they are open to helping to pay for the plan in other ways. The poll asked if people would consider purchasing "Green Victory Bonds" (modelled on the Victory Bonds of the Second World War), and 30% said they would be either certain or likely to buy such bonds, with a further 35% saying they would consider it.

- **Few Canadians had, at the time of the poll, heard of the Green New Deal. But once they learned about it, they liked it.** Unsurprisingly, only 14% of respondents were certain that they had heard of the Green New Deal (GND), and another 19% thought they might have. However, after being given a short description of the GND, 72% responded that they support the key principles of a Green New Deal.
- **Nearly half the public understands that Canada needs to be more open to climate refugees and migrants.** When asked to respond to the statement "As climate change progresses and more people are displaced by major weather events around the world, Canada has a responsibility to accept higher

SUPPORT FOR A GREEN NEW DEAL

Here's a definition of the Green New Deal:

"A Green New Deal is an ambitious vision for tackling the twin crises of climate change and inequality. It would see us cut our greenhouse gas emissions by at least 50% by 2030, while leaving no one behind. It would be a comprehensive plan to massively invest in green infrastructure and renewable energy, and to transform our economy to address the scale of the climate emergency and deepening inequalities. It would see the creation of millions of jobs in the areas of economic/energy transition, affordable housing construction, reforestation, and in the caring economy (education, child care, elder care, etc)."

Based on this description, is this something you strongly support, somewhat support, somewhat oppose, or strongly oppose or do you not have clear views?

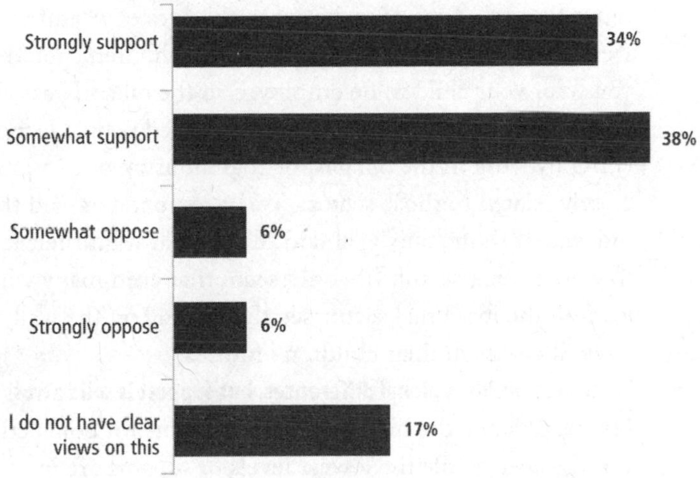

Source: Abacus Data, Summer 2019 Survey of 2000 residents of Canada.

numbers of climate migrants and refugees," 45% agreed that Canada should accept more climate refugees and migrants (14% strongly agreed and another 31% agreed). Only 36% of respondents were opposed to this statement, while 19% indicated they either don't know or had no opinion. While only 45% in support might be discouraging, I expected worse. We are seeing a rise in anti-immigrant views, yet nearly half of us understand that climate change will likely make climate migration a major issue in years to come, and that Canada, in the grand scheme of things, will be geographically lucky and should not respond by pulling up the drawbridge. The strongest level of support for this proposition, at 56%, was among the youngest respondents, those between 18 and 29.

- **Most people don't see a future for their children in the fossil fuel sector.** Survey respondents between the ages of 18 and 65 were asked, "If you have or plan to have children, would you want your child to be employed in the oil and gas industry?" Only 11% said yes. We also asked people if they currently work in the oil, gas, or coal industry, or in a job closely related to those sectors. 5% of respondents said they did, and of those, only 57% said they would want their kids to work in that sector. It would seem that even many who work in the fossil fuel sector see the writing on the wall when it comes to their children's futures.
- **There are notable regional differences, but support is solid across Canada.** Overall, the highest level of support for bold action is in Quebec, while the lowest levels of support are in Alberta. Most of the country falls somewhere in between. But even in Alberta, support for strong policies and action is solid.
- **There are modest but notable differences based on age.** The age cohorts between 18 and 44 years old were generally more supportive of bold action, followed by people over 60. Those age 45–59 tended to have slightly lower levels of support. The fact that millennials (the largest age cohort in

> Canada) are most supportive of bold climate action bodes well for us all. They are just beginning to exercise their political muscle, and what they want and are prepared to hear from our politicians is a harbinger of what will become increasingly possible in our politics.

The trends uncovered by the Abacus survey are confirmed by other recent polling. In spring 2019, climate communication experts Louise Comeau and George Marshall shared results of a months-long opinion research project that had reviewed a large volume of recent climate polls in Canada, as well as undertaking focus group research.[15] They too detected a change in the terrain, with weather events making the reality of climate change more apparent to the public. And they too found a public more prepared to consider and support radical climate solutions.

Comeau also uncovered a disconnect between public opinion and what has been the dominant discussion in Canada. Since at least 2018, the public has been ready to entertain bold, system-change solutions, yet most environmental NGOs and political leaders have been stuck in an earlier mindset, focused on moderate, reformist, and incremental solutions. Indeed, Comeau found that only 5% of mainstream climate-related discussion has been about radical system-change solutions. Some of the fault for that lies with the mainstream media narrowing the scope of debate, but a good part of the blame lies with environmental NGOs themselves (and their funders), who have kept the debate focused on incrementalist options — such as a modest carbon price — and who have spent oodles of money on public opinion polling about those moderate policies. In contrast, until 2019, no polling that I could uncover had tested Canadians' receptivity to climate emergency or wartime-level mobilization frames.

Moreover, the public is often skeptical about solutions that they rightly sense to be overly optimistic or simplistic — people want to hear some frank and hard truths, an acknowledgement that rising to this challenge is going to be hard. And the public is hungry for some *passion* on this subject, even as too many environmental and policy experts have been boring them with dry numbers, economics-speak and making this challenge sound entirely too wonky. (Okay, I may have been somewhat guilty on that score myself over the years.)

Comeau's opinion research, like the Abacus survey I commissioned, also finds particular receptivity for bolder action among youth/millennials, women, and new Canadians. These constituencies want to hear about ethically driven, values-based and caring-based solutions. In the case of newer Canadians, many better understand climate change's global impacts, as they or their family members may have experienced extreme weather events in other countries. And critically, they want to hear a call to action from leaders and spokespeople who look like them, driving home the need to make way for and support more young women and people of colour to lead today's mobilization.

A core takeaway from this opinion research is that our politicians have been underestimating the public. They have failed to take bold action in the face of the climate emergency, insisting the public is "not there yet." But increasingly, the public is ahead of our elected leaders. A solid majority of Canadians are ready to move beyond incremental policies and to entertain truly transformative climate action. Even many of those "in the middle," still wrestling with these ideas, are open to bold leadership. Which is precisely what we need.

RALLYING PUBLIC SUPPORT FOR CLIMATE MOBILIZATION

So, the good news is that the public is closer to accepting wartime-scale mobilization than is frequently appreciated.

But to get there much work remains to be done. Many Canadians are confused and ambivalent. They could be convinced to embrace speedy and transformative action, or not. Many of the questions in the Abacus poll elicited "I don't know/no opinion" responses in the 15–20% range. A sizeable chunk of the public is simply unsure of how ambitious we can be. And so, we need political leadership that seeks to animate the ambition in us all.

A further challenge is that Canadians largely like and believe the "all of the above" approach advanced by the federal Liberals and many provincial governments. For example, a September 2019 pre-election poll from the Angus Reid Institute found 69% of Canadians believe addressing climate change to be a top priority (and more people chose climate change as the top issue facing the country than any other issue). Yet the same survey also found 58% believed oil and gas development should be a top priority for whoever wins the election.[16] And why not? The notion that we can "have it all" — climate action while still benefitting from fossil fuel extraction — is appealing. Turns out the symptoms of the new climate denialism are fairly widespread. The public has, to this point, bought much of the contradiction. What we need now is for our science-aware politicians to stop selling this line.

Polling reveals another core challenge, namely, that the public is increasingly distrustful of and alienated by our politics and public institutions, viewing them as elite-driven and unconcerned with the needs and circumstances of ordinary folks. These are the unsettling yet understandable sentiments upon which right-wing populists prey. A 2019 poll from Ipsos found a majority of Canadians "think politicians aren't concerned with people like them and experts don't understand them." They believe our society is "broken," and 67% agreed that our economy is rigged "to the advantage of the rich and powerful."[17] They aren't wrong. But these feelings make rallying people in common cause under state leadership additionally hard. Consequently, a bold and compelling climate plan faces the additional challenge that it must be convincingly seen as restoring fairness for ordinary people. But as we will see, that's not at all impossible, and indeed is at the heart of the most exciting climate plans now before us.

Ultimately, real leadership means not being overly guided by public opinion research. The point is not to let status quo anxieties and fears define our politics, but rather to see how a transformative politics and inspiring agenda can shift public opinion.

Robert Gifford, professor of psychology at the University of Victoria, in a CBC radio debate with Margaret Klein Salamon, founder of the U.S.-based non-governmental organization The Climate Mobilization, argued the war metaphor is unhelpful. He contends that appeals to sacrifice or dire warnings such as the IPCC report "turn a lot of people off."[18] Klein Salamon, who also has a

Ph.D. in clinical psychology, countered that in the Second World War people were prepared to make sacrifices and accept rationing because "they were for everybody, the rich and the poor got the same amount." She accepts that this task is not going to be easy. It will be disruptive. But she also notes that "the people who lived through World War II on the home front often looked back at it as some of the best years of their lives. They felt productive and meaningfully employed in a cause that was greater than them." She implores us to "tell the truth" about what is now, at this late date, required.

As in the last world war, rising to the climate emergency comes down to a battle for the hearts and minds of the public. How then can support be enhanced for true climate mobilization?

What are the lessons to bring forward from our wartime experience rallying the public? Where is our Climate Wartime Information Board? Where is the government advertising? Where are the stirring speeches? Where's the CBC? And where is the art?

PUBLIC EDUCATION AND ADVERTISING

To begin, we need to increase and improve basic climate education, and we need some sort of public Climate Information Board to lead and coordinate that work. A distressingly large proportion of Canadians lacks basic climate change literacy. For example, in a 2018 survey of 2,000 Canadians conducted by researchers from Lakehead University, only 48% of respondents correctly attributed carbon dioxide and other GHGs as the primary causes of climate change.[19] Too many people do not understand that the combustion of oil and gas in our vehicles, homes and buildings is what causes CO_2 emissions and is thus a major contributor to climate change. They do not realize that a large volume of our electricity in Canada is still produced by the burning of coal or natural gas or diesel fuel, and thus produces GHGs. And they do not know that methane — a particularly potent greenhouse gas — is a major by-product of our agricultural and waste systems and is being constantly leaked in large amounts from our production and distribution of natural gas. The list goes on.

The level of public education on all these matters has been woefully inadequate, and our educational institutions, media and political leaders have not done nearly enough to rectify this. Indeed, in Alberta and Saskatchewan, the public school system is awash in propaganda from the oil and gas industry, while educators who seek to provide lessons on climate change may be chastised.[20] The Canadian public is increasingly worried about climate change, but often people don't understand what actually *causes* climate change. Hence the contradictory polling results in support of pipelines and expanded oil and gas development, even as people rank tackling climate change as a top priority. Nor do most people have clarity about what the necessary solutions are, or that these solutions exist. That needs to be urgently remedied.

Moreover, just as the Wartime Information Board came to appreciate, scientific facts and policy ideas need to be woven together into a compelling vision — something worth fighting for.

The lack of a clear vision holds people back. Those who appreciate the need for climate action are asking people to embrace dramatic change, and yet the picture of our future new life remains quite fuzzy. It is difficult to imagine what our lives and wider society will look like under the kinds of transformative changes needed. And in the current context of economic insecurity, stagnant real incomes and longer working hours, asking people to consider possible sacrifices and further uncertainty is a hard sell.

Given this, a core task to overcome resistance to change is to bring the picture of a new good life into focus, and to answer for the general public some very basic and entirely reasonable questions:

- What will my home and community look like in this new world?
- How will I make a decent living for my family?
- How will I get around?
- Where will our food come from?
- Where will our energy and electricity come from?
- How will I play? (Can I still travel and enjoy my leisure time in a satisfying way?)

- How will we collectively pay for the huge public and private investments that will be needed to get us from here to there?
- What must be sacrificed and why? and
- Who decides? Meaning, will we all be included in a fulsome deliberation about how we achieve the transformation needed, or will the policy agenda be imposed by governments with little appreciation for the economic security of those without political power?

The good news is that answers exist for all these questions, and the picture that emerges of a transformed society can be very attractive, one where we live healthier and more satisfying lives, are more connected to our communities and neighbours and have dramatically reduced poverty, homelessness and inequality. It short, it is a good life, a better life.

Additionally, as climate communications expert Cara Pike told me, we need to show people a clear, step-by-step "action pathway" that will take us from where we are today to that new carbon-free and more just society. And people need to see large-scale examples of the alternatives — massive solar panel arrays, wind turbine fields, banks of EV charging stations on the highways, zero-emission schools and affordable housing buildings.

Of course, all the best information, the clearest answers and the most compelling vision means nothing if people aren't aware of them.

Which sparks the question — where is the public advertising?

Where are the online, TV, radio and print ads from our governments telling us — meaningfully — about their climate action plans? Where are the ads informing us about zero-emission vehicle subsidies and electric heat pump rebates? Where is the public information showing us step-by-step how to convert our homes? Where are the promotional ads inviting young people to train as renewable electricity technicians and installers, building retrofitters, high-speed rail designers, community planners or electric vehicle manufacturers? And where are the public notices advising people that, within a few short years, they will not be able to pipe fossil fuels into their homes or fill up their cars at a gas station, so they can plan accordingly?

During the Second World War, the public was awash in posters and advertisements and notices telling them how to contribute to the mobilization — a coordinated information campaign communicating that a massive collective effort was underway. Where is that today?

Leadership clearly matters in shifting public opinion. But what kind will work for us today?

As I discussed the idea of this book with others, I frequently heard that "people were more deferential to authority back then" — the public was prepared to heed the urgings and orders of its government during the war in a way that may no longer be the case. I suspect there is a fair amount of truth to that view. People feel alienated from our politics today, so leadership needs to come from different sources and take different forms. It needs to be more inclusive and participatory and, well, democratic.

So where are the widespread and innovative community engagements, the citizen assemblies and town halls that invite our fellow citizens to deliberate on the best path forward? People need to feel that they have been able to participate in the transition before us, and not merely had the transformation imposed from on high. There are many models for such innovative engagement exercises, but our governments are not employing them. Indeed, the WIB also came to appreciate this reality in the latter half of the war, and developed ways to involve citizens and soldiers in discussions about how to mobilize, how to support returning soldiers and about the society that would emerge after the war.

We need to do this again today. We need to train a large brigade of well-informed and compelling speakers and educators ready to lead these public conversations in every corner of the country.[21] Our elected political leaders should be leading the charge, but it can't be just them. We need spokespeople who reflect the diverse ethnic and gender makeup of Canada more accurately than our legislatures and government cabinets. Opinion research indicates that people want passionate and charismatic speakers who are seen as ethical, knowledgeable and non-partisan, and who talk less about economics and instead appeal to our collective humanity and values.

And we need different kinds of spokespeople for different audiences. For example, former Ontario Environmental Commissioner Dianne Saxe notes that key to "selling" the phase-out of coal-generated electricity in that province wasn't to pitch the transition in climate terms, but rather to frame it as a health issue, connecting coal power to smog and childhood asthma. The key messengers in that campaign were health professionals.

There are special appeals and messengers that would likely be more effective with small-c conservative constituencies. Conservative-minded people are moved by values of protection and security. Given this, there may be merit in asking military leaders — many of whom have a great understanding of the climate science and its implications — to engage conservative audiences, highlighting the ways the climate crisis is a security threat. And we need faith leaders engaging their constituencies to a much greater extent than we have seen.

But regardless of who these spokespeople are, we need them to communicate a delicate balance of threat and hope, engaging in what my colleague Shannon Daub calls "responsible truth-telling." Our leaders need to speak forthrightly about the emergency because failure to communicate that we are heading towards catastrophe leads to complacency — people won't mobilize if it doesn't appear that our leaders truly believe there to be a crisis. And people need to hear that this transition will involve some hard work. But we also need our leaders and spokespeople to tell us that accomplishing this task is possible — that we have collectively achieved great things before. And that is the power of the Second World War story.

THE ROLE OF THE NEWS MEDIA

Mainstream media outlets and leading political pundits also engage in the new climate denialism.

During the 2015 federal election, journalist and author Linda McQuaig ran as a New Democrat candidate in Toronto. Mid-campaign, she deigned to speak a simple truth about climate and the future of the oil sands. McQuaig said in a CBC interview, "A lot of the oil sands oil may have to stay in the ground if we're going to meet our climate change targets. We'll know that better once we properly put in place a climate change accountability system of some kind." For this great sin she was quickly made the subject of a media onslaught, and her own party distanced itself from her. She was accused of heartlessly ignoring the economic needs and employment anxieties of Albertans. But McQuaig was of course right. The climate science has made it perfectly clear that more than four-fifths of Canada's proven fossil fuel reserves will need to stay underground if we have any hope of keeping global temperature rise under 1.5 degrees.

Similarly, *The Leap Manifesto: A Call for a Canada Based on Caring for the Earth and One Another*, a bold vision to tackle the twin crises of climate and inequality, broadly similar to the Green New Deal,[22] was also released in the lead-up to the 2015 federal election. Over 25,000 Canadians quickly signed their names to the document. But the leading guns of the mainstream media punditry promptly set their hair on fire. The *Globe and Mail* claimed the document calls for "upending of [the] capitalist system." An editorial appearing in the Postmedia chain alarmingly described the manifesto as "a chilling document that suggests pushing Canada farther to the left than has ever been imagined for this country. It is ultimately a plan to completely reject capitalism for something kinder and gentler."

These are but two examples that reveal how the mainstream media and parliamentary press galleries across the land function to narrowly set the terms of allowable debate. And our political leaders, who forever curry their favour, abide it.

The media barrier we face is not merely the space consumed by right-wing outlets and commentators who outright deny the urgency of human-induced climate change (think Rex Murphy, Fraser Institute spokespeople or so much of what finds a home in the *Sun* newspaper chain). It is, more significantly, how the *new* climate denialism finds widespread expression throughout the mainstream media — those who do not dispute the science, but whose unspoken defeatism infects and shuts down the real debate we so urgently need to have.

Mainstream outlets continually amplify the contradictory and incoherent "all of the above" policy-making that has plagued the climate file for years. "No, you don't have to choose. Canada should say 'Yes' to both carbon taxes and pipelines," declared a May 2019 editorial from the *Globe and Mail*. "Canada can lower greenhouse gas emissions while allowing the oil industry to grow," our national newspaper of record assured its readers.[23] The mainstream media largely accepts and promotes the centrist and new denialist contention that we don't have to choose between strong climate policy and continued fossil fuel extraction.

Anyone who has engaged in the mainstream media or on social media on the climate emergency has experienced this dynamic: if you dare to propose bold policy that is aligned with what the science says we must do, you will frequently be chided by some member of the established punditry, who will cynically insist that what you are saying is politically impossible. Speak scientific truth, and get shot down for being politically naive.

The result of this troubling dynamic: the political establishment and punditry become a self-reinforcing circle of diminished expectations.

As all good journalists can agree, words matter. In May 2019, the British newspaper *The Guardian* opened up a fascinating debate within the journalism community. In a welcome move, the paper announced that it was updating its style guide and would no longer be using the terms "climate change" or "global warming," swapping that language for more compelling, urgent and scientifically accurate terminology, such as "climate crisis," "climate emergency" or "climate breakdown."

"We want to ensure that we are being scientifically precise, while also communicating clearly with readers on this very important issue," said *Guardian* editor-in-chief Katherine Viner. "The phrase 'climate change,' for example, sounds rather passive and gentle when what scientists are talking about is a catastrophe for humanity."[24]

In short order, many rightly wanted to know if Canadian media institutions, particularly the CBC, would follow suit. According to a news report on the subject from the CBC itself, "*The Guardian*'s move prompted some discussion at the CBC, and an eventual decision to clarify the public broadcaster's language on the issue. The public broadcaster said use of the words 'crisis' and 'emergency' may be used 'sometimes,' but caution needs to be exercised. 'We never suggested that anyone shouldn't use the words, but we never really articulated their use,' said Paul Hambleton, the CBC's director of journalistic standards. 'The "climate crisis" and "climate emergency" are words that have a whiff of advocacy to them. They sort of imply, you know, something more serious, where climate change and global warming are more neutral terms.' Hambleton said the public broadcaster needs to guard against 'journalism that crosses into advocacy.'"[25]

Of course, we are facing "something more serious," and to fear being seen to imply as much is the new denialism at play. And as for "neutrality," well, in a just war, one generally has to pick a side.

"Surely," I have been asked, "you are not advocating that the news media become the kind of propaganda outlets we saw in the war. And besides, people are more sophisticated today — the kind of propaganda we saw then wouldn't work today."

No, I am not suggesting that our news media become mere propaganda outlets. We want our news to be factual and science-based. But, in the face of a humanitarian emergency and with the fate of civilization as we know it in the balance, there is no virtue in neutrality.

Communications professors Robert Hackett, Susan Forde, Shane Gunster and Kerrie Foxwell-Norton write in their 2017 book *Journalism and the Climate Crisis* that, during the Second World War, media outlets in the Allied countries modelled a form of "patriotic press." "Defeats as well as victories were reported, but there was no pretence of neutrality."[26] They propose a similar approach may be employed when dealing with the climate crisis. Of course, we would not want to see certain elements of wartime reporting — the censorship, the suppression of dissent. But surely we have now arrived at the point where the media should be invoking a sense of urgency, where the volume of coverage should reflect the scale of the emergency, and when distinct news items should be woven "into an overarching narrative" about the climate crisis.

Other experts in this field have urged a similar approach. The American journalist and broadcaster Bill Moyers gave a speech to a conference jointly hosted by *The Nation* and the *Columbia Journalism Review* in 2019 entitled "What if we covered the climate emergency like we did World War II?" "Many of us have recognized that our coverage of global warming has fallen short," Moyers began. He then recalled how, as a child, he used to listen to the radio coverage of the war with his parents, delivered by the renowned journalist Edward R. Murrow and his CBS colleagues. Moyers recounts how, on the eve of the war, CBS headquarters in New York felt there had been "too much bad news" from Europe, and directed Murrow and his European bureau colleagues to produce a song-and-dance feature to lighten things up. Murrow and his fellow reporters chose to ignore those instructions, and instead ended up covering the Nazi invasion of Poland. Murrow's reporting from London during the Blitz dramatically shifted U.S. public opinion in favour of joining the war in support of the Allies. Those reporters were, Moyers noted, "On the right side. At the right time. In the right way — reporting on the biggest story of all, the fight for freedom. For life itself." And we remember Murrow not as a biased advocate or propagandist, but as among the greatest journalists of the 20th century. Moyers implored an audience containing many young and soon-to-be journalists, "Can we get this story right? Can we tell it whole? Can we connect the dots and inspire people with the possibility of change? What's journalism for? Really, in the war, what was journalism for, except to awaken the world to the catastrophe looming ahead of it?"[27]

We desperately need the media, in its climate reporting, to link weather-related events to both their causes (rising CO_2 levels) and solutions. Such reporting has, of late, been on the upswing. But beyond that, should the news media not be advocating for our collective survival and urgent action? Is it not reasonable to expect that the media would seek to rally and mobilize us, as they did in the war? And should climate denialism not be, if not ignored, called out for what it is, including when it takes the form of the new climate denialism — meaning, a policy response that fails to align with what the science says we must do?

Sean Holman, a journalism professor from Mount Royal University in Alberta, penned an open letter to his Canadian colleagues in 2019, urging them to strengthen their climate emergency reporting and to treat the issue as the crisis that it is. Holman was moved to write after news of a new royal baby eclipsed from the headlines the latest dire environmental warning from the United Nations. He beseeched his fellow journalists, "We are responsible for ensuring Canadians have the information needed to make the rational and empathetic decisions that are supposed to underpin our political and economic systems, whether that's at the ballot box or the checkout line. And we are further responsible for exposing public and private institutions when they are harming Canadians with their actions or inactions. By that standard, the climate crisis should be the biggest story of our time."[28] To which I would add, the climate reporting we do get is rarely connected to any sense of *agency* — ideas of what people can do to change the current reality and confront the crisis.

If the public does not, in sufficient numbers, have basic climate literacy — and they don't — then our media have not done some basic work. They should be pressed to do their jobs.

Thankfully, there are now a number of online news media outlets in Canada that have indeed risen to this challenge — the *National Observer*, *The Narwhal*, *The Tyee*, *Rabble*, *Discourse Media*, *Unpointcinq* and *Ricochet* all come to mind. We should all support these publications with our subscriptions and donations, and the government could better support these outlets by making donations to independent non-profit news media tax-deductible. An international consortium of over 250 media outlets that seek to sound the climate emergency alarm have joined forces in an initiative called Covering Climate Now, all committing to boost their climate reporting. Its Canadian partners include the *National Observer*, *The Tyee*, the *Toronto Star*, *Maclean's*, TVO and a number of university papers and other smaller media. But by and large, so far at least, our mainstream broadcasters and print providers are not where they need to be on this defining issue of our time.

In the Second World War, the CBC played a particularly vital role in informing Canadians about the crisis and in mobilizing the public. Today, in comparison, the CBC's coverage of the crisis is weak, although it has improved over the past couple years. The CBC morning radio shows across Canada have *hourly* sports and business reports. Surely then, the CBC can make room for a daily morning update on the climate emergency and efforts to confront it. And just as Lorne Greene and Matthew Halton brought Canadians up to speed on the war effort every night during the Second World War, today's flagship CBC news shows *The National* and *The World at Six* can do so again. And as *The Guardian* has adapted their language, so should the CBC. That is the duty of our public broadcaster, and we should demand it.

In a small but welcome step in this direction, in October 2019, the CBC radio morning show out of Victoria, B.C., began inserting into its stock market reports a "daily CO_2 reading." Each morning they

Weekly averages
11 April 2020: 416.45 ppm
This time last year: 412.67 ppm
10 years ago: 391.12 ppm
Pre-industrial base: 280
Safe level: 350

Atmospheric CO2 reading from Mauna Loa, Hawaii (part per million). Source: NOAA-ESRL

Scientists have warned for more than a decade that concentrations of more than 450ppm risk triggering extreme weather events and temperature rises as high as 2C, beyond which the effects of global heating are likely to become catastrophic and irreversible. Read more about our weekly carbon count

The Guardian

provide the latest carbon parts per million compared to the previous year. *The Guardian* now does the same.

But given how much more fragmented the news media is today — we no longer gather around our radios and TV sets to hear and watch the same shows — we cannot rely upon the CBC alone within the broadcasting universe. The private broadcasters, however, are for the most part missing in action in this urgent task. They too should be pressed into service by their audience and, if need be, by public regulators. The Canadian public regulator, the CRTC, could demand that reporting be scientifically factual (spreading misinformation should not be allowed under broadcast standards). It could also require certain time allotments for vital climate emergency information, as occurs with political programming during elections or other emergency alerts. Public subsidies to the media could be limited to those outlets that provide needed climate emergency information.

Public pressure to improve the climate reporting of private/corporate media is already emerging. In summer 2019, members of the climate emergency group Extinction Rebellion, which focuses on non-violent civil disobedience tactics, staged a "die-in" demonstration outside the *New York Times* that blocked traffic and led to 70 arrests, demanding that the opinion-setting newspaper "tell the truth" and adopt the climate language now used by *The Guardian*.[29]

Let's have the newscasts on the CBC and other networks show us maps each night with the latest climate attacks on our soil, so that we may know the enemy. Let's see the maps of the proposed new fossil fuel infrastructure projects, so we can visualize the scale of what we are up against and the industry's territorial acquisition, and strategize about how to roll it back (see maps on the next page).

Let us also hear much more about what other countries are doing in the battle against the climate crisis — many of which are acting with more determination and focus than is Canada — so that our sense of what is possible may expand. And advocacy be damned — invite us to join the fight to defend the well-being of our fellow humans and their descendants, at home and abroad, in alliance with the millions of other good people around the world who are doing the same where they live.

TOP 10 CLIMATE DISASTERS IN CANADA IN THE 2010s

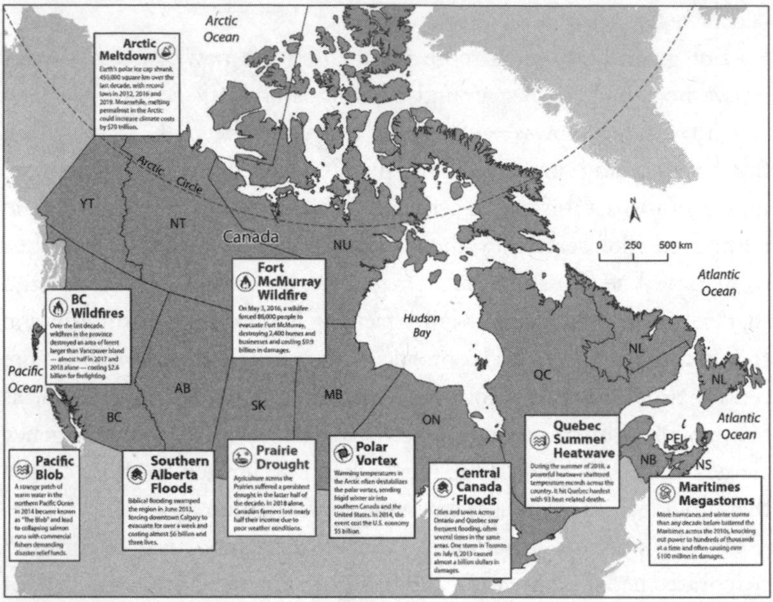

Source: Wilderness Committee.

CANADIAN FOSSIL FUEL PROJECTS

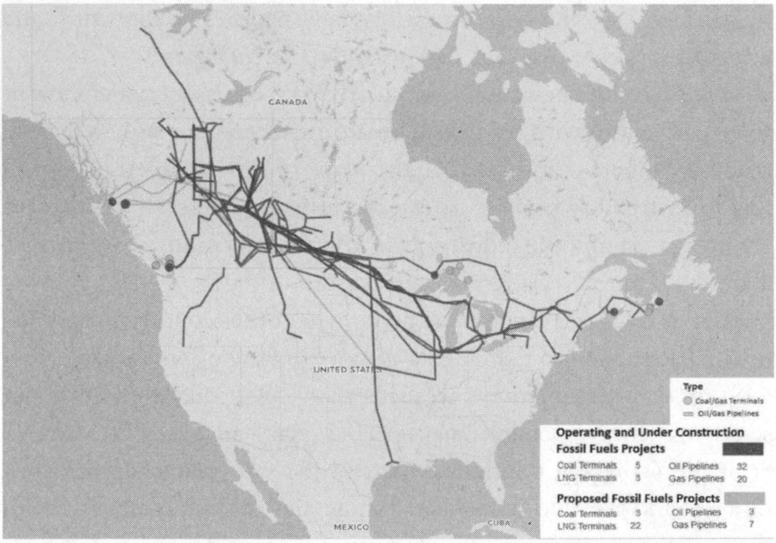

Source: Global Energy Monitor, "Global Fossil Infrastructure Tracker," https://greeninfo-network.github.io/fossil_tracker/.

The frequency and tone with which we see and hear about these matters carries huge weight.

Beyond news programing itself, the major broadcasters and print companies need to change other practices in the face of the emergency, and get their own houses in order. For example, if we are to have confidence that these outlets have chosen the side of humanity, and are not compromised by or complicit with those who seek to block progress, then these media companies should divest from fossil fuels (just as many other institutions and businesses are doing), meaning their portfolios and pension funds should be free of any oil and gas company or related investments. And they should reject any advertising revenues from such companies, as *The Guardian* decided to do in January 2020.

Indeed, as I recently was forced to watch yet another high-production car ad at a movie theatre before the feature film I had paid to see — an advertisement clearly designed to appeal to young people and entice them into the company's attractive domain — I found myself wondering: why are ads for gas stations and gas-powered cars and trucks not banned? We don't allow cigarette ads on TV or radio or in movie theatres, given the known harm these products cause. Why then do we allow ads for the fossil fuel products we know to be a civilizational threat? These ads send a message, even if we don't buy the specific product they are selling. They normalize what must now be wound down and encourage our youth in particular to desire these products. That needs to end.

In their place, we will know that our mobilization efforts are succeeding once we start to see more ads for the things that will speed the transition. But so far, the signs are few and far between, a clear indication that government policies are not yet doing what is needed. Where are the ads for electric vehicles? Or electric bikes? Or electric heat pumps? Where are the campaigns urging us to switch to public transit? I have seen very few. That sends a message too.

ENLISTING ARTS AND CULTURE

Alex Himelfarb began his career as a professor of sociology, but in the early 1980s joined the federal civil service, ended up as a deputy minister and later clerk of the Privy Council — the head of the federal civil service — from 2002 to 2006. Quite an accomplishment, especially considering that Himelfarb was born in a German refugee camp after the war, his Jewish parents having survived the Holocaust, before coming to Canada with his family at the age of one. He has given me, and many others, wise counsel over the years. After reviewing a proposal for this book, he sent me the following note: "I have written previously that great change needs different kinds of actors: poets that explain why the change is needed, what the future could look like and, ideally, inspire us to engage; engineers who design the change, the targets and policies and programs to get there; and moral entrepreneurs both inside and outside government who make it happen. I'd say we have plenty of engineers, many but not enough entrepreneurs, and a serious shortage of poets." Indeed.

During the Second World War, the entire entertainment, arts and culture sector was enlisted to help rally the public. Hollywood cranked out movies to inspire people on the battlefront and home front, and movie stars helped sell war bonds to the public. Likewise, the popular music of the time sought to galvanize support for the war effort. That war had a soundtrack. Indeed, so too did the anti-Vietnam war movement of the 1960s and 70s, and the nuclear disarmament movement in the 1980s.

So where is the soundtrack for today's climate movement? There are a few terrific and contemporary climate-related songs out there, and some bands have made it a focus of their work. But the music industry, like the news media, is more fractured these days. Consequently, there is very little that is truly mainstream. It does not amount to what any of us would consider a popular soundtrack for this emergency moment. Not yet, anyway.

Of course, artists cannot be forced to do anything. They are artists, after all, and produce what they are inspired to produce. But they can be encouraged and supported to create content that speaks to the urgency of this time. The centrepiece of the original New Deal in the U.S. in the 1930s was the Works Progress Administration (WPA). The WPA is mostly associated with huge public infrastructure projects — highways, dams, national parks and other large-scale projects that created jobs for the unemployed during the Great Depression. But another central piece of the WPA was public funding of the arts. As Naomi Klein explains:

> The New Dealers saw artists as workers like any other: people who, in the depths of the Depression, deserved direct government assistance to practise their trade. As Works Progress Administration administrator Harry Hopkins famously put it, "Hell, they've got to eat just like other people." Through programs including the Federal Art Project, Federal Music Project, Federal Theater Project, and Federal Writers Project (all part of the WPA), as well as the Treasury Section of Painting and Sculpture and several others, tens of thousands of painters, musicians, photographers, playwrights, filmmakers, actors, authors, and a huge array of craftspeople found meaningful work, with unprecedented support going to African-American and Indigenous artists.[30]

It's time the climate emergency included a revived public arts program such as that, and our existing arts funding bodies, such as the Canada Council for the Arts, should make climate emergency projects a renewed focus of their support.

Vanessa Richards, a singer, choir leader and community arts programmer in Vancouver, upon hearing of my book idea, thought immediately about the role of arts and culture in rallying public mobilization in the war, and her mind turned to how the arts could once again be pressed into service in the face of the climate emergency. In particular, Richards zeroed in on the potential to marshal young artists. She notes that many young people are extremely tech savvy, skills that could be usefully employed in mobilization efforts today. Young people are already producing vast quantities of viral memes, popular YouTube videos and gaming content, and are increasingly bringing climate-related themes into that work.[31] Richards also laments that too few climate resources have been well-translated for a youth audience.

A body of climate-themed art is slowly emerging. *Globe and Mail* arts reporter Marsha Lederman, in a lovely personal essay, compiled Canadian examples of such art. She found "works of art examining this issue on a deep and meaningful level that may have the power to change minds and provoke action . . . After all the information we have received from scientists, and all the warnings from politicians (well, some of them), is it possible that it will be the artists who can save us? If the science hasn't registered, if the economics haven't resonated — both of which have been laid out in chilling clarity — maybe the arts can be the planet's white knight."[32]

The climate-related art and literature we've seen to date, however, has been mostly dystopian — terrifying and apocalyptic visions of what our world will become after our failure to act. But as Welsh novelist Raymond Williams implores us, "to be truly radical is to make hope possible rather than despair convincing." We need art that expands our political imagination.

That was the spirit behind a compelling short video created and illustrated by artist Molly Crabapple in 2019 called "A Message from the Future," narrated by the inspirational U.S. congresswoman Alexandria Ocasio-Cortez (AOC). Written by AOC and Avi Lewis, and produced by Naomi Klein and the online newspaper *The Intercept*, the video imagines a world two decades into the future and tells the tale of how a successfully won Green New Deal transformed the United States. Within 24 hours of the video's release, it had been viewed over four million times, a testament to people's hunger for hope and for a positive vision of what a new, just and zero-carbon society can look like. If you haven't seen the video, treat yourself to it now.[33]

More of that, please.

Some see hopeful climate poll findings like the Abacus ones, and they worry such positive results may not hold. Those who have been at this for many years have seen support for environmental action ebb and flow over the decades, generally in reverse relationship to the state of the economy.

B.C. Minister of Environment and Climate Change Strategy George Heyman has been engaged with the climate file for many years. He worries, "I remember different times when the environment polled really highly, and then did not. And I think in hard economic times, those [environmental] things drop, jobs rise to the top . . . I totally understand why my neighbour is just thinking: Am I going to have my head above water next week? Or is my plant going to shut down? Or am I going to get a raise that will help me keep up with inflation?"

Yet therein lies the solution.

The answer is not to treat these concerns separately, or to allow them to forever rise and fall in inverse relationship, but to link them and tackle them together. A vital dimension to the recent opinion research is that, in addition to people's growing concern about climate change, they are also increasingly distressed about affordability, employment security and inequality. The Abacus results show that concern about climate change is very high, second only to concerns about the rising cost of living. When asked if taxing the wealthy and corporations, or providing financial support to lower-income Canadians, would increase people's support for bold climate action, a resounding majority said yes. A bold climate program then that is tied to fairness, equality and affordability finds maximum resonance and durability, which helps to explain the appeal of the Green New Deal. We can best motivate and rally people with a big and bold transformative program that tackles the twin crises of climate and inequality.

Which is why, as a next step in mobilization, the next chapter takes a closer look at the role of inequality.

THIS IS OUR STRENGTH
LABOUR and MANAGEMENT
are pooling their strength to give us the means for victory in war · · and progress in peace.

Toronto Public Library

CHAPTER 4

Making Common Cause: Inequality, Then and Now

"I have endeavored to snatch from the exigency of war positive social improvements. The complete scheme now proposed, including universal family allowances in cash, the accumulation of working-class wealth under working-class control, a cheap ration of necessities, and a capital levy (or tax) after the war, embodies an advance towards economic equality greater than any which we have made in recent times. There should be no paradox in this. The sacrifices required by war direct more attention than before to sparing them where they can be least afforded."

— John Maynard Keynes, in the preface to his short book *How to Pay for the War: A Radical Plan for the Chancellor of the Exchequer*, published in early 1940

"If climate policies don't take into account inequalities and differing resources, they will likely make things worse for vulnerable people — those who have done the least to contribute to the problem. Instead, a climate justice approach seeks win-win outcomes spanning employment, health and well-being, and systemic changes that reduce emissions across society."

— Marc Lee, senior economist with the Canadian Centre for Policy Alternatives

WARTIME MOBILIZATION CONFRONTS INEQUALITY

A successful mobilization requires that people make common cause across class and race and gender.

As the Second World War got underway, one of Prime Minister Mackenzie King's foremost objectives was to avoid mandatory conscription for overseas military service. This presented the Mackenzie King government with a formidable challenge — how do you convince hundreds of thousands of people to *voluntarily* enlist, to offer up their lives without duress?

To succeed in such an endeavour, a vital ingredient is social solidarity — a shared belief that *everyone* is in the fight. People need to feel confident that all are contributing according to their means, and making sacrifices in equal measure — the rich as well as the poor. And equally important, people need to know that the society that will emerge after their efforts will be a just and fair one.

The social divisions fuelled by economic inequality, racism and sexism are toxic to such efforts.

The question of income inequality and class figures importantly in the Second World War story. As war mobilization efforts commenced, the governments of Canada, Britain and the U.S. were all concerned with preventing the type of outrageous war profiteering that had marked the First World War and eroded social solidarity generally and undermined military recruitment efforts specifically. It would not do to have some people enlisting to fight and die, while others made a killing from a run-up in prices or from lucrative defence contracts. Britain was therefore quick to pass both an Anti-Profiteering Act and an Excess Profits Tax.

Canada did likewise. The Mackenzie King government instituted both wage and price controls during the war. Far less known is that Canada also imposed controls on profits. The profit margin on all government defence contracts was generally fixed at 5–10% — that's certainly a nice guaranteed margin, but also a cap. And it wasn't just gouging on government contracts that Parliament sought to limit: the First World War highlighted that profiteering could happen in any number of industries, as war produced scarcity and new demands and drove up prices. So, in the interests of ensuring solidarity across classes, *all* profits across the land were capped. Amazingly, under the excess profits tax, for all businesses and corporations in Canada — large and small — the government calculated the average profits during the four years prior to the war (1936–1939, of note, still Depression years), and this became a business's maximum annual allowable profit for the duration of the war. All profits in excess of that cap were taxed at a marginal tax rate of 100%.

The excess profits tax was constructed with care. Corporations could claim depreciation against taxes for plant renovations, machinery acquisition and other expenses. And businesses were promised that 20% of the tax on excess profits they paid would be returned to them after the war, which the government hoped would help with investments needed post-war as plants reconverted to peacetime production. In 1941, Finance Minister J.L. Ilsley said that, "No great fortunes can be accumulated out of wartime profits." Similarly, in a speech to British Columbia lumbermen, forestry baron H.R. MacMillan said, "We must kill off that hangover from the last war — great profits. There can be no profits in this war to capitalists, labour or anyone else. Instead, there will be a sharing of losses."

Looking back some 80 years later, this leadership seems hard to imagine. It is virtually impossible to conceive of the corporate sector abiding such brash measures. Yet as the above quotes illustrate, businesses during the Second World War not only acquiesced to such policies, but some of the country's leading businessmen actively defended them to their peers. While this history is not known to most Canadians today, it is known to some of the titans of Canadian business in the present. My source for this remarkable bit of economic history is a 2005 paper on Canada's wartime production by historian Jack Granatstein commissioned by none other than the Canadian Council of Chief Executives.[1] Clearly, there was a sense of common cause in the war that is not in evidence today. Not yet anyway.

When I've spoken about the Second World War comparison employed by this book, I sometimes encounter the rejoinder "but society was more cohesive back then." Yet interestingly, the state of inequality in Canada (as with our U.S. and U.K. allies) in the years before the war was remarkably high, with over 15% of income going to the wealthiest 1% in all three countries. In fact, as the accompanying chart shows, the share of income going to the top 1% in Canada reached its highest point of 18.4% in 1938, the year before the war. But as also seen in the chart, the war fundamentally jolted income distribution and marked the start of a three-decade period of much greater income equality. Since the mid-1980s, however, all three countries have witnessed a U-turn, with the share of income going to the richest 1% moving back towards pre-war levels (although Canada has seen a slight decline since the 2008 recession).

TOP 1% INCOME SHARES

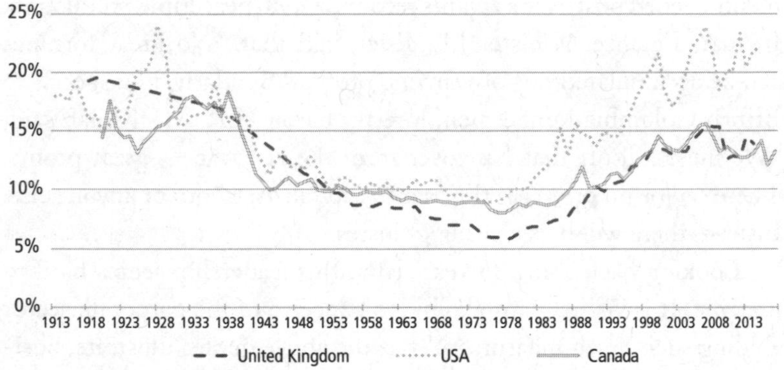

Source: World Inequality Database: Paris School of Economics; for Canada after 2011: Statistics Canada: 11-10-0055-01.

Notes: Pre-1950 U.K. data is limited to only some data points. World Inequality Data based on pre-tax market income. Statsitics Canada Data also based on pre-tax market income (the sum of earnings, net investment income, and private retirement income before tax).

It is striking that in all these countries, early mobilization efforts confronted a reality of stark income inequality — just as we do today — yet all emerged from these transformative experiences as more equal societies.

The Second World War didn't merely shift the distribution of income, it shook up gender and racial relations as well. The war saw a transformation in the lives of women, who poured into the paid workforce by the hundreds of thousands to work in munitions factories.

And while it was primarily men who enlisted into military service, thousands of women did so too, taking on various non-combat roles (the only ones open to them at the time). The war years also saw, albeit only briefly, the introduction of public child care to facilitate the movement of women into needed paid work.

While the war was certainly transformative for women, this should not be overstated. Ruth Roach Pierson and Marjorie Griffin Cohen write of women's labour force experience during the war, "One of the myths about World War II holds that it broke down the sexual division of labour and removed the sexual barriers to occupations." As they explain, even in factory production, clear sexual divisions persisted, with women consigned to distinct roles and provided with a lower level of skills training. And while women's labour was certainly needed in the war, economic planners assumed that after the war women would return to domestic work, even if a majority of the women engaged in paid work during the war had different aspirations.[2] To be clear, the society that prevailed during and after the war remained deeply sexist. But it was also changed.

Canadians from many ethnic backgrounds enlisted. Solid numbers are hard to come by, but Veterans Affairs Canada reports that several thousand Black men and women enlisted in the Canadian military during the war, and it is estimated about 600 Chinese Canadians signed up. Approximately 4,300 "status Indians" enlisted, as did many more Indigenous people who were Métis or non-status.

There is no doubt that racism persisted in the military. In particular, non-white Canadians were effectively excluded from the air force and navy until late in the war and could only enlist in the army. And racial discrimination marked who served as officers in the military hierarchy. Yet those racialized people who chose to sign up often describe experiencing much less racism within the military than in the society they left behind. Indigenous soldiers often recall being subjected to less racism during their time in the military than in civilian life in Canada. Similarly, Chinese Canadian veteran George Chow insisted in my interview with him that "there was no discrimination at all. During the five years I spent in the army, I was just treated as one of the whiteys, you might say." That was likely not everyone's experience. But the reality of serving in combat together was surely equalizing and forged bonds that superseded the intense racism of wider society.

Canada's war experience fundamentally transformed people's expectations of what they were due and what could be accomplished together. A public that had made huge sacrifices was sure as hell not going to return to the conditions that had prevailed in the 1930s. As Granatstein writes:

> The casualties, the sacrifices, the rationing and shortages and hard work had to bring forth something more than a return to the status quo ante. A country that could double its Gross National Product and increase its national budget ten-fold in five years could also look after its citizens. It was the people's war, and the prosperity it brought in its train fed the conviction that victory had to be won at home too. This could be discerned in the opinion polls, in the new electoral strength of the CCF, in the manoeuvrings of politicians poised unhappily on the crest of a wave. "Bring Victory Home!" was one slogan the Liberals tested in 1944, and a popular one. In the end the government went to the people on the motto "Build a New Social Order," a phrase that evoked the same spirit. In this sense the move left by the political parties, the adoption of social-welfare platforms by Liberals and Conservatives, the implementation of major steps towards the welfare state by the Mackenzie King government — all were reflections of the people's war.[3]

The Second World War saw the introduction of Canada's first major universal income security programs. Needing to secure social solidarity across society, the Mackenzie King government instituted unemployment insurance in 1940 and the universal family allowance in 1944.[4]

Throughout the war, the politics of Canada were turning decisively to the left. In by-elections in 1943, the CCF took two prairie seats from the Liberals, and in Montreal, Fred Rose of the Communist-affiliated Labour-Progressive Party defeated an incumbent Liberal. The CCF also made huge gains in the B.C. election of 1941, the Ontario election of 1943, and in 1944 won a decisive victory in Saskatchewan, becoming the first CCF/NDP government in Canada. A national poll in 1943 by the Canadian Institute for Public Opinion found the CCF slightly ahead of the Liberals and Conservatives, with 29%, 28% and 28% support respectively.[5]

Midway through the war, in 1943, leading CCF intellectuals David Lewis and Frank Scott published a book called *Make This Your Canada*, outlining how and why a democratic socialist country should emerge from the war. Academics Roberta Lexier and Christo Aivalis, in a forthcoming paper about that book, note that Lewis and Scott begin with an examination of the wartime economy and the effectiveness of planning to achieve national goals:

> "If we can find the resources and methods to produce tanks, bombs and bullets," they ask, "why can't we find the resources and methods to build homes, schools and playgrounds?" They answer that: "we can, if we have the courage to refashion our society to serve the interests of all the people. Our economic achievements in this war have demonstrated that our country is capable of producing tremendous wealth and that national planning toward a united national objective is indispensable to the creation of that wealth."

During the federal election of 1945, just before the end of the war (which would return the Liberals to power but with a much-reduced majority), the election returns from those in the military fighting oversees indicate that more of them voted for the CCF than for the Liberals, with the Progressive Conservatives a distant third.[6] It was a sure sign that returning soldiers would demand a more just society. Similarly, in the U.K., no sooner was the war over than the British electorate promptly replaced Churchill with the most activist Labour government in that country's history.

The Mackenzie King government, keen to defend itself against the CCF threat, made social welfare reform and a promise of a more equal and caring society a foundation of its post-war promise. Even in the early years of the war effort, planning was already underway for post-war reconstruction. And it was during the war that one of the country's most transformative social welfare plans was written — the Report on Social Security in Canada, more commonly known as the Marsh Report. It was named for its lead author, economist Leonard Marsh, who earlier studied at the London School of Economics under Sir William Beveridge. Both Beveridge and Marsh wrote landmark wartime reports that set out a new progressive social welfare vision for their respective countries. Marsh, who was connected with the League for Social Reconstruction (in some ways, an earlier incarnation of the CCPA), proposed in his report that Canada adopt a comprehensive program of social security, with enhanced unemployment benefits, transition support for returning soldiers, major public works projects to provide employment, child benefits, old-age benefits, maternity leave benefits and health insurance.

Marsh's basic contention was that these programs should be provided and paid for by the country as a whole. He argued that basic benefits were not only important to collective solidarity and family health, but that ensuring income support for people in need would maintain purchasing power and economic prosperity after wartime production wound down. His case for new social programs like child benefits, old-age benefits, and health insurance would become the foundation for the new social policies that would mark the post-war period — the Family Allowance, parental leave, Old Age Security and the Canada Pension Plan, and of course Medicare itself. The experience of collective mobilization would transform our society for generations, even if the actual establishment of these programs did not materialize until some years after the war.

Marsh's report had enormous influence. "The Speech from the Throne on January 27 [1944] was a landmark in the development of the social-security state in Canada," writes Granatstein:

> The "post-war object of our domestic policy is social security and human welfare," the Governor General read. "The establishment of a national minimum of social security and human welfare should be advanced as rapidly as possible." The government pledged itself to guarantee "useful employment for all who are willing to work," to upgrade nutrition and housing, and to provide social insurance. It said that its postwar planning was concentrating on three fields — demobilization, rehabilitation, and re-establishment of veterans; reconversion of the economy; and insurance against major economic and social hazards.[7]

Even the Conservatives saw the writing on the wall during the war years and understood the new reality of how the wartime effort would transform society. Fearing a CCF government after the war, they too adopted major social welfare planks in their platform and, midway through the war, formally changed their party name to the Progressive Conservatives.

The point in recalling all this — as we face today's threat and the need for mobilization — is two-fold. First, to appreciate how inequality serves as a barrier to cross-society mobilization. And second, to understand that effective mobilization isn't merely about building more planes and tanks, or today, wind turbines and solar panels. It requires policies that fulfill a promise that we will better look after one another, that we will offer good jobs and income supports to all, and that people will be treated with dignity and fairness. When you're asking people to share in a great undertaking, that's how you keep everyone on the bus.

INEQUALITY TODAY AS A BARRIER TO BOLD CLIMATE ACTION

There is a common argument made against linking the need for climate action with inequality and social justice issues which goes: "Why make this any more complicated? Getting society off fossil fuels is challenging enough. So why make the task even *more* difficult by requiring our transition plans to rectify the other injustices of the world?" The Green New Deal and other climate justice campaigns frequently encounter this position, often made by climate policy wonks and even some mainstream environmental organizations.

The rebuttal is two-fold. First, it is only by linking these issues that we win over and mobilize broad popular support. And second, these issues are actually deeply intertwined.

Among my core contentions is that inequality itself, along with economic and job insecurity, is a key obstacle to bold climate action. Its corrosive effect operates at both the concrete and psychosocial level.

At a very basic level, inequality undermines trust that "we are all in this together." Many doubt that the task at hand will be undertaken in a manner that is fair. It is hard to rally the public if many believe the rich are merely buying their way out of making change — fortifying their homes, walling their communities or purchasing carbon offsets in the hopes that others will lower their actual emissions. Equally troubling is a cultural narrative that sees climate action as part of an elite project that sees the poor, or those currently working in the fossil fuel sector, as expendable.

But while inequality makes tackling the climate crisis harder, the reverse is also true — less inequality helps to galvanize climate action. Those countries with less inequality, such as Denmark, Norway, Germany and Sweden, are also where we tend to see the strongest climate policies. British epidemiologists Richard Wilkinson and Kate Pickett, in their groundbreaking book on inequality, *The Spirit Level*, highlight that societies with less income inequality are also more open to embracing environmental actions.[8] Wilkinson and Pickett explore how more equal societies are also those with more social cohesion and trust, and warn that "governments may be unable to make big enough cuts in carbon emissions without also reducing inequality."

At a more practical and material level, inequality is interwoven with the climate crisis because the growing income gap exacerbates climate change, while climate change's impacts and many of the policies seeking to tackle it are unequally felt.

The CCPA's Climate Justice Project is a research alliance dedicated to tackling the two inconvenient truths of our time — climate change and inequality. The work of the project has been guided by an overarching question: what does it look like to dramatically reduce our GHG emissions, but to do so in a manner that reduces inequality and ensures that societal and industrial transitions are just and equitable?

A few years ago, the project's director Marc Lee calculated GHG emissions in Canada by household income. Not surprisingly, as the accompanying chart shows, Canadian household emissions are highly unequal; the wealthiest 20% are responsible for almost double the GHG emissions of the poorest 20%.[9] The richer a person is, the greater their carbon footprint — more disposable income means more consumption, bigger homes, more flights, etc.

But consider the implications of this distribution. Even if those in the richest quintile successfully reduce their household emissions by 30% by 2030 — Canada's current official goal — they will still be emitting more than the poorest 20% of households do today. This

CANADA'S GHG EMISSIONS PER PERSON (2009)

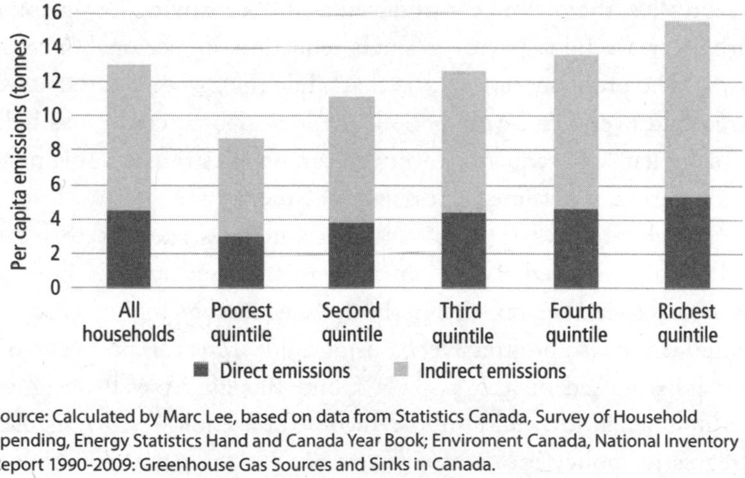

Source: Calculated by Marc Lee, based on data from Statistics Canada, Survey of Household Spending, Energy Statistics Hand and Canada Year Book; Enviroment Canada, National Inventory Report 1990-2009: Greenhouse Gas Sources and Sinks in Canada.

Notes: Direct Emissions include motor fuel and in-home fuel; indirect emissions are those from the production, transportation and use of goods and services that households consume.

surely exposes a fundamental equity challenge and suggests as a matter of justice that higher-income households should bear a greater burden of reducing emissions.

On the flip side, issues of fairness and justice must also be effectively addressed in the design of GHG reduction policies. That's because most conventional climate action policies, in isolation, have regressive impacts — meaning, they hit lower-income families harder as a share of their income. Almost all climate actions have household costs or will increase energy prices in the short- and mid-term, and those costs are harder for low- and modest-income households (not to mention poorer countries) to contend with. That means that redistribution measures — both within and between states — must be at the core of climate action agendas.

This dynamic has also been explored by Lee, who examined British Columbia's experience with Canada's first carbon tax. In 2008, the province introduced a modest carbon tax. The tax started at $10 per tonne of CO_2 in 2008, and then rose $5 each year until it reached $30 per tonne in 2012. (The B.C. Liberal government of Christy Clark then froze the tax for several years, and the B.C. NDP government, as of 2018, has resumed annual $5 increases.) Recognizing that the carbon tax (like any sales tax), in isolation, is regressive, the province introduced an accompanying low-income carbon credit (along with a small reduction in personal income taxes). The problem, however, is that while the carbon tax increased three-fold over a five-year period, the low-income credit was held virtually flat. Consequently, over the carbon tax's initial implementation period, it became regressive over time.[10]

The good news is that policy design flaws such as this can and should be fixed. Lee's paper on this subject models how to construct a carbon tax that is both considerably higher ($200 a tonne) while also progressive, by using a generous carbon tax credit that is patterned on the Canada Child Benefit. A well-designed credit aimed at low- and middle-income households can transform a regressive policy into a progressive one — a policy that simultaneously reduces emissions and redistributes income from wealthier to lower-income people.

The key is to bring an equity lens to the design of all climate actions and to hardwire into our plans a commitment to tackling inequality.

This approach shouldn't be limited to a tax credit. We need to think ambitiously, just as the Marsh Report did back in the war years, and as the Green New Deal invites us to do today. We have a vision with maximum public appeal when our post-carbon society also includes a comprehensive plan to tackle poverty, a bold national affordable housing strategy that ends homelessness, a commitment to living wages and the introduction of the next generation of social programs — universal public elder care and child care, public pharmacare and dental care, tuition-free post-secondary education and free public transit — programs that effectively socialize these costs and removes them from the affordability anxieties of Canadian families.

For years now, various public opinion polls have highlighted a paradox: respondents indicate high levels of concern regarding climate change and strong support for policies that would reduce GHG emissions, yet low levels of willingness to personally pay for climate action.

But there is no great mystery to this mixed response. As the Abacus poll revealed, affordability is the top concern of Canadians, ahead of climate change. That only stands to reason; it is a more immediate concern to most people. Hence the powerful need to link these issues.

When we do so, the appeal is dramatic, far surpassing the levels of support for any one political party. As the Abacus poll showed, when Canadians were given a short explanation of the Green New Deal — describing it as an ambitious vision for tackling the twin crises of climate change and inequality — it proved immensely popular, supported by 72% of Canadians surveyed.

The same trend is evident in the U.S. After Democratic congresswoman Alexandria Ocasio-Cortez first popularized the Green New Deal in late 2018, a poll by the Yale Program on Climate Change Communication[11] found that 81% of Americans, a substantial majority of registered Democrats and Republicans alike, supported it. And similarly, when AOC shared in January 2019 that her main proposal for how to pay for the Green New Deal would be a new top marginal income tax rate of 70% for the wealthiest Americans, a poll, reported by Fox News no less, found that 59% of Americans supported the idea.[12]

Conversely, asking people who don't have their basic needs covered to make more sacrifices or to boldly leap into a new carbon-zero future is often a non-starter, not to mention unreasonable. Failure to link these issues results in resistance, which at the margins now takes a particularly hateful form.

The election of Donald Trump reinforces this important warning when it comes to climate action. If governmental climate actions consign many working-class people to the economic scrapheap, if their economic security is seen as expendable by "elites" making the decisions, if it feels that "not everyone is doing their fair share," then climate action risks heralding the same spectacular political dysfunction we have seen playing out in the U.S. under Trump and in places such as France, where the Yellow Vests movement first appeared, sparked by a climate-linked fuel tax that many rightly saw as unfairly designed. When the Yellow Vest symbol was imported to Canada it morphed into an ugly mix of defence of the oil sands with white nationalism, racism and xenophobia.

We are witnessing a distressing return of fascism, and the emergence of political leaders both in Canada and elsewhere who are happy to give public expression to our ugliest tendencies. And like the 1930s, the rise of the far right emerges from a toxic stew of racism and xenophobia played out against and catalyzed by a backdrop of growing inequality and economic insecurity.

Historically, when we have seen the emergence of far right and neo-Nazi parties, it is frequently the product of austerity and neoliberal policies.

The great economist John Maynard Keynes warned of the dangers of economic abandonment one hundred years ago. In 1919, Keynes, then a youngish economist (and years before authoring his groundbreaking *General Theory of Employment, Interest and Money*) was part of the British delegation sent to negotiate the Treaty of Versailles at the end of the First World War.

But Keynes did not stay. He was so appalled by what he witnessed — by how ordinary German people were being made to pay for odious reparation debts that would cripple their economy — that he quit the meeting in disgust, returned home and penned a short book called *The Economic Consequences of the Peace*. In it, he warned the treaty would not allow the German economy to recover post-war, leaving a population resentful and vengeful. With chilling prescience, Keynes predicted another war. In the years that followed, Hitler's Nazi party capitalized on the humiliations and austerity that resulted.

In the 1980s, the U.K. saw an upswing in neo-Nazi movements under Prime Minister Margaret Thatcher's public spending cuts. More recently, we've seen this dynamic at play in numerous European countries. In Greece, the neo-Nazi Golden Dawn Party made its largest political gains (winning 7% of the vote in 2015) in the face of punishing austerity imposed by Greece's debt holders.

It is worth noting how debt figures centrally in these examples, debt that ordinary people have been made to repay despite stemming from economic choices not of their making. Rising levels of household debt should serve as a warning — a proxy for the economic anxiety weighing upon so many families — here in Canada and elsewhere.

Our governments certainly aren't helping matters when they exempt some of the worst GHG emitters from carbon-pricing policies, or when some of the largest transition subsidies go to some of the wealthiest corporations in Canada. One egregious example was the federal Liberals' 2019 decision, under its clean energy program, to extend $12 million to the hugely profitable Loblaw corporation, owned by the Weston family, one of the richest in Canada, to assist with upgrading the company's refrigeration systems at stores across the country.[13]

And so, as we redouble our climate action efforts, we must also redouble our commitment to climate *justice*.

Inequality is not only a moral problem — it is a practical barrier in getting to carbon-zero. Runaway wealth is associated with runaway emissions, while poverty and inequality — in all their dimensions — will be exacerbated by climate policy, unless this reality is explicitly recognized and tackling inequality is built into the design of climate action measures. High levels of inequality undermine social cohesion and promote social divisions, rather than building the social and political trust needed to chart a future based on a sense of shared fate. If climate policies are not perceived as fair, public support will not be sustained, and political determination will shrink accordingly. The more a robust climate action plan is linked to an exciting plan to tackle poverty and inequality, along with a hopeful and convincing jobs plan, the more we maximize public support. Plus, it's the right thing to do.

Photo of Rt. Hon. W.L. Mackenzie King voting in the plebiscite on the introduction of conscription for overseas military service. / Library and Archives Canada

CHAPTER 5

Confederation Quagmire: Regional Differences, Then and Now

"I said to him once, 'Why don't Canadians love Canada the way Americans love America?' And he said, 'It's not a country you love, it's a country you worry about.' It's true!"

— author Mavis Gallant, recalling a conversation about nationalism with Robertson Davies, in *Canadian Forum* magazine, February 1987

"In wartime the provinces were forced to concede their jurisdiction over provincial resources to the central government. Canada passed in the twinkling of a pen from a free enterprise system regulated by ten jealously competing sovereignties to a centrally directed economy regulated by the government's perception of the needs of the war."

— historians Robert Bothwell and William Kilbourn, biographers of Second World War Minister of Munitions and Supply C.D. Howe

Climate mobilization in Canada cannot succeed if undertaken by a strong central government alone. It needs to be a society-wide endeavour, and all levels of government have an important role to play — the federal government, each province and territory, each Indigenous nation and each municipal government. Unfortunately, some jurisdictions have been digging in their heels on climate action.

The challenges of federalism are always high on the agenda of any Canadian prime minister, and its trials often feel intractable. Progress on any number of files needs support and cooperation from the provinces and territories, and at any given time, provincial leaders may seek to make political hay by picking a fight with the feds.

Of course, the reality of confederation is now understood to be about more than just federal-provincial relations. Growing public and legal recognition of Indigenous title and rights adds important new complexity to the more than 150-year-old question of how decisions are made in Canada and by whom, particularly with respect to land and resources.

Regional economic differences and the realities of confederation are among the political and economic barriers to bold climate action in our country. The federal government has been reticent to undertake actions it fears would alienate one or more regions and foreclose on political support there. Further complicating climate action is the reality that natural resource policy falls, for the most part, under provincial jurisdiction in Canada's constitution (Indigenous rights and title aside).

Were Canada a unitary state, acting in the face of an emergency might be easier. But for better or worse, we aren't. Indeed, Canada is arguably one of the most decentralized federations in the world. Which is why in this chapter we explore whether we can secure sufficient collaboration across the various levels of government to confront the climate crisis.

The Second World War required that the Mackenzie King government negotiate deep and troublesome regional differences and rifts. Foremost among these, the declaration of war in 1939 threatened to split open one of the country's longest-standing political divides. Opposition to the war among French-Canadians in Quebec was particularly strong. Mackenzie King's core policy for keeping Quebec onside was a promise at the war's outset that he would not impose mandatory military conscription for overseas service. But many in the province distrusted Mackenzie King's assurance, and there were mixed signals from Ottawa. In a remarkable assault on civil liberties, under the War Measures Act, the mayor of Montreal, Camillien Houde, was imprisoned for four years for speaking out against the war and conscription.

Two weeks after Canada declared war, the premier of Quebec, ultraconservative Union Nationale leader Maurice Duplessis, dissolved the provincial legislature and called an election. "The war seemed to offer Duplessis the opportunity to renew his mandate," writes historian Granatstein. "He could, he obviously believed, capitalize on the very great unease in the province, turn his election campaign into a triumphant referendum against the war . . . and in announcing his decision for an election, Duplessis spoke of Ottawa's long campaign to destroy provincial autonomy and of the way in which 'le pretexte de la guerre declarée par le gouvernement fédérale' had been used to foster 'une campagne d'assimilation et de centralisation.'"[1]

Eighty years later, virtually the same words in English have been intoned by the Conservative leaders of Alberta, Saskatchewan, Manitoba, New Brunswick and Ontario in opposition to federal carbon pricing. They too have sought to make political gain by targeting federal climate policies. The main difference is that this strategy has often proved effective for today's Conservative leaders, whereas the tactic backfired for Maurice Duplessis and he lost the subsequent Quebec election. Given that Quebecers understood, by this point, that war was inevitable, they were more inclined to vote for Quebec Liberals who would maintain better relations with Ottawa and hopefully hold the Mackenzie King government to its non-conscription promise.

Mackenzie King did indeed do everything he could to honour his pledge, avoiding mandatory enlistment for overseas service as long as possible. In 1944, he even lost his minister of national defence, J.L. Ralston (a decorated First World War veteran), who Mackenzie King fired from cabinet after Ralston's strong protest of Mackenzie King's refusal to support overseas conscription. When, in the wake of the Normandy invasion and mounting losses, the Mackenzie King government did ultimately decide to send conscripts overseas in late

1944, Quebecers were largely prepared to abide the decision. Having seen Mackenzie King's efforts, most Quebecers were ready to accept the move as necessary. Mackenzie King thus managed to navigate what was potentially the most dangerous quicksand of the confederation quagmire with limited political casualties.

The Mackenzie King government also faced a squad of provincial premiers during the war that included some notoriously big political personalities — William "Bible Bill" Aberhart in Alberta, Mitch Hepburn in Ontario and Thomas "Duff" Pattullo in B.C. None of these men were keen to relinquish power to Ottawa, all had major disagreements with the prime minister (Pattullo had tried unsuccessfully to push the Mackenzie King government to adopt more ambitious New Deal-type projects during the Depression), and the first two strongly disliked or distrusted Mackenzie King. If and when Justin Trudeau and William Lyon Mackenzie King get to meet in the hereafter and argue about which of them had to contend with the more troublesome band of premiers, it is not clear to me who would rightly win the debate.

Nevertheless, the challenges of financing the war forced a rethink of federal-provincial financial relations. While most provincial premiers refused to entertain permanent changes to the division of taxation powers, they were all prepared to accept that, during the war, the federal government should be empowered to do what was necessary. A new Dominion-Provincial Taxation Agreement Act was passed in spring 1942, with which, until the war's end, the provinces forfeited to Ottawa all powers over income and corporate taxation in exchange for a federal transfer, so that Canada's dollars could be mobilized for a total war effort to maximum effect.

Imagine that. Premiers — many of whom made their political fortunes attacking Ottawa — relinquished provincial powers of taxation to the federal government for the duration of the war. Quite a contrast with the present. It is remarkable what becomes politically possible once all are agreed that an emergency exists and that cooperation is needed to ensure an all-embracing national effort.

HOW DO YOU SOLVE A PROBLEM LIKE ALBERTA?*

The oil and gas industry's importance to the overall Canadian economy is often overstated. The industry represents 8–9% of Canada's GDP, and only about 1.5% of national employment (the gap between these two statistics gives some indication of how capital intensive the industry is, and how much of our oil and gas is exported in low-value, unprocessed form). From a trade perspective, oil and gas account for about 5% of the total value of Canada's exports. But as many economists have noted over the years, oil and gas exports are a double-edged sword — they inflate the value of the Canadian dollar, thereby hurting the competitiveness of other non-energy exports. These numbers should provide reasonable confidence that we can plan for a two- to three-decade wind down of the industry.

But while the overall Canadian economy is not highly dependent on the extraction and export of fossil fuels, the same is not true of certain regions, notably Alberta and Saskatchewan, but also north-east B.C. and parts of Atlantic Canada.

In Alberta, for example, fossil fuels are the single largest sector of the provincial economy, accounting for 31–34% of provincial GDP in recent years and employing over 160,000 people (or about 7% of total provincial employment). In Saskatchewan, fossil fuels represent 21–23% of provincial GDP and just under 3% of employment. In Newfoundland and Labrador, now home to a massive offshore industry, oil and gas is responsible for about 25% of the province's GDP about 3% of their employment. And while oil and gas represent only between 1 and 2% of British Columbia's employment, in north-east B.C., the sector is of much greater significance, representing about 9% of regional employment.

The reality of these regional economic distinctions is reflected in public opinion variances. While public support for climate action is solid across Canada, it ranges from highs in Quebec to lows in Alberta. Compared to the Second World War, the willingness to act

---

* With apologies to *The Sound of Music.*

in the face of an emergency has been turned on its head, with Alberta assuming the oppositional role that Quebec had played.

But it's important not to paint the Alberta public with too broad a brush. Opposition to climate action among Albertans is often overstated. In the 2019 Abacus poll I commissioned, the level of support in the province for climate action was heartening, notwithstanding Alberta's current economic reliance on fossil fuels. For example:

- 58% of Albertans report they either think about climate change often and are getting really anxious or think about it sometimes and are getting increasingly worried. That's lower than other regions of Canada, but still a majority who are worried.
- 47% of Albertans believe climate charge is now an emergency, or will likely be one in the next few years.
- 67% of Albertans agreed climate change represents a major threat to our children and grandchildren.
- Surprisingly, 50% of Albertans support or can accept phasing out the extraction and export of fossil fuels over the next 20–30 years.
- 51% support or can accept banning the sale of new gas-powered vehicles by 2030.
- 64% of Albertans support or can accept requiring all new buildings and homes to heat space and water without fossil fuels by 2022.
- 62% of Albertans support our governments making massive investments in new green infrastructure, such as renewable energy (solar panel fields, wind farms, geothermal energy, tidal energy), building retrofits, high-speed rail, mass public transit and electric vehicle charging stations, as well as reforestation.
- When given a definition of the Green New Deal, 56% of Albertans support it (and only 21% oppose it).
- Only 18% of Albertans say they would want their children to be employed in the oil and gas industry.

As Edmonton-based climate justice activist Emma Jackson pointed out upon digging into the Abacus results, Albertans' response to the question "To what extent have you or someone close to you experienced the effects of climate change?" was comparatively low. Only 68% of Albertans replied positively to this question (the lowest level in Canada), despite the province experiencing some of the worst flooding and wildfires in the country in recent years. This speaks to the media's and others' failure to connect natural disasters with the climate crisis. Indeed, a chill has often been placed on any attempt to do so. In both mainstream and social media, it was frequently viewed as "insensitive" or somehow in poor form when the 2016 fire forced the evacuation of Fort McMurry to note any connection to climate change, let alone the irony that the city at the centre of the oil sands was being subjected to this climactic assault.

Interestingly, in the Abacus survey, Atlantic Canada, despite also being quite reliant on fossil fuel extraction, tended to record some of the highest levels of support for bold climate policies in the country (frequently coming in just behind Quebec and occasionally ahead). There is likely a lesson to be found there about how these discussions have unfolded in eastern Canada compared to the prairies.

Our caricatures of Alberta aside, there is in fact a vibrant climate movement in the province, although more so in Edmonton than in Calgary. Climate Justice Edmonton is very active, along with many young 350.org/Our Time activists. In June 2019, I attended a Green New Deal event in Edmonton that filled a church with about 400 people. The event happened to be the day after Trudeau approved the Trans Mountain pipeline expansion for the second time, and everyone in that church was strongly opposed (a very welcome sight for a British Columbian interloper such as myself). The global September 27, 2019, climate strike protests drew a crowd of approximately 5,000 in Edmonton. And when Greta Thunberg came to Edmonton on October 18, 2019, to join a #FridaysForFuture climate strike, 10,000–12,000 Albertans came out to join her in front of the provincial legislature.

There has been important Indigenous leadership in Alberta's climate justice movement, and there are numerous Indigenous activists, particularly from lands that have been severely damaged by oil sands development, such as those of the Lubicon Cree and the Beaver Lake Cree, who are committed to blocking fossil fuel expansion and to moving towards energy transition. And increasingly, there is a new generation of labour leaders and activists who are also keen to see Alberta transition away from fossil fuels.

The point being that, notwithstanding the current politics of the Kenney government with its "war room" and "enemies list" of those who do not support oil sands expansion, finding common cause across our confederation needn't be viewed as an insurmountable task, particularly as millennials' political influence increases.

Nonetheless, concerns about regional opposition haunt and complicate political considerations.

When I interviewed former NDP MP Libby Davies, she expressed the view that having an NDP government in Alberta between 2015 and 2019 constrained what the federal NDP was willing to say about climate and the oil sands. B.C. NDP support for LNG development has similarly resulted in the federal NDP failing to outright oppose such projects. This is a particularly thorny issue for the NDP, as the provincial and federal wings of the party are formally connected (unlike for other parties), a link Davies believes should be ended.

When I spoke with Liberal MP and cabinet minister Joyce Murray, she stressed that, "We are a federation. So the idea of having an agreement [the 2016 pan-Canadian climate plan] that was signed by all the provinces and territories is pretty powerful. The idea of going it alone and then having all of the provinces and territories attack us for jurisdictional overreach is probably not as effective." She underscored that a federal government needs to navigate the federation challenges "with as much good faith as possible, and continue to seek cooperation from all of the players, whatever the politics, whatever the political stripe, and be optimistic that, over time and in the short-term, all provincial governments of every political stripe will understand the urgency."

And yet, how did that work out? The ambition in the pan-Canadian climate plan was greatly eroded to win wide agreement, particularly from Alberta, only to have the Kenney government elected in 2019 and renounce the plan.

Notably, these regional interests have also been reinforced by Bay Street in the heart of central Canada. As the Corporate Mapping Project has highlighted, Canada's largest financial corporations are heavily invested in the oil sands.[2] Two weeks after the 2019 federal election, the Canadian Press reported, "CIBC's chief executive Victor Dodig rallied support for Canada's energy sector [at a speech in Calgary], saying it's the country's 'family business' and that the shortage of pipeline capacity represents a 'critical threat' to our economy . . . Dodig added in his comments at the Economic Club of Canada that Canada not only needs to maintain its position as a leader in 'responsible energy development,' but grow it for 'the benefit of Alberta and all Canadians.'"[3] So the federal government hasn't been held back merely by oil patch interests; they are also being encouraged in their policy contradictions by some of their long-time friends on Bay Street. That dynamic may, hopefully, be shifting, with some investment funds starting to divest from fossil fuels. But as Dodig's comments indicate, some old habits die hard.

Let us take Alberta to be a proxy for all oil-and-gas-dependent regions of Canada, and consider these core questions: Is it possible for bold leaders to have an honest and respectful conversation with Albertans about the climate emergency and the future of the oil and gas industry? And can such an engagement produce a hopeful vision for the province — one people can accept and perhaps even embrace?

Alberta Federation of Labour (AFL) President Gil McGowan's understanding of the Alberta economy and the global oil industry runs deep. Recent Alberta governments of both the left and the right have promised working families in that province that they can bring back "the good times" and restore jobs in the oil patch, and have falsely told Albertans that new pipelines to tidewater would result in greater wealth and revived investment. McGowan, in contrast, has been trying to tell working-class Albertans a more complex and challenging story.

"We have seen the last of the boom years," McGowan told a just transition conference in Edmonton I attended in the summer of 2019. As he explains, a trifecta of factors is fundamentally impacting Alberta's oil sands. First, there is an oversupply of oil in the market, driven by the fracking revolution in the U.S. Second, automation is transforming the oil patch, such that even a return of some capital investment will not bring back lost jobs. "For example," notes McGowan, "you have drilling rigs that used to need fifteen, sometimes twenty people, to operate. The new drilling rigs can be operated by two people with laptops." And third, climate change and the global policy response mean that demand for Alberta bitumen and investment dollars will be in long-term decline. The upshot is that we will not see a return of large-scale employment to Alberta's oil patch, regardless of who is in government. Instead, McGowan and the AFL have sought to advance a vision for what they call "The Next Alberta," in which they urge the provincial government to make the energy transition a top priority.

But that's a hard message for people to hear. And so, in the face of the resulting economic insecurity and anxiety, many are instead drawn to the scapegoat politics of the Jason Kenneys. To some extent, such a response is understandable and speaks to a failure of climate change activists and progressives to ensure those facing deep anxiety feel heard and understood. "We haven't addressed their angst," says McGowan.

"Politics is all about narratives, especially during election campaigns," McGowan told me in an interview. "And [in the 2019 Alberta election] Jason Kenney had a narrative. It was very simple. And I guess it was kind of attractive in its simplicity. Basically he said, 'Alberta was rich, they elected the New Democrats, and then we were less rich. They destroyed the economy. Let us get rid of them, and bring back the Alberta Advantage and bring back our prosperity.' It was complete horseshit and based on a gross misreading of what is actually happening in our province. But it worked because it is what people wanted to hear."

McGowan clearly takes pride in the fact that the Alberta Federation of Labour has been talking about climate change and developing just transition ideas since the late 1990s. "We think it is imperative that as a society and as a province, we start to prepare. Because if we are not prepared, the change is going to hit us much harder in terms of lost jobs and lost economic opportunities."

I asked McGowan if there was a scenario in which the Notley government could have been re-elected, particularly after the political right was reunited by Kenney in the new United Conservative Party. He insists:

> I continue to believe that there was a path to re-election for the Notley New Democrats, but it was a very narrow one and it depended on them embracing an alternative story about Alberta's economic future, one that acknowledged what we describe as energy transformation and the move away from fossil fuels. The work that we as a labour movement here in Alberta have done — on climate change and green jobs and the energy transformation — is partly driven by values. We really believe that everyone, including the labour movement, has a responsibility to help address the unfolding climate emergency. But it is also based on our understanding of economic policy, in particular what we think would be best for working Albertans and the Albertan economy. And we have come to the firm conclusion that sticking our heads in the sand and pretending that change is not happening is not good for Alberta workers and, over the long-term, it is not good for the Albertan economy. You can pretend that change is not happening but it does not stop it from happening . . .
>
> One of the things that frustrates me about the conversation that has unfolded in Alberta over the last three or four years, including in the recent provincial election campaign, is that our politicians, regardless of their political stripe, are not being straight with Albertans about what is really happening with the global oil and gas economy.

> They are not telling them the truth about the move away from fossil fuels. Especially when it comes to how quickly that is going to happen and what it is going to mean for employment here in Alberta.

McGowan remained publicly supportive of the Notley government's campaign to win new bitumen pipelines, arguing that the industry will still be around for a few decades and will need new investment in transportation capacity, a contention I certainly don't accept. But neither does he hold with the view, advanced by many pipeline advocates, that Alberta's economic problems all stem from being "land-locked" and unable to get its product to new markets. He describes the Conservative campaign to blame Alberta's woes on foreign-funded attempts to block new pipelines as "tinfoil hat crazy stuff."

McGowan makes the interesting observation that, over most of the last two decades, as much of the world was learning hard lessons about the false promises of neoliberalism and how its policy agenda was leaving working people less economically secure, high oil prices, until 2014, meant that Albertans were largely shielded from this reality. "We [in Alberta] were insulated from all these negative trends that were dragging working people down in almost every other jurisdiction around the world. We were insulated, safe in an oily bubble. And so the working class and the middle class here in Alberta were prospering in a way that was not the case almost anywhere else in the world." That's the shiny rear-view mirror picture many in Canada's most climate-resistant province are still holding on to.

Almost 40% of Canada's GHG emissions are from Alberta (a staggering reality), and thus, climate action needs Alberta. Given this, McGowan worries that his province is on a collision course with the rest of the country. He hopes that, as we seek to tackle the climate imperative, other Canadians can approach his fellow Albertans with understanding. "It is hard to divorce people's attitudes from the structure of the economy. And our [Alberta] economy has been a very successful economy. It has delivered robust middle-class lifestyles for millions of people for very long periods of time. But it is all based on fossil fuels."

Can a collision be avoided? McGowan believes it's possible. He says we need "leaders at both the provincial and federal level who understand the nature of the problem and who recognize that Albertans will have to do a lot of the heavy lifting and pay the heavier price, and there needs to be some validation for that."

Beyond rhetorical validation, McGowan believes rethinking federal transfers can help to break the logjam. That is surely an idea worth pursuing, although not through the Equalization Program (as Kenney and Saskatchewan premier Scott Moe have demanded). That program, which seeks to ensure all provinces have equal capacity to fund equivalent levels of public services, is more or less working as it should. According to Alberta's own 2019 provincial budget, if Alberta taxed individuals and corporations at rates comparable to other provinces, it would have "at least $13.4 billion more in taxes."[4] Re-opening the equalization formula would be a messy and complex proposition.

Instead, I would propose an entirely new federal transfer program to the provinces — a Climate Emergency Just Transition Transfer. Such a transfer could be specifically linked to funding green infrastructure projects that would create thousands of jobs, along with training/apprenticeships. It could be a mechanism to renew confederation while rising to the climate crisis.

The transfer's distribution could be based on a formula linked to recent GHG emissions in each province (but fixed from that point onward, so that it does not perversely incentivize continued high GHGs). Doing so would recognize that jurisdictions such as Alberta, Saskatchewan and Newfoundland and Labrador face a more challenging task and will have to do more of the heavy lifting to transition their local economies. As Alberta currently produces about 40% of Canada's GHG emissions, it would receive about 40% of the transfer money. Alternatively, the formula could be linked to how many people are currently employed in the oil and gas sector; again, the bulk of the support would go to the regions where it is most needed.

I don't think, however, that we should just hand over the money to provincial governments. Instead, we should establish new just-transition agencies — one in each province — jointly governed by the feds, provincial and local governments, and, vitally, Indigenous nations from that province, with civil society representatives too from labour, business and academia/NGOs. This would ensure the transfer money isn't simply absorbed into provincial budgets or used to displace other infrastructure or training funds. It would ensure the money is used for its intended purpose. There are already models for a joint structure like this in Canada, such as the Port Authorities. We might want to require that provincial co-governance be linked to cost-shared contributions.

There is a long list of worthwhile projects such a transfer could fund. The key is that this transfer would represent *real* dollars for *actual* transition and new jobs (not vague assurances and the historic false promises of just transition). It could include a good jobs guarantee for energy workers. And there should be other conditions tied to the transfer: ending fossil fuel subsidies; minimum apprenticeship placements for women and Indigenous people; minimum royalty rates charged to fossil fuel companies; and no new fossil fuel infrastructure. It could also, over time, be tied to demonstrated reductions in GHGs. An innovation such as this could be a linchpin within an overall transition plan that is fair and just.

That said, in such a "new deal," Alberta will be obliged to revisit its own taxation policies. Alberta is currently highly undertaxed. The province is alone in Canada in refusing to institute a provincial sales tax, made possible only because, for decades, Alberta has coasted on oil income. Alberta will need to raise its own taxes, otherwise a new federal climate transition transfer would be seen as the rest of Canada subsidizing Alberta's under-taxation.

Ricardo Acuna is the long-time executive director of the Parkland Institute (the CCPA's sister think tank in Alberta). Parkland has a long history of taking dissenting and sometimes unpopular views within the institute's home province, and they were early critics of oil sands development. When the Notley government was elected, Parkland supported many of the government's moves to reform social policy, health care, education and labour laws, but they were critical of the NDP's continued support of oil sands and pipeline expansion.

And like McGowan, Parkland has valuable insights into how to navigate the challenging intersection between climate action and the confederation quagmire.

Acuna expresses some sympathy for the pressures an NDP government in Alberta inevitably faces. "They wanted from the beginning to play nice with industry and build a big tent. And I think in many ways that is a political imperative in Alberta. The first day after the election, oil stocks plummeted," and Notley sought to quickly reassure the oil majors. Acuna believes the threat of a capital strike loomed large.

Acuna wishes the NDP had adopted a "go big or go home" approach. In particular, he believes the government should have substantially raised oil royalties, as well as income taxes on wealthier individuals and corporations, and he argues those increased royalties should have gone into a long-term public savings plan to be used for economic diversification.

In an interview with Acuna and Parkland researcher Ian Hussey, Hussey notes that, while many companies are active in the oil patch, just five companies control about 80% of the industry, and they have continued to pull in substantial profits. He thinks the Notley government should have hiked royalties, and then only offered reductions to those corporations willing to provide community benefit agreements — a commitment to boost local employment, improve opportunities for female workers in the oil patch and fund the reclamation of abandoned oil wells, which represent a massive and expanding public liability. Both Acuna and Hussey lament that Alberta's governments have not bargained harder with the industry over access to this public resource. Acuna believes the failure to do so stems from fear of the industry, particularly in the context of rising unemployment and lower oil prices.

Acuna and Hussey also worry about a coming confrontation between Alberta and the rest of Canada. But they hold out hope that if Indigenous efforts to block pipelines are successful, and if governments elsewhere successfully implement climate policies that drive down demand for oil, then in time the sources of these confederation conflicts with Alberta will simply recede — the oil sands will slowly be forced to wind down, regardless of Alberta's policy choices.

As our leaders seek to have an honest conversation with Albertans about what the climate emergency demands of us, some approaches may find a better reception than others. And the wartime frame might help.

In early 2019, a communications initiative called the Alberta Narratives Project issued their final report and recommendations.[5] The project, supported by 75 organizations, convened 55 dialogues across the province, investigating how to constructively talk about climate with Albertans — what rankles and what resonates, both with the population at large and with various constituencies?

The Alberta Narratives exploration revealed a public that desires an acknowledgement that this transition is going to be hard. People all understand that oil and gas currently represent a huge portion of Alberta's economy. But they also get that diversification is necessary, they share a love of the outdoors, they understand that the weather is changing, and they take pride in a culture of helping one another (as the nation has witnessed when Albertans faced recent fires and floods).

The narratives project also warns, "Climate campaign messaging that blames the fossil fuel industry closes down the conversation. Many of those who might share concerns about climate change feel that they, too, are being attacked." This speaks to the industry's capture of Alberta's political culture. It remains my contention, however, that the climate emergency requires that we confront the oil and gas industry — the corporations who have actively blocked progress on climate. The key, therefore, is to carefully and consistently distinguish between the corporations and its workers, and to ensure that those employed in the sector do not feel under attack. That ought to be doable. After all, in the years since the collapse of oil prices in 2014, the major oil companies have continued to record massive profits, even as they laid off tens of thousands of people and abandoned communities from which their wealth was amassed. The Parkland Institute has done a commendable job of highlighting these trends.[6] Hussey believes the years since the oil price crash have already produced a shift in public perception within Alberta towards the industry: "There is less trust of the industry. There are literally thousands of workers who got laid off during the latest price crash that have left the industry, have found jobs in other sectors, and are not coming back." He believes most Albertans understand that there is no going back to the heydays of the 2004–2014 boom. All the more reason, then, to embrace this time to confront the oil and gas corporations. And if that means a fight, then, in the words of President Roosevelt when he faced down the wrath of corporate America in the 1930s, we should "welcome their hatred."

The Alberta Narratives report emphasizes common values such as "pulling together," helping each other in times of need, and "loyalty to the wider community" — precisely those values to which a wartime-scale effort seeks to appeal.

The narratives project also points to ways in which it may be possible to animate the pride of Albertans: "In regard to Alberta's oil and gas industry specifically, they were proud of its pioneering origins, and of how it had overcome major geological and technical issues. They saw this as a mark of an entrepreneurial culture in Alberta, a culture of innovation, problem solving and can-do attitude." Hussey similarly highlights that Albertans "have respect for doing hard things. Albertans are not afraid of hard work. The whole stereotype about rolling up your sleeves and all of that is true." Those are exactly the "we can do it" wartime values that need to be tapped as we mobilize for the climate emergency. If Albertans, or Newfoundlanders for that matter, can build the wonders of engineering and ingenuity that are the oil sands and the offshore oil industry, if they could build whole new cities in the space of a couple decades, then surely we can do what is needed to now transition our economy off fossil fuels.

According to the project, younger people are the most likely to accept the science of climate change. Climate justice activist Emma Jackson (who works with both Climate Justice Edmonton and 350.org) similarly contends, "There is a massive generational distinction that I have seen. I think that young people in Alberta really, deeply understand the climate crises and are hungry to see their province taking far more ambitious action." She wants to see training and alternatives that allow younger Albertans to envision a different future.

An Albertan conversation about this climate crossroads moment can also remind Albertans about what a raw deal they are getting, not from confederation, but from the oil companies. Alberta has historically played an aggressive role in asserting provincial jurisdiction over oil. But in the early days of the oil sands, under Premier Peter Lougheed, the province was fierce in its negotiations with Ottawa and the industry alike, insisting that the Alberta public was the owner and ensuring that returns to the treasury in resource rents/royalties were sizable. Subsequent Alberta premiers from Getty and Klein to Stelmach, Redford, Prentice, Notley and now Kenney remain committed to the battle with Ottawa, but unlike Lougheed they have been much more acquiescent to the demands of the oil industry. (Although Stelmach briefly tried to boost royalty rates, he then caved and ultimately resigned in the face of a swift backlash orchestrated by the industry.) As author Kevin Taft explains, Alberta has been keen to assert its authority, and then hand the winnings off to the oil giants.

The story of Alberta's oil and gas royalties is a tragic tale. Only under Lougheed were royalties — the returns to the public for the right to extract these resources — at appropriate levels. Lougheed used those earnings to create the Alberta Heritage Fund in 1976, a public savings fund whose earnings could help diversify the economy and meet the collective needs of Albertans into the future, long after the oil ran dry. This public innovation became a global model, most notably for Norway. After oil and gas was discovered in the North Sea, Norway (whose population is only slightly larger than Alberta's) sent bureaucrats to Alberta to study the Heritage Fund, leading to the creation of Norway's sovereign wealth fund in the mid-1990s. How these two jurisdictions have managed their oil wealth makes for a fascinating comparison, and was the subject of a 2015 study by Bruce Campbell, the CCPA's former national executive director. Campbell writes:

> Shortly after the discovery of oil in Norway, a strong consensus emerged among the political parties and across Norwegian society about how to manage oil wealth . . . Based on the view that multinational oil companies needed

> to be controlled, the Norwegian state took on the central role as both regulator and producer . . . There was also consensus that its petroleum wealth should be appropriated by the state and distributed equitably within Norwegian society. In Norway, the state has always been in the driver's seat in determining petroleum development, owning 80% of oil and gas production and controlling the transportation infrastructure. In Canada, private interests — foreign and domestic — have dominated Alberta's petroleum sector.[7]

After Lougheed's retirement, his successors began hacking away at the province's royalty rates and corporate taxes — they now rank among the lowest among petro-states — and no subsequent Alberta premier added anything of note to the Heritage Fund endowment. Its value today stands at about $18 billion, equivalent to a little over $4,000 per person in Alberta. In contrast, the Norwegian government has continued to plow earnings from its oil development into its collective savings fund, and their sovereign wealth fund now sits at about $1.4 trillion Canadian dollars — a remarkable amount that is equivalent to about a quarter million dollars for every individual in Norway, giving that country an extraordinary capital pool from which to fund a transition off fossil fuels. That's what it might have looked like, had Alberta's government truly stewarded this resource in the public interest. Instead, the oil and gas companies made off like bandits.

COOPERATIVE FEDERALISM AND NATIONAL SECURITY

University of British Columbia political science professor Kathryn Harrison has studied federalism and Canadian environmental policy since the late 1980s, and she's passionate about the need for climate action today. When we talked in the summer of 2019, I asked her perspective on what to do about the barrier confederation poses to climate progress.

"We have a massive challenge," says Harrison, "because we have a regionally diverse country and the scale of difference in per capita emissions within Canada is at the outer limits of the international scale. Saskatchewan and Alberta's per capita emissions are higher than those of OPEC states. So, we have [GHG emission levels within much of Canada] like Western Europe and something above Bahrain within the same country."

Harrison believes one of the barriers to progress on the environmental file in Canada is that, "we have assumed we need federal-provincial consensus. And because there are only ten provinces (compared to fifty states in the U.S.), the first ministers or their environmental ministers can all sit around a table. They sit around a table as equals and there has been this assumption that we have to have consensus. What this means is that, in practice, different provinces have exercised vetoes at different times. And Alberta has exercised the veto over Canadian climate policies since the early 1990s. For a long time, Ontario did it over the regulation of the auto industry."

The last point is worth highlighting. Many in Alberta express frustration that the climate debate in Canada is so focused on the oil sands, while the auto industry seems to escape scrutiny. Why do plans to build a fossil fuel car plant or a new highway to move those cars not receive the same climate attention as a pipeline? If there is to be some sort of confederation quid pro quo in a cooperative national plan, the speedy transformation of the auto industry in central Canada needs to be on the table too.

Harrison does see some hope in what transpired in the creation of the pan-Canadian climate plan. "Something important happened. The federal government created this idea that there is a 'backstop.' The federal government effectively required carbon pricing, and said to the provinces, 'If you do not act, we will do it.' And that is a really important precedent. The problem, however, is they built Alberta's level of ambition as the foundation of the Canadian plan. And it is not enough."

Harrison notes that a useful comparison is Medicare. Health care is generally provincial jurisdiction. And public health care was initially a provincial innovation, led by the CCF's Tommy Douglas when he was premier of Saskatchewan — just as carbon pricing was initially a provincial innovation. But once there was enough provincial and public support, Medicare became a federal policy, and later the Canada Health Act (introduced in the early 1980s) linked federal funding transfers to the provinces for health care to a requirement that the provinces adhere to core Medicare principles regarding public and universal access to care. The provinces sometimes balk at what they perceive to be federal infringement into provincial jurisdiction, but given strong public support for Medicare, no province is willing to push back too hard. If such a model can be made to work for Medicare, why not climate, energy and resource policy as well, despite claims of provincial authority?

In June 2019, as tensions mounted between Ottawa and Conservative premiers over climate and energy policy, *Globe and Mail* columnist John Ibbitson wrote, "This Prime Minister's approach to federalism — which can be summarized as 'do as I say, or else' — has premiers from the Bay of Fundy to the Rocky Mountains to the Arctic Ocean in open revolt."[8] Ibbitson contends that the Trudeau government, like many previous federal governments, wrongly believes that Ottawa has "the right and duty to act in the national interest, even if it meant intruding in areas of provincial jurisdiction." Ibbitson praised the approach of the Harper government for maintaining relative peace with the provinces. Under Harper, he claimed, "any pan-Canadian initiatives, such as working toward a national securities regulator, were voluntary at the provincial level." Ibbitson's solution? "Canada must act to combat global warming. But the provinces must take that action. Ottawa can encourage, co-ordinate, even administer if asked. But it must never impose."

But arguments such as this will see progress on climate forever stalled. The whole point of understanding this crisis as an emergency is that we must reject "voluntary" participation. And the federal government does indeed have a duty to act.

While I was born and raised in Montreal, in a province that fiercely defends its jurisdictional powers, I have now lived for more than half of my life in British Columbia. I am grateful to be a Canadian. But I also now feel a strong sense of British Columbian identity, and I have been known to mutter (particularly in the face of efforts to force a pipeline down our throats) that the powers-that-be in central Canada "don't understand the west coast." When I first moved to B.C. from Toronto (where I attended university and briefly taught) in the early 1990s, I was struck by the fact that people's provincial attachment here in the west was much stronger than it was in "the centre of the Canadian universe." The mainstream media conversation here is much more dominated by provincial issues, compared to Ontario.

Even the organization for which I worked for 22 years — the CCPA — has a fascinating federated structure, some of it governed by bylaws and some by historic convention, like many a constitution. Each provincial office of the CCPA operates with a huge amount of autonomy, which it fiercely and rightly defends, and concerns itself primarily with provincial policy matters. Yet each office also sees itself as part of a national whole, with a shared sense of national purpose and a commitment to cooperate in the advancement of a shared mission. Because in the struggle for a more just and sustainable world, the forces we are up against don't operate in just one province, and many of the needed solutions cannot work in one province alone.

There is no doubt that many of the climate actions we urgently need come under provincial (and in some cases municipal) jurisdiction. Many climate mobilization efforts will achieve their maximum benefit if they are the result of federal-provincial cooperation. It is also vital that our climate emergency plans at every level — federal, provincial and municipal — be developed and implemented in collaboration with Indigenous communities and in a manner that complies with the United Nations Declaration on the Rights of Indigenous Peoples.

But it is also unquestionably the case that, as we face down the threat of climate breakdown, we need the national government to be able to take forceful action. And the necessary mobilization we now need cannot be prevented by the foot-dragging of some provincial governments.

The good news is that most of us appear to concur. The 2019 *Confederation of Tomorrow Survey of Canadians*, a major survey of our attitudes towards confederation, offers hopeful results. Andrew

Parkin, summarizing the survey's findings in *Policy Options* magazine, writes:

> The current debate on how best to respond to climate change . . . unfolds against the backdrop of a federation in which few citizens are seeking more power for Ottawa, and in which residents of provinces such as Quebec and Alberta are particularly protective of their provincial government's prerogatives over energy resources. Viewed from this perspective, the deck on climate change policy is not necessarily stacked in the federal government's favour.
>
> It is all the more striking, then, that on the specific issue of climate change, there is relatively weak public support for a province-first approach. Only 12% of Canadians trust their provincial government more to make the right decisions when it comes to addressing climate change. More than twice as many (29%) trust the federal government more, and almost three times as many trust both governments equally. About one in five (19%) trust neither government.[9]

Even in Alberta, only 14% of respondents favoured their provincial government to make the right decisions on addressing climate change, and the same percent in Quebec. Only Saskatchewan stood out, where 25% most trust their provincial government to lead on climate. The survey found Canadians are more likely to support a federally led national climate policy, and this is even true in a province like Quebec, where people generally support decentralization. Only in Saskatchewan, and by bare margins in Alberta and PEI, did a plurality (but not a majority) of respondents favour the provinces setting their own climate policy.

There are ways to fairly accommodate and constructively engage Alberta and other fossil-fuel-reliant regions as we ramp up an ambitious climate plan and to bring these places into the fold.

That said, the Canadian constitution, from the days of the British North America Act, is clear on one matter — national security is federal jurisdiction. That is why, in the Second World War, the provinces were prepared to relinquish immense powers to the federal government. And the climate crisis, as we have established, is clearly an urgent matter of national security.

PART THREE

MOBILIZING ALL OUR RESOURCES

WikiCommons

CHAPTER 6

Remaking the Economy, Then and Now

"Limiting global warming to 1.5°C would require rapid, far-reaching and unprecedented changes in all aspects of society."
— Intergovernmental Panel on Climate Change, Special Report, 2018

"So far as it is possible, Canada's effort in this war must be a planned and concerted national effort . . . In order to have the tremendous quantities of supplies available at the right time, and in the right place, it is imperative that the economic life of Canada be reorganized, but not disorganized. The economic forces of the country require to be mobilized, just as the armed forces are mobilized. This task can be performed, in the main, only by the national government. Its adequate performance, however, demands the co-operation of provincial and municipal authorities, as well as of business, labour, the farmers and other primary producers, and of voluntary organizations of all kinds . . .

Within a few hours of the outbreak of war, the government established the War-Time Prices and Trade Board to prevent hoarding, profiteering and undue rise in prices of necessities. The duties and powers of the Board are extensive. It confers with manufacturers, wholesalers and retailers, with a view

> *to enlisting their co-operation in ensuring reasonable prices, adequate supplies and equitable distribution of all necessaries of life . . .*
>
> *The War Budget, although necessarily burdensome, was founded upon the very just principle of taxation — ability to pay. Upon those making profits from the war, we have placed a heavy excess profits tax . . ."*
>
> — radio broadcast by Prime Minister Mackenzie King, October 1939 (one month after Canada declared its entry into the Second World War)

Climate science forcefully tells us that we have one decade to at least halve our GHG emissions and three decades to fully wean ourselves off fossil fuels. Consequently, most of the known fossil fuel reserves in Canada must remain in the ground and the industry wound down. But the challenge is much greater than that. Our entire economy and society — our homes, communities, food, travel, leisure and more — are currently deeply tied into the use of oil and gas. For many of us, this feels like a daunting task, the scope of which, at times, seems overwhelming and impossible.

Yet, as the Mackenzie King quote above reminds us, we have fundamentally retooled our entire economy before, and done so in far less time.

TRANSFORMING THE CANADIAN ECONOMY FOR WAR

The scope and speed of Canada's economic transformation during the Second World War was spectacular. No one could have predicted the scale and rapidity of the conversion, and the story of how this was achieved should rightly serve as a source of tremendous inspiration.

The importance of Canada's wartime production on the home front cannot be overstated. Whereas the First World War was characterized predominantly by tens of thousands of soldiers facing each other across trenches and fighting with guns and small arms, the outcome of the Second World War was strongly determined by the quantity and quality of military equipment and machines — the planes, naval ships and new technology weapons.

Canada's wartime manufacturing was nothing short of stunning, all the more so because most of the production capacity had to be built from scratch. Between 1939 and 1945, Canada produced:[1]

- Over 800,000 military transport vehicles (jeeps and trucks) and 50,000 tanks;
- Over 40,000 field, naval and aircraft guns;
- Over 1.7 million small arms;
- 348 merchant cargo ships and 393 naval war ships; and
- Over 16,000 military aircraft — at its peak, aircraft manufacturing was producing 300 planes a month.

Historian Jack Granatstein, in a 2005 paper prepared for the Canadian Council of Chief Executives, drove home the scale of the effort this way: "On June 12, 1943, the *Globe and Mail* printed a chart showing one week's production from Canada's factories. Each week, the newspaper noted, 900,000 Canadian workers, men and women, made at least six vessels, 80 aircraft, 4000 motor vehicles, 450 armoured fighting vehicles, 940 heavy guns, 13,000 smaller weapons, 525,000 artillery shells, 25 million cartridges, 10,000 tons of explosives, and at least $4 million worth of instruments and communications equipment. It was not until 1944 that Canada reached its peaks in production, so there was more to come. That almost none of these weapons, ammunition, and equipment had been produced in Canada in 1939 — that very few in fact were even capable of being produced — is an indication of just how effective Canada's wartime industrial mobilization had been."[2]

"This was," concludes Granatstein, "an astonishing feat of production and organization, a massive effort by every sector of the Canadian economy and by Canadian workers and business leaders."[3]

The private sector had a key role to play in wartime production. But critically, it was not allowed to determine the allocation of scarce resources during the period of economic transition. Rather, private factories were told what to produce, because in an emergency, we don't leave such important and urgent decisions to the vagaries of the free market.

Remarkably, the Canadian government — under the leadership of C.D. Howe, dubbed the "Minister of Everything" — established 28 public crown corporations during the war to meet the supply and munitions needs of the war effort. And the government carefully tracked and coordinated all the key inputs needed for wartime production.

One cannot talk about the fundamental retooling of the Canadian war economy without talking about the central role played by C.D. Howe. He brought to the task a fierce sense of purpose. And so the story of his leadership is worth telling.

C.D. Howe: Minister of Everything[4]

There is, for me, a rather delicious irony in my newfound appreciation for C.D. Howe. He was by many accounts the most powerful and important minister in Mackenzie King's wartime cabinet. But today, if people are familiar with his name, it is likely because the neoliberal, Toronto-based think tank — the C.D. Howe Institute — is named for the man. Having spent 22 years with the CCPA, we were from time to time in intellectual battle with the free-market think tank. And to be clear, C.D. Howe himself was a free-market guy. He came from the private sector, and his social circle was populated by the business elite of the country. Howe's biographers, historians Robert Bothwell and William Kilbourn, write, "Certainly he loved the company of the rich and powerful, and tended to accept uncritically many of their social views."[5] Within the awkward coalition of business-types and social progressives that has long been the Liberal Party of Canada, Howe certainly sat on the right flank of the party.

Yet paradoxically, during the Second World War, C.D. Howe was, for all intents and purposes, the minister of state economic planning.

Howe, as Minister of Munitions and Supply during the war, was happy to give contracts to the private sector. But he was also an engineer in a hurry. If the private sector was not willing or able to rapidly meet a wartime production need, Howe was content to have the government step in and get the job done. In this regard he was remarkably non-ideological and amazingly free of shackling economic assumptions about what is and is not allowed.

Clarence Decatur Howe was an American by birth. He was born in the town of Waltham, Massachusetts, in 1886 to a family that had lived in New England for generations. His father was a house builder and local alderman, his mother a local teacher. They were conservative and traditional people and lived what was then an upper-middle-class life.

With both a father and grandfather who were carpenters, Howe likely inherited a love of building, and chose to pursue engineering at the Massachusetts Institute of Technology. In 1908, at age 22, he moved to Canada to become a professor of engineering at Dalhousie University in Halifax. But he did not remain in academia for long. In 1913, now in his late 20s, Howe left Dalhousie when an opportunity arose via a former colleague to help develop grain elevators on the prairies for a new government initiative. Howe became chief engineer of the Canadian Board of Grain Commissioners (a body that would ultimately evolve to become the Canadian Wheat Board in the 1930s).

After doing that work for three years, and sensing business opportunities during the First World War, "he decided that he knew enough to take himself and his acquired expertise into private practice," writes historian Robert Bothwell. He established the C. D. Howe Company, "prospered, and Howe-built elevators soon dotted the prairies."[6] The company later branched into pulp mill design as well.

Howe did not enlist during the First World War. He was doing work that felt important, a necessary contribution to meeting wartime food supply needs. He was becoming a very wealthy man at the intersection of the private sector, government contracts and working with farmer cooperatives. Along the way, he was also building a strong reputation as someone who could get big and challenging jobs done quickly, and making business and political relationships across the country. Now with strong ties and a deep connection to Canada, Howe applied to become a British subject and Canadian citizen.

Biographers Bothwell and Kilbourn describe a man who was well-liked and who could talk with equal ease with businessmen, politicians, and farmers. By the late 1920s, driven by strong growth in Canada's agricultural sector, the Howe company had 175 employees and had built grain elevators from Prince Rupert, B.C., to Ontario.

Then the Depression hit and work dried up. "Canada's shrunken wheat exports needed no more terminal space," write Bothwell and Kilbourn. "As the Howe Company's prospects dwindled, so did its staff. By the end of 1933 only the three partners remained, along with a secretary and a junior engineer."[7] Then his partners both left and it was time to close up shop. Time to try something new.

Howe was not partisan. His social networks among the business elite were populated by Conservatives and Liberals alike. On economic matters he leaned right. He was a "pull yourself up by your bootstraps" kind of guy.

But by 1933 people had growing doubts about Prime Minister R.B. Bennett and his Conservatives and their ability to pull the country out of economic despair. That year the Liberal Party approached Howe to run in his home riding of what is now Thunder Bay (then still known as Port Arthur). The Liberals' hope was that Howe could take the seat from the Conservatives, who had long held the riding.

Howe accepted the Liberal nomination in 1934 and was elected to the House of Commons in the October 1935 federal election that saw Mackenzie King's Liberals win a large majority. Despite his rookie standing, Howe was named to the new cabinet, the only engineer among them, and became Minister of Transport.

During his first term in office, in 1936, Howe introduced the motion in the House of Commons establishing the CBC (although the creation of the CBC was more driven by Mackenzie King himself and built upon ground that had been laid by the previous Bennett government). By 1939, when war was declared, 85% of Canadians were conveniently reachable by the new public broadcaster. Howe's other early and notable crown corporation creation pre-war was the establishment of Trans-Canada Airlines (what would later become Air Canada). Having moved from the private to public sector, Howe seemed keen to bring his entrepreneurial spirit to the public realm and to not let the private sector secure all the benefits of emerging domains such as broadcasting and commercial air travel.

In early 1940, the Liberals were returned to office with an even stronger majority. And now, with the war effort moving into high gear, war production was given its own ministry — the Department of Munitions and Supply — and Howe was made its minister. He was also brought into the War Committee, the small cabinet subcommittee that oversaw the wartime effort.

Howe was now 54 years old, and he was firmly in charge of his portfolio. As Bothwell writes, the Department of Munitions and Supply "looked like nothing else on the bureaucratic map, and it did not function as an ordinary government department."[8] Unlike in a typical government ministry, Howe created his own executive team of "dollar-a-year men" he recruited from the private sector to help him organize war production. "The deputy minister did not ask to make decisions, nor was he allowed to. In Munitions and Supply it was the minister, not the deputy, who was the 'executive head.'"[9]

With the fall of France in summer 1940, only the Commonwealth was left to face the Nazis (as the U.S. and Soviet Union would not join the war until the following year). That's when Canadian war production moved into high gear, and notably, with little concern for cost. After all, as Howe famously stated — and with deep resonance today — "If we lose the war nothing will matter."[10]

Howe was requisitioning supplies and opening factories all over at a frantic pace, and his Wartime Industries and Control Board was ensuring the supply chains. The Department of Munitions and Supplies was frequently providing substantial capital assistance to private contractors as an incentive to move them into war production and to help them retool for the new task. Howe's frequent presence at the opening of factories across the country was also an exercise in political communication — it let the public know that the government was constantly upping its activity in the face of the emergency.

In some cases, these factories were importing components from the U.S. but, for the most part, the production was all in Canada. For his head of war production, Howe chose Harry Carmichael, formerly the vice-president of General Motors Canada.

As production ramped up, Howe began to establish new crown corporations that could ensure supply needs, taking on the task of bulk purchasing for all defence contractors' key inputs. As Bothwell and Kilbourn explain:

> The Munitions and Supply Act was amended to permit the minister to set up crown companies using the existing Companies Act. Overnight, C.D. Howe became an industrial tycoon . . . The minister could buy and repair, mobilize and construct, requisition and order anything he felt necessary to the production of munitions. There was no barrier to his powers, not even a constitutional one, for in wartime the provinces were forced to concede their jurisdiction over provincial resources to the central government. Canada passed in the twinkling of a pen from a free enterprise system regulated by ten jealously competing sovereignties to a centrally directed economy regulated by the government's perception of the needs of the war.[11]

But the central planning was being led less by senior civil servants and more by Howe's recruits from the top ranks of the private corporate sector. As Granatstein writes:

> He began to look to Canadian business for executives who could step in to organize and galvanize war production and allocate scarce commodities. He expected their employers to pay their salaries, and he offered nothing beyond a dollar a year, only expenses; many of those he brought to Ottawa declined to take their expenses at all. The "dollar a year men," as they quickly became known, were the cream of Canadian business, men like H.J. Carmichael of General Motors, R.C. Berkinshaw from Goodyear Tire and Rubber, Henry Borden, a powerful corporate lawyer from Toronto, E.P. Taylor, a Toronto businessman and brewery owner, H.R. MacMillan, the British Columbia lumber giant, and W.C. Woodward, the west coast department store owner.[12]

By early 1941, the government had recruited 107 of these dollar-a-year men to head up the crown corporations being established, to run various departmental production branches, or to oversee the supply of key inputs, and more would join their ranks as the war proceeded. In the face of today's climate emergency, as one contemplates what leadership — and non-leadership — we have seen thus far from Canada's corporate sector, it is remarkable to recall how many businessmen were prepared to leave their private sector work for the war and contribute their time for nominal pay to help operationalize vital wartime production. "The recruits — financiers, industrialists, and lawyers — brought in talents that had previously been in insufficient supply or completely unrepresented in Ottawa. '. . . the challenge,' Howe reflected in 1960, 'to me was tougher. But I stuck to my system. I asked myself, "who can do this particular thing better than I can?" and I got 'em.'"[13]

As 1940 drew to a close, Howe's team determined it needed to undertake a detailed inventory of both Britain's and Canada's war production needs, and then match that with a careful accounting of Canada's production capacity — both what was available and what could be quickly built. This necessitated Howe and a delegation of advisors going to Britain in December of that year, travelling by ship from New York City.

On his way to London, sometime after midnight on December 14, disaster struck. The ship they were travelling on was torpedoed and sunk by a Nazi submarine some 300 miles off the coast of Iceland. Those aboard had to scramble to lifeboats, where they spent hours in the rough seas, before being found by a Scottish merchant ship that defied orders to avoid submarine infested waters and responded to the SOS issued by Howe's ship. Almost everyone in Howe's party survived except one of his closest advisors, Gordon Scott, who died during the ordeal while the rescue was underway — his lifeboat, hurled by waves, crushed him as he attempted to climb a rope ladder onto the merchant ship. According to Bothwell and Kilbourn's account, Howe was tough and remained defiant throughout the event. The rescue ship took them to Scotland, where Howe bought a new suit and quickly made his way to London for his meetings.

When told that Howe's ship had gone down, and still unaware if those aboard had been rescued, Prime Minister Mackenzie King wrote in his diary, "Throughout the day I have been turning in my mind possible men to take his place but I can think of no one"[14] — a clear indication of the esteem Mackenzie King now held for Howe and what he was accomplishing.

Over half of Canada's war production was fulfilling contracts for Britain. Under heavy bombardment, Britain was in little position to undertake much of the manufacturing for its own wartime supply needs. A key problem for Canada's finances, however, was that neither was Britain in an immediate position to pay. Over the course of the war, Canada extended about $3 billion in financial aid to Britain, mostly to cover British purchases of Canadian orders.[15] This put considerable pressure on Canada's finances. But this problem found some modest relief in 1941 when Howe and Mackenzie King secured a deal for the U.S. to send $200 to $300 million in war contracts to Canada, providing a welcome boost to Canada's cash flow.

By early 1942, Howe's department had grown to 4,000 employees coordinating the nation's war production, and would ultimately peak at 5,000. A key resource in short supply was skilled managers. But these too were being quickly produced thanks to Howe's style of delegation. By this stage in the war, "Howe had been conducting a school for managers for two and a half years, training men [and given the deep-seated patriarchy of the day, and at great loss of potential capacity, they were essentially all men] to think for themselves, decide for themselves, and create for themselves, without referring to a distant head office or wait for orders from a telephone."[16]

Reading Bothwell and Kilbourn's biography of Howe, certain qualities of his leadership stand out. As a minister, he was a quick study, he made decisions decisively, and he was fiercely committed and single-minded when necessary. He was an astute negotiator and politically skilled. Howe could be tough and curt (he had a temper that sometimes exploded), but also charming. And he was a good delegator. When it came to the management of the crown agencies he created, Howe was hands-off, unless a problem developed, such as a production drop off, a labour dispute, or a sharp increase in costs.

Howe would emerge during and after the war as one of the most powerful people in Canada and was widely seen as having a key hand in the direction of the Canadian economy. Yet few would have predicted this prior to 1940. He was an impatient pragmatist, keen to get the job done in whatever way made most sense. In comments he made at a dinner in his honour just before his death in 1960, "He was not, he claimed, obsessed by doctrines like socialism or conservatism. In general, if private enterprise could do the job, it should be allowed to do so; where it failed, or 'where some things needed to be done by the nation . . . we did them.'"[17]

Bothwell speculates that perhaps it helped that Howe was an engineer by training and not a lawyer (like so many politicians of the time). He was imaginative, keen about technology and willing to take "well-calculated risks."

Howe was a federal cabinet minister for all the 22 years he served in elected office, from 1935 until 1957, when he was defeated and left politics.[18] He died three years later at age 74.

Wartime Production: Ships, Planes and Vehicles

While one man, C.D. Howe, was central to Canada's wartime economic mobilization, it was the creation almost overnight of a vast network of workers, managers and factories that created our wartime capacity.

On June 18, 1940, the Canadian government introduced into the House of Commons the National Resources Mobilization Act, which Mackenzie King said would "confer upon the government special emergency powers to mobilize all our human and material resources for the defence of Canada." The Act brought in male conscription, although for service only on Canadian soil or in Canadian territorial waters. But as Mackenzie King went on to explain, military service was only one part of what was required. "The skilled worker in the factory, the transport worker and the farmer . . . are as essential to the effective prosecution of the war." And Mackenzie King noted that under the act, the government was empowered to "call property and wealth, material resources and industry to the defence of Canada."[19]

Immediately prior to the war, as Granatstein writes:

> Most heavy industrial goods were imported, and even the auto sector relied heavily on American motors. Moreover, there were almost no munitions plants in that last year of peace. There was only a small federally owned arsenal in Quebec City (that primarily made limited quantities of small arms ammunition) and a subsidiary plant in Lindsay, Ontario, reopened in 1937. The British government just before the war started had placed a small contract with Marine Industries Limited of Sorel, Quebec, to make one hundred 25-pounder artillery field guns. There were a few tiny aircraft manufacturers that produced airplanes on an almost piecework basis.[20]

But by spring 1940, major contracts for war supplies started pouring in from Britain and emanating from Ottawa itself. Decisions were made and contracts awarded with remarkable speed. Production efforts were often started from scratch and local skills were in short supply. If domestic technical expertise was absent, the government brought in American or British experts to get production facilities up and running, and to train the local workforce. Harry Carmichael, Howe's head of production, drawing upon his experience at General Motors, introduced a nationally coordinated system of subcontracting and sourcing for various parts and supplies that involved factories in every corner of the country.[21]

In some cases, the government turned to private corporations to meet its production needs, such as General Motors, De Havilland, Massey Harris and Boeing, and many smaller companies to produce parts. In other cases it created new crown corporations to get the job done.

Jeremy Stuart, in his MA thesis on the war's dollar-a-year men, explains the process thus:

> The Department of National Defence set the armed forces' requirements, and passed the requirements to Munitions and Supply to decide how and where to meet these needs. The tendering process generally went as follows: first, the

> production branch within Munitions and Supply which specialized in the item decided where it should be produced. If capacity did not exist in Canada, and production likely could not be induced, then the item was given to a Crown Corporation. If the item could be produced by private industry then the tendering process began.[22]

By April 1942, Stuart's research reveals, the Department of Munitions and Supply and its team of dollar-a-year men were reviewing an average of 411 tenders *a day* from potential contractors. Howe's department also housed its own team of lawyers (separate from the Department of Justice) and even started a night shift of lawyers (yes, you read that right) so that they could expedite the negotiation and signing of contracts with production firms. And when bottlenecks developed, they began to issue "go-ahead" letters or preliminary contracts, so that work could begin while the details of contracts were finalized.

This is what emergency operations looked like. As we think about the urgency of the transition needed today, it is fascinating to note the ways in which Howe and his senior advisors structured the Department of Munitions and Supply in a manner that would speed up production and insulate their work from the rest of government in the interests of maximizing swiftness and flexibility. They recruited outside leaders who were not weighed down by the usual bureaucratic norms of doing things. They set up separate legal teams that, unlike conventional government lawyers who are frequently hardwired to tell elected leaders what they *cannot* do, were specifically tasked with making things happen quickly. And they used orders-in-council to expedite work.

The logistics and sourcing challenges were immense. Every ship and plane required countless parts. The coordination of supply chains was a monumental task. New engineering challenges emerged constantly, requiring ongoing modifications and innovations, all on tight timelines.

What was accomplished was, without doubt, an extraordinary feat. Yet by what logic, with our enhanced productivity, education and technology, do we think we could not undertake a similar task today?

While Canada had historically been home to a large shipbuilding industry, it had atrophied over the Depression years. The few remaining shipyards mostly undertook repair work, and a few small ships were constructed for coastal purposes. During the whole of the 1930s, Canadian shipyards built only 14 steamers exceeding 46 metres in length. All that changed with war, and the number of workers employed in the industry expanded tenfold.

At the outset of the war, the Canadian Navy had six destroyers and four minesweepers. By the end, our Navy had 471 ships. But we weren't just producing for ourselves — with supply and battleships being regularly sunk, Britain needed replacements too. So, over the course of the war, Canada produced a remarkable 348 ten-thousand-ton merchant cargo ships (using a design imported from Britain, these ships were called the North Sands, Park, Fort and Victory Ships in Canada and were known as Liberty Ships in the U.S.) and 393 naval warships (primarily corvettes, minesweepers, frigates and destroyers), employing a total of 84,000 people mainly in B.C. and Quebec, but with sizable workforces in the Maritimes and Ontario as well. Over 300 companies across the country were involved in supplying parts for the shipbuilding effort.

As of 1941, the man tasked with overseeing wartime shipbuilding was H.R. MacMillan, the B.C. forestry magnate, who had been recruited as one of the dollar-a-year men. MacMillan claimed to have improved upon the original British design for the Victory cargo ships and was able to produce them more cheaply than the Americans. (As an aside, my grandfather spent most of the war building Liberty Ships, first in the shipyards of Los Angeles and then in his hometown of Newark, New Jersey. He was an artist who started painting ships but later became a welder. I still have his ball-peen hammer from those shipyards, with his initials welded into the head. While he was an animator by trade — he had helped to lead a long strike at the Walt Disney Studios just before the U.S. entered the war — his favoured art form in retirement became metal sculpture, using the welding skills learned in the wartime shipyards.)

Warships were produced mainly along the St. Lawrence River and in the Maritimes. The cargo ships were produced primarily on the west coast. British Columbia's shipyards produced over 300 ships during the war, about 250 of which were the ten-thousand-ton merchant marine cargo ships. As Rod Mickleburgh writes in his history of the B.C. labour movement, "From fewer than a thousand employees in 1939, Vancouver and Victoria shipyards grew to more than thirty-one thousand workers, turning out hundreds of merchant ships and naval vessels at breakneck speed."[23] Indeed, so large did that workforce grow that Local 1 of the Boilermakers and Marine Workers Union of Canada, representing workers in the shipyards of Vancouver and North Vancouver, became the single largest union local in Canada.

That Vancouver shipbuilders local was headed up by a feisty organizer named Bill White. He is the subject of a wonderful oral history called *A Hard Man to Beat* by Howard White (no relation).[24] White, who hailed from a poor farming family on the Prairies, had been an RCMP officer in the Arctic who had quit after refusing to crack down on unemployed workers in the Depression. He tried to enlist when war was declared, but at age 35 and married, the military turned him down. Instead, he landed a job at one of the Vancouver shipyards. After witnessing frequent health and safety violations that caused horrendous accidents and death — these were "par for the course, you see," remembers White — and hardwired to intervene, White quickly became a shop steward, was recruited into the Communist Party and before long was the local's president. Under wartime legislation, wages were frozen. "We were under tremendous pressure not to interrupt war production," recalled White. "Conditions was the only issue left, and these old-time bosses, oh Jesus, they wouldn't budge an inch on conditions." So that's where the union put its focus, and over the wartime years, employing some dramatic showdowns and sit-down strikes, they secured new rights, bonus pay for working in particularly challenging conditions (dubbed "dirty jobs") and new health and safety protections.

The cargo ships produced for the merchant marine played an essential role in the war. In the face of constant attack by German U-boat submarines, the merchant marine convoys carried vital supplies to the Allies in Europe. Extensive repair work was another major focus of Canada's shipyards, so that disabled ships could be returned to service.[25]

Sad to think that today, B.C.'s shipyards are rarely able to land a contract to build a single publicly owned and operated ferry. And in the face of the climate emergency, and despite a large ferry and cargo fleet that now needs to be converted from diesel fuel to electric power, we have not yet mustered the imagination to activate our own shipyards to get that necessary work done.

Before the war, about 40 airplanes a year were manufactured in Canada in eight small plants. By the end of the war, the Canadian air force had grown to become the fourth largest in the world. And during the war years, Canada produced over 16,000 military aircraft (in particular the Hawker Hurricane and Mosquito fighters, Lancaster bombers, various training aircraft and a variety of reconnaissance planes). At its peak, aircraft manufacturing was employing 120,000 people — over 30,000 of whom were women.

The Department of Munitions and Supply's Aircraft Production Branch sought orders from private producers, but it also established two crown corporations, Federal Aircraft Ltd. and Victory Aircraft Ltd., to meet its production goals. Victory Aircraft, based in Malton, Ontario, produced mainly Lancaster bombers. "From the first blueprint to the first test flight took only sixteen months, an impressive accomplishment. The workforce escalated from 3,300 in 1942 to 9,521 in 1944, most of them initially unskilled and about a quarter of them women."[26] Once production got up to speed, the plant produced one plane per day, and 430 Lancasters by war's end.

The Canadian Car and Foundry Company (Can Car) plant in what is now Thunder Bay landed the contract to build the Hawker Hurricane fighter plane, the plane credited with shooting down more enemy fighters in the Battle of Britain than any other. Production of the Hurricanes occurred under the direction of chief engineer Elsie MacGill from Vancouver, then only 33 years old and the first female aeronautical engineer in Canada. MacGill became something of a national superstar — there was even a comic about her, albeit a deeply sexist one. By the end of the war, the plant had produced 1,450 planes.

On Lulu Island in Richmond, B.C., and in Vancouver's Coal Harbour, Boeing set up plants producing airplanes. And like elsewhere — because in an emergency, this is what is done — the government built over 300 homes to accommodate the needed workers.

As for vehicles, Canada produced more military transport vehicles during the war than Germany, Japan and Italy combined. Granatstein writes that these trucks, mainly Canadian Military Pattern vehicles produced by Ford, GM and Chrysler, were "arguably Canada's biggest industrial contribution to victory."[27]

Of course, a similarly extraordinary level of war production was unfolding south of the border. Even a year before the U.S. entered the war, in December 1940, President Roosevelt declared that the U.S. "must become the great arsenal of democracy." He was preparing the American population for the inevitability of U.S. entry into battle, but already making it clear that the economic task was on. By the time the U.S. formally declared war in December 1941, behind the scenes, officials had already developed a "Victory Program" to mobilize millions of men into the armed services and a plan to spend $64 billion on military production, an amount that would eventually triple.[28] Economic advisors had spent the better part of the previous two years planning for how to quickly convert the U.S. economy for wartime production.

The year before the U.S. entered the war, the American auto industry had sold 4.6 million civilian automobiles. Four days after the Japanese attack on Pearl Harbor, the U.S. auto manufacturers were ordered to cease civilian production,[29] and by February 1942 — a mere two months later — the last civilian car rolled off the assembly line. Take note. In the United States of America, ground zero of auto culture and free enterprise, for the balance of the war, the production and sale of the private automobile was illegal. Same in Canada. All those car manufacturing factories and their workers were still plenty busy, but now focused on military production. That is what happens when people understand they are in an emergency and refocus their resources accordingly.

Not everything about Canada's war production was perfectly designed or implemented. Occasionally war production efforts landed in trouble or controversy. There were some complaints about graft, preferential contracting to friends of the government and about some of the dollar-a-year men being in positions to benefit their own companies.

There were occasionally conflicts between the dollar-a-year men and the career civil servants. And there were conflicts between Howe and Minister of Defence Ralston over "manpower," namely, how many people to recruit into the military versus how many to assign to war production. As the armed forces grew, industrial production of munitions started to experience labour shortages. The solution, of course, was staring them in the face — bring in the women.

With respect to private production, because all the government defence contracts were structured as "cost-plus" arrangements (either cost-plus a set fee per unit or cost-plus a guaranteed percent of 5–10%), somewhat perverse outcomes could result. Whatever the cost of production, the shipyard or factory owners were assured a given profit. Given the lack of experience in munitions production in Canada in the early years of the war, this arrangement was understandable. The problem was that the cost-plus contracts created no incentive for owners to control costs — indeed, quite the opposite; the contractors were often content to see wages escalate, hire unneeded workers or even turn a blind eye to theft of supplies. As local union president Bill White recalled, "The deal was, you see, the shipyards got 10% on every dollar that was spent, so the more waste there was, the richer they got. The more cost the more plus."[30] One of the solutions to keep this dynamic in check was the creation of crown corporation competitors.

Crown Corporations and State Planning to the Rescue

While Howe favoured private enterprise, when the private sector couldn't meet a need quickly, his solution was the creation of new crown corporations. As Granatstein writes (ironically, in his paper prepared for the Canadian Council of Chief Executives): "There was a shortage of rubber? Set up a Crown company to produce synthetic rubber. Wood veneers for aircraft were in scarce supply? A Crown corporation could do the job. Machine tools? Howe's Citadel Merchandising could get them and make sure they went where they were most needed. In all, 28 Crown corporations came into being during the war, some manufacturing, some purchasing and distributing, others supervising and controlling. The establishment of Crown companies, operating with great flexibility outside the usual bureaucratic restraints, allowed for efficiencies."

Let that sink in — the war effort saw the creation of 28 crown corporations to get the job done. Today, there are 47 federal crown corporations in total, although ten of these are national museums or arts and culture organizations.

"The government was ideologically supportive of crown corporations and they were established in situations where public ownership was a more effective policy instrument than regulation," wrote York University professor of public administration Sandford Borins in 1982.[31] In some cases they did what the private sector could not, while in others they served as "a yardstick competitor," ensuring private sector munitions producers weren't bilking the system. "The characteristic of crown corporations making them preferable to regulation was that they were more effective at monitoring the private sector and ensuring policy coordination between the public and private sectors."

Of the 28 crown corporations established in the war, Borins distinguishes between "production" ones and "administrative" ones.

PRODUCTION CROWNS	ADMINISTRATIVE CROWNS
Defence Communications Ltd.	Aero Timber Products Ltd.
Eldorado Mining and Refining Ltd.	Allied War Supplies Corporation
Machinery Service Ltd.	Atlas Plant Extension Ltd.
National Railways Munitions Ltd.	Citadel Merchandising Company Ltd.
Polymer Corporation Ltd.	Cutting Tools and Gauges Ltd.
Quebec Shipyards Ltd.	Fairmont Company Ltd.
Research Enterprises Ltd.	Federal Aircraft Ltd.
Small Arms Ltd.	Melbourne Merchandising Ltd.
Toronto Shipbuilding Company Ltd.	North West Purchasing Ltd.
Turbo Research Ltd.	Park Steamship Company Ltd.
Victory Aircraft Ltd.	Plateau Company Ltd.
	Veneer Log Supply Ltd.
	War Supplies Ltd.

ADMINISTRATIVE CROWNS
Wartime Housing Ltd.
Wartime Merchant Shipping Ltd.
Wartime Metals Corporation
Wartime Oils Ltd.

According to Borins, when choosing to go the crown corporation route, the reasons included "ensuring secrecy, providing continuity of production, coordinating complicated projects, taking risks that would have been unacceptable to the private sector, responding to the initiatives of other public sector agencies, and preventing inefficiency."

For example, in the interest of secrecy, Research Enterprises was established to work on radar technology, and Turbo Research was established to work on jet engines. Defence Communications and Small Arms Ltd. were established at the request of the Department of National Defence, which wished the work of these corporations to remain secret. In mid-1942, Howe was approached by the Americans about the top-secret "Manhattan" atomic bomb project and asked to help supply uranium for the endeavour. Uranium was then mined by the Eldorado Gold Mine company near Great Bear Lake in the Northwest Territories. Consequently, Eldorado Mining and Refining was expropriated — nationalized for national security reasons — and made into a crown corporation.

Toronto Shipbuilding, Quebec Shipyards and Victory Aircraft had all been private companies prior to the war but were expropriated and nationalized during the war when the government determined they were not operating at adequate efficiently. In other cases, as historian and postdoctoral researcher at Trent University Alex Souchen explained to me, crowns were newly established when the private sector was reluctant to make major capital investments in an activity that was inherently temporary.

The Polymer Corporation was an example of how a crown corporation was used to coordinate the work of a sector: "Due to the shortage of natural rubber after Japanese victories in the Pacific, the government decided to build as rapidly as possible a synthetic rubber plant which would be large enough to achieve economies of scale. The construction was a massive undertaking, involving a great deal of complicated engineering. The completed plant was managed by a consortium of six companies" all of which benefitted from the coordination of a neutral third party — the government.[32]

In some cases, the government created crown production firms simply to ensure competitive bidding. Given that very few firms were in a position to bid on any particular contract, an appropriate contract fee was otherwise hard to determine. The government tended to own and operate at least one firm within each major production field — shipbuilding, aircraft manufacturing, munitions production — so that it had a window into each industry and could ensure the private operators were operating efficiently and fairly.

As for the "administrative" crown corps, these were mainly to ensure effective cooperation across a sector when needed. Often the firms were created at the initiative of the relevant controller at the Wartime Industries Control Board, to ensure adequate supply of a given resource at the best price possible, such as Aero Timber Products, Veneer Log Supply, Fairmont (supplying rubber), Melbourne Merchandising (supplying wool) and Plateau (supplying silk). In other cases, the collective purchasing was of higher-value manufacturing inputs: Citadel purchased machine tools for other production companies, and Cutting Tools and Gauges Ltd. did so for various manufacturers. Centralized bulk purchasing benefitted everyone, ensuring lower prices for inputs and smooth supply chains (rather than various manufacturers competing with one another for inputs) — something well worth considering today in many areas. Might not the same logic exist for the manufacturing of heat pumps, wind turbines and solar panels?

Wartime Housing, established to build housing for soldiers and war production workers, also offers an interesting lesson for today. It acquired land, handled development, local permitting and planning, and hired architects. All these functions benefitted from economies of scale. Doing this through a crown corporation also meant that housing was built where it was needed, close to war production plants or military bases. The building of the housing itself was contracted out to private local builders, as that is where the expertise lay.

Wartime Housing morphed into what is now the Canada Mortgage and Housing Corporation (CMHC), but CMHC is sadly no longer in the business of building new supply, despite a housing crisis. But given how costly land and development has become today, might the Wartime Housing corporation not be a model for how we could build an adequate supply of carbon-zero affordable housing today?

Early in the war, the government understood it faced two major economic challenges. First, all the wartime manufacturing and construction required a major expansion and coordination of basic inputs. And second, to control costs and to avoid wartime profiteering, inflation had to be kept in check.

Consequently, when war was declared, the Canadian government established two new agencies: the Wartime Prices and Trade Board (WPTB), tasked with controlling both wages and prices for key items like fuel, sugar, wool and, interestingly, rents,[33] and the Wartime Industries Control Board (WICB), which controlled the supply of key resources needed for wartime production including oil, coal, gas, steel, metals, chemicals, power, rubber, machine tools and timber.

At the WICB, "controllers" were put in charge of all these major branches of the economy. As Newfoundland academic Dennis Bartels, citing A.F.W. Plumptre's description from the time, recounts, "each Controller had sweeping power: 'For instance, the Machine Tool Controller [had] authority over machinery of every kind and could enter on, take possession of and utilize any land, plant, factory or place used or capable of being used for making or storing machine tools.'"[34] The regulations governing the WICB were compiled into a 1944 book over 500 pages in length.[35] The controllers were not only empowered to direct where resources and supplies went, but also (in collaboration with the WPTB) to set minimum and maximum prices for those supplies, so as not to permit excessive markups or profiteering. The controllers could demand account books from any company doing business in their field and were even allowed to enter and search business properties to secure those accounts if they felt it warranted.

The oil controller was a man named George Cottrelle. "The task would have exhausted a lesser man," recount Bothwell and Kilbourn, "but Cottrelle reduced his worries by applying a simple maxim: no exceptions. Cottrelle's inspectors spread out across the country, investigating, justifying and occasionally prosecuting, sustained by widespread public approval and a considerable network of private information. Those who violated the law were hauled in to explain themselves; those who wasted gas found their rations suddenly reduced."[36]

To meet the need for more electricity to power wartime production, there was a huge ramp-up in hydroelectric power investment and the construction of new dams. By 1945, hydroelectric generating capacity had expanded over pre-war levels by 40%.[37] In Quebec, the provincial government nationalized the electric companies and formed the new crown corporation Hydro-Quebec.

Farmers in Canada expressed grievances during the war that they were not getting a fair price for their produce, that they were denied the right to sell into the U.S. market for a higher price, and were thus being asked to shoulder an unfair burden. In response, the WTPB created another corporation, the Wartime Food Corporation, "to buy all the beef cattle needed to adequately supply the domestic market, by paying producers the price they would have received in the U.S. The Corporation would then sell that beef to packers (or the military) at a price that would not break the price ceiling and absorb the loss."[38] The model worked. Food production in Canada increased by almost 50% during the war, while the prices received by farmers grew by about 60%, whereas prices paid by consumers were held steady.[39] The increase in production occurred despite thousands of farmers and farm workers enlisted into the military, although this was offset by the fact that relocated Japanese Canadians, prisoners of war and conscientious objectors who refused conscription were ordered to work on farms.

Once again, what is notable is that, in wartime, neither private sector actors nor the supply-and-demand price signals of the market were allowed to determine the allocation of scarce resources that were needed for the war effort. Rather, the state ensured that wartime production needs took priority, and supply chains were organized to meet those needs. Effectively, Canada's wartime economy was a planned economy.

Household Management: Canadian Families Called to Do Their Part

It wasn't only major industry that was subject to planning in the war, or whose use of scarce resources was controlled. The Greek etymology of the word "economics" is the science of "household management." And households were indeed managed.

The war saw household consumption dramatically transform. Quotas — or rationing — were applied to various goods such as meat and fuel, recycling was strongly urged, families grew "Victory Gardens" and people dramatically switched their transportation from private automobiles to public transit — coincidentally, all actions that also reduced GHG emissions, even if that was not their intent. In a prescient nod to the need to retrofit our homes today, Canadians were exhorted "to minimize [energy] waste, people were instructed to clean furnaces regularly, seal windows, close off unused rooms and not heat garages."[40]

From personal and family experience, Harold Steves knows a lot about how household consumption changed during the war, and more importantly, how families adjusted their behaviour to "do their part."

Steves, born in 1937 and now in his 80s, remains a vigorous environmentalist and political activist. He was first elected as a city councillor in Richmond, B.C., in 1969 and has now served consecutively as a Richmond councillor since 1977, most recently re-elected in 2018. During a brief interlude, he served as an MLA in the B.C. provincial NDP government of Dave Barrett (1972–1975), and while not a member of cabinet, he is credited as the driving force behind that government's creation of the Agricultural Land Reserve, the groundbreaking legislation that has helped preserve farmland in British Columbia.

While Steves is an honorary lifetime member of the NDP, he's politically hard to pin down. He describes himself, with unique specificity, as "a Green New Democrat from 1972," and on social media is actively critical of the current B.C. NDP government for approving

LNG and the Site C dam. His entry into the world of politics was environmentalism. In 1968, he helped to found one of the first environmental organizations in Canada — the Richmond Anti-Pollution Association.

Outside politics, Steves has been a farmer for most of his life, and he continues to live in the same family farmhouse where he grew up. His ancestors were pacifists who immigrated from Germany to the east coast in the 18th century in the hopes of finding a more peaceful life. But in an ironic twist, the family's Richmond farm, which sits strategically at the intersection of the Salish Sea and Fraser River, was expropriated and turned into a military base for the Royal Canadian Artillery during the Second World War. The family had to move a few houses down the road. At age four, young Harold Steves became the base's mascot — the soldiers even had a tiny army uniform tailored for him, which Steves still has — and his mother opened up a canteen to offer food and beverages to the troops who now occupied their farmland. The family kept all their wartime artifacts from that period, and Steves now, quite literally, has a small Second World War museum in his farmhouse basement.

I went to visit Steves and his museum on June 6, 2019, coincidentally the 75th anniversary of D-Day. His collection includes wartime posters and photos, cans of foods and bottles of drinks sold at the time (including local salmon canned in Richmond but mostly shipped to Britain during the war) and various and sundry paraphernalia. Many of the items capture what families in Canada were encouraged to do during the war. Steves still has the ration books his family used — booklets containing stamps of various colours and shapes that households had to use when purchasing their quota of core necessities such as meat, fuel, sugar, butter and rubber. The creation of these ration books during the war was a huge undertaking. Ration card applications were sent to every household and "volunteers then went door-to-door to collect the completed applications, which were then turned over to transcribers [female volunteers] who used the information to fill out the ration cards [by hand]."[41] Nearly 12 million ration books were distributed in only 14 days.[42] While wartime rationing was not without its rule-breakers, in the face of the climate emergency, it is notable that the public has previously been willing to ration its consumption of both meat and fuel, although ironically, overall meat consumption actually rose during the war, no doubt because this more equitable model allowed poorer families to consume meat that had previously been unaffordable.

Some old posters on display urge people to buy Victory Bonds, while others heavily promote recycling of all manner of items, for conversion into wartime supplies. Their messages — such as "Waste Is Unpatriotic" — pulled no punches in the name of subtlety. "Bomb 'Em with Junk!" declared another, imploring people to collect and turn in anything with iron, steel, copper, brass, aluminum, zinc, lead and tin. "There Are Bombs in Your Barn . . .There Are Guns in Your Attic!" one recycling poster beseeches. The government also collected hundreds of thousands of pounds of both bones and fat, for use in industrial glue manufacturing and as a weapons lubricant. It is notable how many of the things people were urged to do and conserve were precursors to what today we view as "good for the environment," even if the purpose was different then.

Of special interest to Steves, given his farming background, was the drive for people to grow Victory Gardens. While farms in Canada operated at full-tilt during the war, much of what they produced was needed for export to war-torn Britain and to feed those in military service. And food was not being imported. Consequently, families were urged to grow as much as they could of the food they needed themselves. And grow they did. In 1943, the government estimated that 115 million pounds of vegetables had been grown in Canadian Victory Gardens.[43] (In the U.S. it is estimated that 40% of the food consumed by households during the war was self-grown in Victory Gardens, but a comparable Canadian statistic is not available.) The Steves family took that commitment up a notch. They had established a mail-order seed company before the war and were expert growers. To help others with less experience, Steves's father and grandfather used a horse-drawn plow to dig up the local Richmond boulevards, so that non-farming neighbours could plant their own gardens on public land.

Adherence to these public calls — from recycling to growing gardens — was extremely high during the war. I asked Steves what he thought motivated such compliance. "People really believed in what they were doing," he contends, "and that is what is missing today in terms of climate change."

Delivering the Goods

"Canada delivered the goods," Granatstein wrote in his 2005 report to Canada's major corporate leaders, "producing 40% of Allied aluminum and 95% of the nickel. It mined 75% of the asbestos, 20% of the zinc, 12% of the copper, and 15% of the lead. Very simply, without the aluminum provided by the Dominion, the Royal Air Force could not have fought the war . . . Canada produced an array of military equipment, its war production overall ranking fourth among the Allies, behind only the United States, the United Kingdom and the Soviet Union. For a nation of just 11 million people, this was little short of amazing."[44]

All in, "Canadian wartime industrial production was valued at more than $9.5 billion in 1940s dollars (in today's dollars, more than $100 billion). Another $1.5 billion was spent on defence construction and the expansion of war plants, all paid for by the government. For a nation that had begun the war with a Gross National Product of $5.6 billion, this was incredible. That Canada's GNP in 1945 was $11.8 billion, more than double the total six years before, is accounted for in large part by the extraordinary production of the nation's war factories and mines."[45]

In fact, the transformation of the economy in the war years had to happen twice: once to ramp up war production, and again a few years later when the economy needed to be reconverted again to peacetime.

This latter effort also required careful planning, and again, C.D. Howe was tasked with overseeing this effort. As of 1944, the Minister for Munitions and Supply was also made minister responsible for post-war reconstruction. And Howe began to plan again. His team considered how many returning soldiers could be reabsorbed into various sectors — farming, forestry, mining and construction industries — and they assessed what factories could be fairly easily converted from wartime into peacetime production. And they understood too that a comprehensive system of supports would be needed for returning soldiers, covering income, training and housing.

Nevertheless, the immediate post-war years were challenging, as government wartime contracts dried up. The economy struggled to revert to peacetime, marked by a period of flat economic growth and labour unrest. In that respect, we may have an advantage in the present, in that we do not need to convert the economy twice, as they did in the war years, but only once, on a permanent basis.

The wartime production effort and other home front efforts were not without faults. Howe and his team made errors. There was inefficiency and waste. There were black markets and graft. Some of this may be inevitable, as a society seeks to move quickly and experiment with new approaches. Also, Howe effectively created a parallel public service, but his "work-arounds" that sought to expedite production eliminated the checks and balances that normally prevent such problems. The lesson for us today, then, is not to end-run the public service, but rather to transform it.

As we contemplate the economic transformation now before us, it is also worth considering: was the war production effort more public or private? Ultimately, it is hard to say because it was all so integrated, with government sending contracts to the for-profit sector, and in turn recruiting senior managers into the public realm. Granatstein writes, "Howe and his advisers believed that private enterprise was inherently more efficient than government-run operations. The Second World War made the government — or at least C.D. Howe's part of it — operate much like a corporation." Tommy Douglas, then still a CCF Member of Parliament, expressed the concern in 1942 that "instead of government taking over industry, industry has taken over government."

But that was surely overstated. Unlike the way in which our mostly market-based economy functions today, the private sector in wartime did not get to *decide* on the allocation of resources. Rather, the economic mobilization was coordinated and of course paid for by the public sector — by public servants who planned the overall effort, orchestrated the supply chains, regulated economic conduct including prices and profits, and directed massive public investments into realizing this economic transformation. And in both Canada and the U.S., wartime production was not only done by private for-profit corporations (under the direction of the public authorities). A huge amount was undertaken by newly established public enterprises specifically created to meet the wartime effort.

The takeaway is that meeting the wartime effort involved both harnessing the entrepreneurial enthusiasm of the private sector but also realizing that the overall coordination and investment of the state was essential. The conversion of our economy was thus a product of both socialism and capitalism — not purely public or private, but planned.

While Howe was a private sector-oriented guy, he was also a practical man in a hurry. As Borins concludes, "Howe and his associates had an ideology consisting of the following elements: winning the war was the over-riding priority; cooperation between business and government was essential to the war effort; and institutional innovation was justified if it served the war effort."[46] That's the focus we need today.

REJECTING ECONOMIC STRAIGHTJACKET THINKING: HOW CONVENTIONAL ECONOMICS LIMITS OUR ACTIONS TODAY

My point in recounting all the planning and economic controls employed during the Second World War is not to say that we require all of that again. Rather, it is merely to blow open our sense of what we might consider to meet the urgent task now before us, and to highlight what becomes possible once we recognize an emergency.

What is notable about Canada's wartime economic policies is that our leaders then were not bound by the straightjacket of neoliberal economic thinking.

Thus far, however, our leaders today very much are. So much of our climate-response policies to date can be summarized in four words: incentives, rebates, carbon pricing. We are fiddling at the margins while the planet burns, hoping that market-based signals can sufficiently alter household consumption and business investment. They won't.

The fundamental point of an emergency is that action is not voluntary. For example, if a wildfire is nearing a community (as occurs with increasing frequency in this era), and the government declares a state of emergency, people are not told "we *encourage* you to leave." Rather, they are ordered to evacuate. It requires good collaboration with local governments and Indigenous communities, clear communication and trust in our public institutions and first responders. But we treat such events as a true emergency. Choosing to stay put is rightly viewed not only as individually reckless but also as collectively irresponsible — to do so puts emergency responders needlessly at risk.

The climate crisis, unhelpfully, moves in slower motion. But let there be no doubt, the flames are licking at our heels.

Yet as one surveys the federal, provincial and municipal climate plans across Canada to date, what stands out is the overwhelmingly voluntary nature of almost all the policies. In the face of clear knowledge that we have to stop driving fossil-fuel cars and switch our home heating off oil and gas, governments are offering incentives and rebates and "encouragement," but with a few notable exceptions, what they are decidedly not doing is *requiring*.

Stubborn neoliberal economic assumptions have gotten in the way of serious climate action and foreclose on options that must now be considered. This dynamic plagues not only the mainstream discussion, but also progressive elected leaders and some environmental non-governmental organizations. Whether it is the need for strong regulatory measures, effective economic planning or large-scale public investment, the dominant economic paradigm of our time has prevented us — and our governments of all political stripes — from consideration and adoption of the economic policies the crisis demands. Instead, almost all the political oxygen has been consumed by "market-based solutions."

Before continuing, perhaps it's best to offer a simple definition of what I mean by neoliberalism. Firstly, neoliberalism is a set of public policies that privilege and prioritize the "free market" and the interests of for-profit corporations. These include: tax cuts (particularly those benefitting the wealthy and corporations), public spending cuts (which generally follows from tax cuts), privatization of public services and public enterprise (or building new public infrastructure as a "public-private partnership" that bestows huge benefits to large for-profit consortiums), deregulation of industry (for example, getting rid of environmental and safety regulations that hamper private businesses) and "free trade" agreements that aim to maximize the mobility of investment and corporations while limiting the ability of governments to set policies — including climate policies — in the public interest. Neoliberalism fetishizes the goal of balanced budgets and austerity, even when there is no genuine economic rationale for doing so. But neoliberalism is also an ideological orientation, one that fundamentally does not believe in the wisdom or efficiency of delivering services and accomplishing great tasks collectively through our governments, and that disparages and undervalues the public sphere. Neoliberalism is not so much a response to socialism or communism, as it is to Keynesianism, the economic framework that prevailed during the Second World War and the three decades that followed, which held that, even within free-market economies such as ours, there was much that government can and should do to boost employment, meet our collective needs and improve overall well-being and equity.

Neoliberalism has now dominated our politics and economic debates for four decades. Many of its core tenants are embedded not only in conservative parties, but in the policies of *all* our mainstream parties (including the Liberals, the NDP and the Greens). And it has another legacy that haunts us now. It has bred a cynicism about government and our collective capacity to act, and has left us feeling more isolated and powerless to effect change. "There is no such thing as society," U.K. Prime Minister Margaret Thatcher famously said in her iconic elucidation of neoliberalism. "There are individual men and women and there are families." That's a sentiment that severely undermines what we need to accomplish together today.

Why have our governments, even progressive ones, been so reticent to undertake large-scale investments in green infrastructure and renewables? Why have they been reluctant to consider ambitious new taxes or higher oil and gas royalty rates to fund such investments? Why have they shied away from an exciting and truly ambitious green jobs plan? And why have they felt unable to use the regulatory power of the state itself to simply mandate necessary changes?

The answer is that ultimately, they accept a core — yet false — neoliberal assumption that only the profit-seeking private sector truly creates wealth and jobs, and secondly, they falsely believe that only the resource and manufacturing sectors produce "real" jobs. These governments have been unwilling to oppose new fossil fuel projects because they have assimilated the proposition that without these private-sector investments in major resource projects, rural communities in particular will have no other prospects for economic security and employment. They have internalized that these are the only game in town. And they fear retribution — in the form of capital strike or flight — from the fossil fuel sector (and perhaps the larger corporate/financial sector) if they were to truly confront these powerful interests.

I recall being in a meeting with a former B.C. NDP MLA (before they were government), and when I recommended substantially raising royalty rates for natural gas, he got wide-eyed and replied, "Look what happened when Ed Stelmach tried to modestly raise royalties in Alberta! The oil and gas companies moved their rigs over the border to the B.C. side until the Alberta government cried uncle." His point was not that these corporations play hardball and that we must steel our resolve to confront them. Rather, he was already resigned to what governments purportedly cannot do. (My counter-advice was that a progressive government should take out an ad in *The Economist* seeking a CEO for a new natural gas crown corporation and see if the private industry doesn't cry uncle.)

Similarly, in a meeting with a senior civil servant responsible for provincial climate policy, I once asked why the government doesn't simply order companies providing district-wide central heating that burn natural gas to switch to renewable energy by a given date, and if they refuse, expropriate the company at a fair price. The response was a blank stare — today's civil servants, unlike those of the Second World War, are just not wired to contemplate such things.

Former MP Libby Davies spoke of an assumption in Canada about what constitutes "good governance." We have this idea that we have to "appease" the corporate sector and private interests. Social democratic governments don't feel this any less. Indeed, arguably, they feel it more — they are afraid of being red-baited and therefore seek to prove they aren't scary. Hence the Alberta and B.C. NDP try to show industry that they are friendly, even when that means adopting policies that conflict with climate science. Davies also believes the NDP is captive to insiders who are, at their core, centrist and deeply cautious on economic policy.

There is a deep paradox at play: neoliberal policies have deepened inequality, and economic and job insecurity, yet the very people who advanced these policies now capitalize on these very conditions; having made people less secure, they use this fear as a pretext for furthering fossil fuel development.

Even the Green Party of Canada, which in most respects has, as one might expect, the strongest climate policies, still continues to campaign on promises to balance the budget — a sop to neoliberalism of no actual economic value, but which severely hamstrings the party's capacity to advance significant and necessary green infrastructure spending.

In economics sessions I have, on rare occasions, conducted with progressive politicians, I'm struck by how many good elected folks fundamentally lack confidence on economic matters and harbour the fears and assumptions just mentioned. Consequently, when these progressive people win election into government, they permit vested interests within the civil service and powerful corporate interests to tell them what is and isn't "allowed" when it comes to economic policy-making. This dynamic plagues otherwise progressive people who lack confidence in economics, and it is heightened when senior civil servants remain in place after a change of government — the same people giving the same advice as always.

In contrast, in popular economics workshops I taught over the years, I always felt my core task was to challenge the hubris of economics and the straightjacket thinking of mainstream economic assumptions in order to bolster people's confidence to engage in economic debates and to imagine different possibilities. The goal was to encourage people to reject the harmful notion that only the profit-seeking sector undertakes real job and wealth creation, inviting them to appreciate that, in fact, wealth is created whenever natural resources, human ingenuity, human labour and finance capital are combined to add value (including the production of services, not only goods). The for-profit private sector is certainly effective at doing this (although not always in the most socially and environmentally helpful ways), but it does not hold a monopoly on such activity. The public sector does it, as do the co-op sector, Indigenous nations, municipalities, worker-owned businesses, crown corporations and credit unions. Once we are liberated to recognize that all these sectors can create common wealth and employment, we can start to imagine more democratic and decolonized economic practices and paths. And, as we contemplate what we must undertake to confront the climate crisis, we can start to envision all these sectors engaging in valuable job-creating work.

Neoliberal ideas and assumptions are like zombies. Not only do they die hard but sometimes experience and evidence must kill them many times over before we finally leave them in the trash heap of this unequal and unsustainable era, in which our economy treats our shared environment and atmosphere like a free toilet, leaves so many behind and sees so many wrestling with precarity and unaffordability. Yet in the face of the climate emergency, many environmental organizations and governments of all political stripes continue to waste precious time entertaining only neoliberal policy options.

It was my good fortune to have a few economics teachers who broke the conventional mould of the discipline. One of those was Marjorie Griffin Cohen, a professor emeritus at Simon Fraser University and the founding chair of the CCPA's B.C. office. During the early debates over B.C.'s carbon tax, Cohen was presciently among those warning of the limited benefit of carbon pricing. She always clearly understood that if we want a change to occur, we need to use the power of regulation to make it happen.

"Relying primarily on a market solution to deal with climate change has been a huge strategic mistake," Cohen shared with me. "The argument centred initially on which type of market mechanism would be best — carbon tax or cap-and-trade — neatly shifting the discussion away from the core problems and strong regulation. Carbon pricing as the main approach for reducing GHG emissions is not just inadequate, it also takes the power structures and all the inequalities of the existing economic system as given."

"Many supported carbon pricing because economists were almost universal in championing this approach (rather than outright corporate regulation)," she observed. "It was considered to be more efficient than anything else. Even those activists who recognized its likely inadequacy thought it was at least a 'step in the right direction.' This was mostly because corporations and governments had resisted dealing with carbon emissions altogether, but when forced to choose some type of action, carbon taxes were their preferred approach. But ultimately, carbon pricing is tinkering with a monumental problem that requires direct and concerted action, through governments, to control the behaviour of corporations in the production and use of carbon. Instead, we are fighting to rely on small, incremental increases in carbon prices to direct individual behaviour, with the hope that these collective actions will trickle up to ultimately influence the behaviour of corporations. This hyper-focus on one solution — and the very hard work to even achieve carbon pricing — means giving up on corporate regulation and giving up on bringing about a serious change in the balance of power in our economy."

B.C. is now lauded for its introduction of a carbon tax in 2008, and I support the tax. But a distressing truth is that B.C.'s GHG emissions in 2018 (the last year for which we have data) stood at about 66 megatonnes, four megatonnes higher than in 2007, the year before the carbon tax was introduced. True, emissions might have been higher still without the carbon tax. But that's ten years with precious little progress to show. Ultimately, the planet does not care if our GHG emissions are relatively lower than might have occurred under status quo conditions. It's time to dramatically bend the curve.

Here then are core lessons we should take from our wartime experience retooling the economy:

- Spend what needs to be spent — in infrastructure, in training, in new economic institutions and firms, and in contracts — to get the job done.
- Recognize that an emergency means shifting from voluntary incentives to mandated changes — at the household, community and industrial level. Use regulatory fiat as needed to require changes that must happen.
- Set clear and ambitious targets, both for the overall economy and for various sectors.
- Conduct a national inventory of conversion needs, so that those needs can be matched with production, distribution and training capacity.
- Establish and empower new agencies and crown corporations as needed to get the job done. In particular, consider creating at least one crown corporation in each major sector needed for the transition, to ensure a public competitor exists to control costs and prices.
- Prioritize and coordinate the use of scarce resources and key inputs for the task of producing what is needed.
- Centralize power and coordination as needed, but also liberate and empower local and sectoral leaders to do their part.
- Limit household consumption of items as necessary for the transition.
- Galvanize and inspire citizens, workers and business leaders, so they rally to the urgent task at hand.

Now we consider what these lessons might look like, applied to today's emergency.

MOBILIZING THE ECONOMY TODAY

The task before us now involves nothing less than the wholesale transition of our economy, industry, communities and homes. The IPCC, in its pivotal 2018 report, notes that limiting global warming to 1.5°C will "require rapid and far-reaching transitions in energy, land, urban and infrastructure systems (including transport and buildings), and industrial systems." In short, we need to decarbonize and electrify everything, while also ensuring that we are no longer generating electricity by burning fossil fuels. And we need to do it all in a hurry.

Planning for a New Carbon-Zero Economy

Some may contend that war-style planning is no longer possible, that in today's free-market and globalized economy, the days of being able to coordinate society-wide economic transformation are over. True, technology may make some forms of planning more challenging. Globalization and digital realities may make it easier for both companies and households to hide income and expenditures, or to escape government regulations. But it is also true that today's technology makes planning easier.

Corporations themselves are already showing us what is possible. As Leigh Phillips and Michal Rozworski argue in their 2019 book, *People's Republic of Walmart: How the World's Biggest Corporations Are Laying the Foundation for Socialism*, major multinational corporations like Walmart and Amazon, as well as government operations like the U.S. Pentagon, are proving modern technology allows for planning and supply-chain logistics coordination on a scale that Second World War planners could only have dreamed of.

Phillips and Rozworski write, "For many progressives, the story of logistics and planning seems musty and old." But they invite us to re-engage with the need and potential for economic planning to meet the core challenges of our age. They outline how "big data" makes planning possible in entirely new ways. And they rightly point out that companies like Walmart and Amazon, despite being larger "economies" than many nation-states, carefully manage their global supply chains in much the same way that Canada's WICB did during the war. "It is already the case," they write, "that great swaths of the global economy exist outside the market and are planned." Indeed, according to the United Nations Conference on Trade and Development, about 80% of global trade takes place in "value chains" that are already managed by transnational corporations, and much of this is intra-corporate, meaning the movement of goods and services between international subsidiaries and branches of the same corporation.

The problem with these corporations, emphasize Phillips and Rozworski, isn't with their size per se or with planning, it is that these larger-than-state economies aren't democratic. The real issue, as we contemplate the energy transition before us, isn't "Can we have sufficient planning to move us off fossil fuels?" We can. The true challenge is — can we have more democratic planning? Will the transition need to be top-down like it was in the war, or can we embrace models of collective economic planning and organizing that are more participatory? This is indeed a challenge, given the short time we have, and one marked by an inherent tension — we know this transition needs to be state-led and fast, but can it somehow resist the top-down logic of the state? "Nothing about us, without us," as the social movement slogan, first coined by the disability movement, says.

An overall plan for economic transformation aligned with the climate emergency needs to include eight core elements.

First, like C.D. Howe did, we need to conduct a "national needs inventory" of all the items required to electrify everything with renewable power. How many heat pumps will be needed? How many electric buses, cars and trucks? How many electric charging stations and where should they be located? How many solar panels should we install on homes, public buildings and commercial buildings? How many additional solar fields will be needed, and how much power do we need them to generate? How much wind power is needed, and how many turbines will we need to deploy to produce that? Where should tidal and wave power be produced? What is needed to electrify our ferry fleets and our government vehicles? Which communities will be best served by geothermal power, and which neighbourhoods can be serviced by community renewable energy utilities? Et cetera.

The good news is that many of these core estimates have already been done. Stanford University's Mark Z. Jacobson and his team have not only calculated how every U.S. state can meet its energy needs through renewables (and without relying on new mega hydro dams, new nuclear power, biofuels or carbon capture), but they have also very kindly done so for Canada and 138 other countries. An engineering professor (as C.D. Howe was in his early career), Jacobson's work is both hopeful and practical. For each jurisdiction his team takes on, they meticulously calculate how much energy is needed, how much can be produced by each of wind, water and solar power, how many wind turbines, solar arrays and, to a lesser extent, tidal/wave turbines will be needed, how much it would cost to produce this new power, what materials and land will be needed, and approximately how many jobs will be created producing this new capacity. It's a remarkable body of work, showing that we already have the technological capacity for the world to get to nearly 100% renewable energy. (There remain some technological challenges to meet the final 5–10% of our energy needs, mainly with respect to air travel, shipping and a few industrial processes, but new technology is quickly coming online to meet these needs too.)[47]

For Canada, Jacobson and his team calculate that we can attain 100% renewable energy by 2050 through a combination of:

- 27.5% onshore wind power;
- 22.9% offshore wind power;
- 14.5% hydro power;
- 9.8% concentrating solar plants;
- 9.1% commercial and government rooftop solar power;
- 6.9% solar plants;
- 5.3% residential solar rooftop power;
- 2.2% from wave devices;
- 1.7% geothermal power; and
- 0.2% tidal turbines.

Were we to adopt this plan, Jacobson and his team calculate we would generate 315,138 construction jobs, and 367,889 operating jobs (with each of these jobs representing 40 years of continuous employment). They estimate all this new power generation will require only 0.25% of Canada's land mass. They also contend that overall demand for energy would be 38% less than under status quo conditions, as fossil fuel extraction and combustion is much less efficient than renewable electric technology. And they estimate this energy shift will result in $110 billion in health cost savings (due mainly to reduced air pollution).[48]

Similarly, the David Suzuki Foundation's Clean Power Pathways project is producing a detailed roadmap towards 100% renewable energy for each province in Canada, with a high level of detail.[49] And at a global level, the Drawdown initiative and its exhaustive book, edited by Paul Hawken, outlines 100 global solutions to reverse global warming. It too provides a guide for what we should be pursuing.[50]

Second, informed by the inventory requirements outlined above, as occurred in the war, we must coordinate the mass production of the equipment needed to realize our new GHG reduction targets. We will need factories to produce solar panels, wind turbines, electric heat pumps and electric buses at a mass scale. It is worth appreciating, however, that the technical challenge of such production is far less than building complex planes and warships. And the technology needed already exists. As much as possible, we should seek to manufacture these items in Canada.

Third, we must develop a clear wind-down pathway for all fossil fuel extraction in Canada, guided by a robust just transition plan for existing fossil fuel workers and communities that currently rely on these industries.

An essential starting point is to declare that there will be no new fossil fuel infrastructure. While we have two to three decades to wind down the fossil fuel industry, it makes no economic or ecological sense to spend billions of dollars on new projects, particularly when that infrastructure has a working lifespan and expected financial returns that well exceed 30 years. Decisions we make today about infrastructure such as pipelines and new processing plants matter greatly because they tend to lock in place a certain development path. If we know we need to set out on a path towards fossil fuel wind down, then the time is past to be investing in new infrastructure that supports ramp-up.

Fourth, in place of fossil fuel infrastructure, we need to develop a massive green public infrastructure plan, involving all levels of government. We need to invest billions of dollars in renewable energy, building retrofits, high-speed rail, mass public transit, electric vehicle charging stations and methane capture from farms and landfills. And we need it at a scale well beyond what any political party has campaigned on to date. This would represent the foundation of a wartime-level employment plan.

And because even under the best-case scenario a certain amount of global warming is already locked-in, we also now need to undertake major investments in climate adaptation and resilience infrastructure, with a focus on ensuring that vulnerable communities are better protected from climate disasters and related events (extreme weather events, forest fires, extreme heat events, flooding, etc.). For example, we need to construct and reinforce dykes that threaten vulnerable communities. And we need to significantly invest in forest management that will lessen wildfire risks to rural and Indigenous communities, while providing thousands of sustainable jobs in resource-based communities.

Fifth, we need a large-scale program to repair and enhance Canada's natural climate sequestration systems — helping nature suck carbon from the atmosphere. That includes an extensive reforestation program, and of course the preservation of existing old-growth forests (the world's intact forests, of which Canada is home to a large share, represent a vital element of slowing climate change).[51] We also need to encourage value-added wood production and building (which locks away carbon), and develop new sources of fibre such as hemp and bamboo. There is scope for extensive job creation in this area. We should adopt an ambitious goal, such as planting five or ten billion trees.

Sixth, as advocates of a Green New Deal have made clear, we need more than just direct climate infrastructure investments — we also need large-scale investments in social infrastructure and the caring economy if we are to realize the equality goals outlined earlier and keep everyone on the bus. That means investments from all levels of government in zero-carbon public and non-profit housing — a bold commitment to build hundreds of thousands of new units of non-market housing. And it means federal and provincial funding for universal, public, accessible, quality child care and home care for seniors and people with disabilities. These are public services that are already virtually carbon-free and would represent a major enhancement to household affordability.

Seventh, and perhaps most controversially, like the rationing we saw in the Second World War, we will likely need to adopt a system of carbon quotas that decline over time for all households and businesses. This radical option could be held in abeyance for a few years, to be employed if we find by mid-decade that we are not on track to meet our 2030 emissions target.

Eighth and finally, we need to set in law and regulation clear dates by which certain things *must* happen. For example:

- Cutting Canada's GHG emissions by at least 50% by 2030.
- Ensuring Canada achieves net-zero GHG emissions by 2050. (Both these targets require interim benchmarks that must be hit along the way, and we should establish rolling and declining three-year carbon budgets for the country, each province and major industrial sectors to ensure we stay on a pathway to meet our targets.)
- Banning the sale of all new fossil fuel-combustion vehicles by 2025.
- Phasing out the business licenses of conventional gas stations, with all of them expired by 2040 (unless they have converted to electric-vehicle charging stations).
- Prohibiting the use of natural gas or any other fossil fuel in all new buildings by 2022.
- Regulating that, by 2025, if a furnace or boiler needs to be replaced in any type of building of any age, it cannot be legally replaced with a fossil-fuel-fired unit.
- Cutting off *all* natural gas (and any other fossil fuels) to homes and buildings by 2040. Meaning, by this date, all pre-2022 buildings will need to have been retrofitted so they are no longer using a fossil fuel heating or energy source. This is important, because more than half the existing stock of buildings today will still be standing in 2050, when the IPCC says we must be off fossil fuels entirely. Given this,

energy retrofits cannot be voluntary by this point — they need to be mandated.

- Ending the use of all fossil fuels for electricity generation by 2030. Canada is already on track to do this for coal-generated power, but this must also be extended to natural gas–fired electricity generation. And we need a plan to get all remote and Indigenous communities off diesel-powered electricity.
- Phasing out the use of hydrofluorocarbons, HFCs (a highly potent greenhouse gas), as Europe is doing.
- Ending the extraction and export of all fossil fuels by 2050, if not 2040.

These targets could be combined into a single overall Climate Mobilization Emergency Act (modelled on the National Resources Mobilization Act of 1940).

Clear targets such as these — embedded in law and well publicized — will send a much stronger signal to the market than any form of carbon pricing. They communicate to businesses and consumers that they must reorient their plans accordingly. If effectively enforced, these targets will push manufactures, builders, installers and extraction companies to make investment plans that align with these dates. And what are now small industries (such as electric heat pump or geothermal installers) will quickly scale up to meet the demands that will result from such regulations. Many low- and middle-income families will need financial help to get their homes off fossil fuels, and that must be part of a just transition plan.

For some perspective on what the transition before us requires, the chart on the next page shows total GHG emissions in Canada broken down by major sector. A wartime-level mobilization is needed if we are to eliminate all these sources of GHGs in the next three decades, and preferably sooner. There are technical reports available that outline the mechanics of how to decarbonize each of these domains. And ultimately, the details of such efforts should be developed through

CANADA'S GHG EMISSION BY SECTOR (2017)

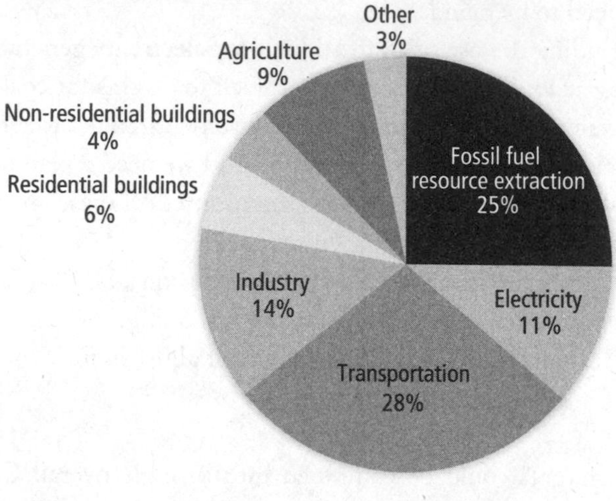

Source: Environment and Climate Change Canada: A-Tables-IPCC-Sector-Canada
Notes: Fossil Fuel Extraction includes oil, gas and coal; agriculture includes some forestry; and "other" includes waste, construction and mining.

transparent and inclusive consultation, collaboration and partnership with frontline and vulnerable communities, Indigenous organizations, labour unions, civil society groups, academia and businesses. But next I offer some broad elements of how the decarbonization of each of these sectors can occur.

Transforming Transportation

Anthony Perl is a professor of political science at Simon Fraser University, the former director of the university's Urban Studies Program, and an expert in transportation. Perl started writing about high-speed rail back in the 1990s. He is co-author (with Richard Gilbert) of the 2010 book *Transportation Revolutions: Moving People and Freight Without Oil* in which he argues that, rather than thinking about incremental shifts to our transportation, we need to embrace a "revolutionary" approach that substantially changes how goods and people move within a few years. "Transportation revolutions take society to a different place than incremental change," he writes. It involves disruptive thinking and a new vision.

As the chart above shows, transportation is currently responsible for 28% of Canada's GHG emissions. One piece of changing this reality is shifting how we drive — getting people out of gas vehicles and into electric ones and expanding the use of car-sharing. But we must be go well beyond that. We need to fundamentally rethink how we get around, and how our communities are organized. As Perl explains, we must also make much greater use of rail and water to move people and goods, rather than roads and air. And vitally, we must use more collectively managed travel over isolated and individualized travel — meaning much greater use of public transit.

Getting our transportation sector to carbon-zero thus requires:

- expanding public transit, including a plan to make public transit not only more accessible and convenient, but also dramatically more affordable (minimally, that means free public transit for lower-income people, but could well involve making transit a "free" publicly paid service, just like health care);[52]
- converting all our public transit to electric buses and trains;
- expanding the use of electric car-sharing and bike-sharing services;
- launching high-speed rail between our major cities and across the U.S. border, and the electrification of freight rail;
- building out public electric-charging infrastructure at every highway rest stop;
- improving the efficiency of ships and moving them to electric battery power;
- subsidizing electric bikes (as we do now for electric cars), which have the potential to substantially reduce car use, and dramatically improving urban and suburban biking infrastructure and safe paths;
- improving the efficiency of airplanes and moving them to renewable fuels and electric engines;
- ramping up the use of video-conferencing and telecommuting for work and business;
- rethinking city planning to move our communities from

a car-centric culture to "complete communities" — where people of all incomes can live within easy walks, rides and transit from the services they need and from where they work and play;[53] and

- banning the manufacture and sale of new gas-powered vehicles within a few short years, financial support for vehicle owners to retrofit their vehicles (swapping internal combustion engines to zero-emission alternatives) and offering generous "scrap it" rebates to everyone who turns in their fossil fuel vehicles (getting those cars, trucks and vans off the road entirely and recycling their materials).

Our governments are starting to declare climate emergencies. Yet they continue to allow the manufacture, sale and advertising of new fossil fuel–burning cars. We are clearly still in cognitive dissonance about what this emergency means.

The B.C. government has passed a law banning the sale of new fossil fuel burning vehicles as of 2040. It's good to see clear regulatory dates being set. But a single province doing this just invites imports from elsewhere. Why aren't all governments in Canada — not least the federal government — bringing in such laws? And why isn't the ban date for new vehicles 2030? Or 2025? Is this an emergency or isn't it? If we could entirely cease the production of civilian vehicles during the war, why can't we act accordingly today?

Other than B.C.'s 2040 ban, all other Canadian government strategies for zero-emission vehicles (ZEVs) are incentive-based — offering partial rebates (as the federal government now does, in addition to B.C. and Quebec), or financial incentives for installing level 2 charging stations, or access to special traffic lanes or exempting ZEVs from bridge tolls (as Quebec does). Such measures are all helpful, but they miss the point of an emergency — in an emergency we don't merely encourage people to change, we mandate it.

With respect to electric vehicle manufacturing, if the private sector does not indicate its preparedness to move swiftly towards the manufacturing of electric cars, buses, trucks and vans, then like C.D. Howe did, why not create a new crown corporation that will? For example, workers from GM's auto plant in Oshawa, which the company closed at the end of 2019 after more than 100 years of operations, have called on the federal government to take over the company as a publicly owned enterprise and make a $1.4 to $1.9 billion investment to convert the plant for electric vehicle manufacturing. The group leading the call is Green Jobs Oshawa and involves the union representing the 2,500 workers at the plant. "What could make more sense than the government stepping in and saying, 'We will ensure that we will keep jobs here, that we develop the technology here and that we build the vehicles of the future here?'" said Tony Leah, chairperson of Unifor Local 222's political action committee.[54] The workers hope a converted plant would focus on the production of electric vehicles for government fleets, such as those of Canada Post. In many ways, these workers are simply calling for what should have happened back in 2008, when the government gave GM and Chrysler a $13.7 billion bailout in the wake of the financial crisis but then failed to leverage that equity stake to drive a climate-inspired change.

Electric high-speed rail technology has been around for decades, and high-speed rail is the norm in much of Europe, Asia and parts of North Africa. In North America we have been laggards. Does Canada have the population base to support high-speed rail? "Sure we do," says Perl. He recommends starting with routes between Montreal, Ottawa and Toronto, as well as between Calgary and Edmonton, and between Vancouver, Seattle and Portland (talks are already underway between the provincial and state governments for that latter route in the Pacific Northwest).

Air travel is currently responsible for about 5% of global GHG emissions, and the sector's emissions continue to escalate sharply. In addition to changing how we travel, eliminating aviation emissions will require some new technology still in development. For short-haul air travel — which represents the majority of the world's flights — electric motor options are already coming on stream. Perl notes that China is now making wide use of such technology. And in March 2019, B.C.-based Harbour Air, one of the world's largest operators of seaplanes, currently operating over 40 planes and running over 30,000 flights a year, announced a bold plan to partner with the magniX electric aviation company to convert its entire fleet to all-electric.[55] But the majority of GHG emissions still emanate from longer flights, and that remains a major challenge. If flights were capped at current levels, efficiency improvements already occurring would help to bring GHG emissions down. But beyond that, there is no imminent magical solution other than that we must significantly reduce such air travel.[56]

Marine transport needs to move to electric batteries. A Richmond, B.C.–based company, Corvus Energy, has become a world leader in the production of advanced batteries for marine vessels. Its technology is being used by governments in Scandinavia to convert public ferry fleets to electric. But ironically, this Canadian company has few Canadian contracts. Where are the plans to convert our public ferry fleets? Perl recommends this technology be supplemented by "sky sails," significantly reducing a ship's draw upon its engines.

Overall, aviation and marine travel will likely need to move to some combination of electric, hydrogen and bio or "renewable" fuel sources.

At a fundamental level, we need to stop building new infrastructure — wider roads, bridges and airport expansions — that we know will be obsolete in 30 years. That is not, in an emergency, how we should be deploying our capital spending or engineering talents. Those resources are now needed elsewhere.

And reaching back to our war experience, Perl also reminds us how "leadership has revolutionized transportation in hard times before." The Second World War saw the rationing of fuel and tires, and massive public campaigns urging people to use transit and to carpool. One U.S. government campaign poster even employed the subtle slogan "When you ride alone, you ride with Hitler: Join a car-sharing club today!"

Transforming Our Homes and Buildings

Our homes and buildings currently produce 10% of the GHG emissions in Canada, mainly from burning natural gas for heating and hot water. We have the technology for this to end.

For new homes and buildings, we already know how to build

passive structures that need virtually no energy for space heating. We now need updated building codes that make the employment of these techniques mandatory within the next couple of years. There are already prefab passive solar homes coming on the market, which can be used to quickly provide secure and healthy low-income housing that is carbon-zero. According to the Energy Mix (an electronic newsletter and website that amalgamates climate and energy news), Ontario-based Quantum PassivHaus claims it can build one of these prefab housing units in their shop in seven to eight days, and then install it within a day and a half.[57] That's the kind of speed we need to scale up.

For existing homes, better insulation, programable thermostats and other building retrofits have the potential to save households considerable money on their monthly bills. What's more, it's well established that improved energy efficiency and conservation is where we get the best bang-for-the-buck returns when it comes to energy investments. Modern electric heat pumps are so much more efficient than traditional baseboard electric heating that monthly operating costs are comparable to low-cost natural gas. However, the upfront cost of converting one's home heating to electric heat pumps can be steep. So, low- and middle-income families need financial help making the switch. But government action and mass adoption of heat pumps can also significantly reduce their cost. Heat pumps and their installation in much of Asia, for example, is much cheaper because they are the norm and economies of scale and building codes make their use much less expensive.

Commercial and apartment buildings also need to fuel-swap, either at the building level, or at the neighbourhood utility level. In Vancouver, the new housing developments in False Creek are all required to join a publicly owned neighbourhood energy utility that derives most of its heat by extracting it from waste water.[58] Similarly, all new buildings in downtown Richmond, B.C., must now tie into the city's public district energy utility, which uses geothermal heating.

Societies have quickly transitioned home heating sources before. In England, prior to the 1960s, most homes used what was called town gas — a coal-derived gas that was particularly volatile and noxious — delivered by municipal pipe. Town/coal gas contained high carbon monoxide levels and its use was associated with distressingly high suicide rates. The dangers associated with town gas convinced the government that it should be phased out and replaced with safer natural gas. That decision necessitated replacing all home appliances, a conversion achieved in a ten-year period.[59]

Transforming Industry, Agriculture and Electricity

Industry (including steel, aluminum, cement, pulp and paper, wood, chemical and other manufacturing) currently accounts for 14% of Canada's GHG emissions, agriculture and logging another 9%, and electricity production 11%. Fossil fuel extraction and processing is responsible for 25% of Canada's GHGs, but as previously argued, the 20- to 30-year plan for this sector should not be transformation but rather wind-down.

Some provinces such as Quebec, B.C., Manitoba and Newfoundland and Labrador already derive virtually all their electricity from hydro power. But in other jurisdictions across Canada, much of the electricity is still produced by burning coal and natural gas. That needs to end by 2030.

Some key industries still rely heavily upon the burning of fossil fuels, or use specific production processes that heavily emit GHGs, such as concrete, steel and aluminum production. In these cases, clear and mandated timelines will be needed for these industries and manufactures to swap their GHG-emitting activities for new technologies (such as renewable energy, non-GHG replacements for cement, electric arc furnaces,[60] etc.).

The agricultural sector also needs a GHG wind-down plan, collaboratively developed with farmers and farm communities. Many agricultural greenhouses use natural gas as a heating source, and they will need to switch to biofuels, geothermal or electric heat pumps. Intensive industrial livestock production is a massive source of methane, a particularly potent GHG, and consumes a disproportionate share of land and energy. Farmers will need financial support to move off fossil fuel machines and nitrogen (another GHG) fertilizers, deploying methane capture technology and replacing much of their industrial livestock operations with non-meat protein production. Farmers will need help making other adjustments too, as global warming itself impacts production in numerous ways. We should increase our support of local and organic farmers, and can use public purchasing (by schools, hospitals and other public institutions) and promotional campaigns to sustain local farming over industrial production and imports.

Ultimately, every industry and business will need its own plan to eliminate its GHG emissions. Sectoral transition boards should be established to plan and oversee the conversion of each major industry as it eliminates its GHGs. As for the individual business level, U.S.-based The Climate Mobilization, in its very detailed and much-recommended Victory Plan,[61] proposes that every private firm and non-profit organization with over $10 million in annual revenues be mandated to produce a "zero-emissions plan" for their operations, outlining their pathway to decarbonization.

Transforming Our Public Institutions and Services

In the transformation before us, we need our government to lead by example and quickly align the public sector with the climate emergency. For all the activities and institutions within the public domain, governments should model what needs to occur and use their purchasing power to drive change.

For example, our public school boards, post-secondary institutions, public health authorities, hospitals and clinics, municipal community centres and libraries, and government service offices should be: converting their vehicle (car, van and bus) fleets to full electric; ensuring all their parking lots are equipped with electric-vehicle chargers; bargaining eco-transit benefits for those staff who don't use a car to commute; rapidly phasing out natural gas for heat and hot water from all public buildings, immediately ensuring that no new buildings tie into gas and putting solar panels on all public roofs; and requiring that public cafeterias (such as those in schools and hospitals) only purchase local food, cutting down on transportation emissions.

The same is true for all existing crown corporations. There are innovative and exciting models for how our crown companies can become beacons of change. For example, the Delivering Community Power campaign developed by the Canadian Union of Postal Workers and the team at The Leap has proposed an inspirational and total transformation of Canada Post, calling for the corporation's vehicle fleet (the largest in Canada) to be quickly electrified and for all Canada Post buildings (there are more Canada Post outlets in Canada than there are Tim Hortons) to put solar panels on their roofs and to offer electric vehicle chargers. Additionally, marrying these climate goals with socio-economic ones, Delivering Community Power has urged Canada Post to offer community postal banking, so that rural and lower-income people can access affordable banking services, and avoid the predatory lending and usurious interest rates of the payday-loan industry.[62] Indeed, to take this a step further, postal banking could become a local network of public investment banks, specializing in offering financing to local social enterprises, co-ops and others wanting to develop climate-friendly ventures.

A similar creative reimagining of all existing crown corporations is now needed. The task of our public utility corporations is now obvious — ensure all our power needs can be met through renewables. Crown liquor corporations need to think inventively about how their purchasing can drive changes in agriculture, in addition to converting their vehicle fleets. Public housing authorities need to develop clear timelines for retrofitting all their existing buildings and aggressively build out the kind of carbon-zero low-income housing described above. And as noted, our public ferry corporations must quickly convert their ships to electric power. BC Ferries, the publicly owned corporation that operates B.C.'s extensive ferry fleet, currently has plans to electrify only its smaller ferries. It is seeking to convert some larger ships from diesel to liquified natural gas, a move that modestly reduces GHGs but is surely not the answer — the days of thinking of LNG as a transition fuel are past. Blair Redlin, hired by the B.C. government to conduct a review of BC Ferries, writes in his 2018 report, "The steps that have been taken are positive and are making a difference but are not yet aggressive enough given the seriousness of the climate change crisis and the targets in the provincial Climate Leadership Plan." Redlin notes that other public ferry corporations are taking bolder action. For example, "In 2014 Sweden was the first country to operate an electric passenger-only ferry, followed in 2015 by Norway with the first electric vehicle ferry. Results for the Norwegian electric ferry have been impressive with a 95% reduction in CO_2 emissions and an 80% reduction in costs. The battery system on the Norwegian ferry is provided by Corvus Energy of British Columbia. Based on its initial success, the Norwegian Parliament recently directed that by 2026 all ferries operating in its fjords must be electric."[63]

First Nations governments have also been passing climate emergency motions, and they too need to be taking the next step and developing climate action plans that apply to all the assets, homes and businesses that come under their authority.

Our governments should be widely adopting videoconferencing and telecommuting for their employees, helping to cut down on vehicle and air travel. And our legislatures should adapt their schedules and explore technology options for virtual parliamentary engagement so that MPs and MLAs can reduce their travel and spend more time at home and in their constituencies.

We also need our governments to rethink and model a new form of budgeting. Canada has developed a strong and robust tradition of financial budgeting and accountability within government. Legislative finance committees conduct annual consultations, collecting public input into budget priorities. Most governments adopt three-year financial budget plans. Treasury boards have practices and procedures for government ministries and agencies to stay within their budget allocations. Auditors General review government spending and highlight wasteful or inefficient practices. And a federal Parliamentary Budget Officer offers expert advice on government fiscal room and projections.

Now we need to develop and implement similar procedures and practices for the climate emergency. Our governments need to adopt a "carbon budgeting" framework to plan for and track GHG emission reductions, with processes similar to financial budgeting: public consultations, federal-provincial-territorial-Indigenous meetings to determine carbon-budget sharing, rolling three-year carbon budgets that decline as needed to meet our 2030 and 2050 targets and that allocate GHG allowances to various jurisdictions and industries, and a federal Carbon Budget Officer to independently oversee these commitments.

And finally, while this book has rooted its mobilization ideas in the positive lessons from the Second World War, it is also the case that the military itself is a huge consumer and combustor of fossil fuels. While we will surely need the services of the military at home and abroad in the coming years as communities confront climate disasters, the military must also convert and electrify its vehicles and ships and develop an ambitious plan to end its use of fossil fuels.

Transforming Household Consumption

"Dig on for Victory!" declared wartime posters urging people to plant Victory Gardens. The changes in household consumption and recycling during the war years weren't only important to save scarce resources — they also gave everyone a sense that they were part of the collective war effort.

That history of how households rallied to the cause can inspire us today. Modern urban agriculture movements have found a muse in Victory Gardens. The urban farming group Action Communiterre in Montreal's Notre-Dame-de-Grâce neighbourhood has a Victory Garden network. Knowing how wartime families shifted their transportation habits and conserved energy and fuel can motivate us to rethink our practices today.

But if we are unable to dramatically bend the curve on our GHG emissions very soon, more radical measures will be needed, modelled on the household rationing of the war.[64] We might well need to implement carbon quotas. Under this approach, we would calculate Canada's fair allotment of the world's remaining carbon budget, allocate a certain share to government (which in turn would be divided among the various levels of government), a certain amount to various industries and the remaining share to households, to be equitably apportioned.

Unlike in the war, rationing would not be needed for specific goods such as natural gas, vehicle fuel, meat or air travel. Instead, households and companies could be allotted annual "carbon quotas" for purchasing any fossil fuel. Those quotas would decline each year, in line with our society's overall GHG reduction targets. A model like this still gives households choices about how to "spend" their carbon quotas — if air travel is important to you, you can concentrate your quota there and eat less meat; if you really love meat, you can travel less; etc.

Such a system is technically possible. The carbon content of fossil fuels is already well known and reflected in the carbon tax one pays when filling up a car with gas or paying a home utility bill. But businesses can also calculate the carbon embedded in the goods and services we purchase, and these could be reflected next to the sticker prices we have long known (or simply be embedded in overall prices). Households or individuals could be granted carbon debit cards to operationalize the system, effectively resulting in a parallel carbon currency.

I first encountered this model in British writer and climate champion George Monbiot's book *Heat: How to Stop the Planet from Burning*, in which he outlines how a system like this could work.[65] As Monbiot highlights, a carbon rationing system isn't only valuable for reducing GHGs — it also strongly communicates that everyone is doing their equal part, as occurred in the war. It helps to build the sense of social solidarity discussed earlier.

And a carbon quota system has an additional benefit over a carbon tax/price. As we've seen, carbon pricing in itself is regressive — the cost impacts lower-income households more as a share of their income, while the rich can simply continue to emit as much carbon as they wish and pay the tax (which the wealthiest would hardly notice). The regressive nature of carbon taxes can be mitigated by way of a lower-income carbon credit. But a carbon quota avoids this dilemma entirely — all households are given their fair and equal allotment of carbon rations, regardless of income. A carbon quota/rationing system thus has the benefit of both equity and simplicity.

Monbiot contends that this model would preserve more individual "freedom" than a purely regulatory approach that tells people what we must do and monitors our choices and behaviours. Moreover, he argues, "the market created by carbon rationing will automatically stimulate demand for low-carbon technologies, such as public transport and renewable energy." But even in this equitable model, lower-income households that live, for example, in poorly insulated homes, would still need financial support to make needed transitions.

Any way you cut it, though, as the use of fossil fuels is forced down and the commodity becomes increasingly scarce, there will be some form of rationing. The only question is whether we will have rationing determined by who can pay the escalating price, or some more equitable approach.

New Crown Corporations: Let's Get This Done!

When interviewing Alberta Federation of Labour President Gil McGowan, I asked what he thought the Notley government should have done when it was in tough negotiations with the oil and gas companies about declining employment, notwithstanding the meagre royalty rates they are charged. "I would have created an energy crown corporation," replied McGowan without missing a beat. McGowan says such an option was in fact discussed by the Notley government, but rejected. McGowan thinks such a new crown corporation could have helped to move the industry up the value chain — "instead of just ripping and shipping our raw resources" — and ultimately to expedite the transition off fossil fuels and into renewables.

When interviewing B.C. Minister of Environment and Climate Change Strategy George Heyman I asked if the B.C. government had considered creating any new crown corporations to help meet its climate goals (and I noted that the provincial NDP governments of the 1970s and 1990s had established a number of new crown corporations, including to help with economic transition, such as Forest Renewal BC). "Not specifically," Heyman replied. "We did not think in those terms. What we did was to make *Clean BC* a whole-of-government initiative and ensure that in every mandate letter giving direction to ministries, crown corporations and agencies, working to help meet government's climate targets was explicitly stated."

It is true that a whole-of-government approach is needed. It is unfortunate, however, that neither of these contemporary NDP provincial governments seriously considered how new public corporations could be employed to expedite climate action. The same is not true for social democratic parties elsewhere, who in some cases are not only re-embracing the need for state enterprise but are reimagining what modern public ownership can look like and how it could be more democratically and locally governed.[66]

A central dilemma facing transition is that oil and gas companies continue to put billions of dollars on the table for new fossil fuel infrastructure (pipelines, oil sands projects, LNG), while investment in new economy infrastructure on the scale required has not been forthcoming (notwithstanding the fact that the jobs produced per dollar invested in renewables, public transit and building retrofits exceed those in oil and gas). The problem, as Phillips and Rozworski remind us, is that "what is profitable is not always useful, and what is useful is not always profitable." How then might public enterprise (at all levels of government), or Indigenous development corporations, or worker or community cooperatives, or social enterprises help to break this logjam?

When I have shared the story of C.D. Howe and the 28 crown corporations he established during the war, I am frequently asked: what would those crown corporations be today? It's an excellent question.

First, recall some of the reasons why Howe and others have felt the need for public corporations in the past:[67]

- To secure supply chains for needed production, while also gaining the cost benefits that come from bulk purchasing some inputs or from the economies of scale that come from mass production.
- To ensure a public competitor operates within core sectors, to keep the private players honest and prices/costs down, and to avoid profiteering in the face of an emergency.
- To develop a technology that the private sector either won't or cannot develop, perhaps because the endeavour is too risky or costly or sensitive or short-term.

- To provide a good or service that the private sector won't because there is not enough profit to be made, or to ensure that the good or service will be universally provided at an affordable price.
- To ensure that needed technology is made widely available, rather than tightly held as a private proprietary asset for the purposes of maximum profit.
- To ensure that the public gains maximum return from the development or commercialization of a public resource.
- And lastly, as was certainly the case with Howe, to make something happen quickly at a mass scale when the private sector proves unwilling or unable to do so.

Given such considerations, here is my list of possible candidates for new crown corporations for meeting the climate emergency:

- A federal high-speed rail corporation. The original railroads were foundational to the creation of Canada. Why not keep the next-generation ones public? If our population base has failed to make this proposition sufficiently attractive to the private sector, let's just do it ourselves.
- Nationalize a GM auto plant (or others) to expedite the production of electric vehicles, with a focus on electric vans, trucks and buses to replace public and commercial fleets.
- Renewable energy corporations, likely in every province, that focus both on building large-scale renewable projects (solar fields, wind farms, wave/tidal power generation and neighbourhood geothermal/heat exchange systems) and on the widespread installation of solar and geothermal systems at the individual building level. Much like the development of large hydro dams in our past, there is a logic to doing this work publicly. These projects will frequently require public land. And, starkly different from oil and gas, once a project is built, the energy itself is free. Nature, in its bountiful generosity, provides the sun's rays and the wind and the warmth of the earth, day after day, at no monthly

charge to the user, making these activities less attractive to massive for-profit corporations. Oil has produced the most profitable corporations on the planet. That's never going to happen with solar.

- New municipal and/or provincial public corporations focused on the mass installation of electric heat pumps. This would simplify what is currently a very complicated exercise for many homeowners (I say based on personal experience), keep the price down by bulk purchasing/importing heat pumps and eliminate profit margins from the installation work.
- If the mass importation of heat pumps proves too challenging or costly, then consider a new crown corporation to manufacture heat pumps for the millions of homes and buildings across Canada.
- Similarly, if we no longer wish to import most solar panels and wind turbines, then we should consider new crown corporations that would mass produce them in Canada.
- A crown corporation to expedite the development, manufacture and deployment of new non-fossil fuel agricultural equipment (freeing farmers from the costs of oil). And recalling how well the management of food production worked during the war, for both consumers and farmers, we should restore and expand the use of public supply management boards (jointly governed with farmers and communities).
- Federal and/or provincial crown corporations to oversee the decommissioning of abandoned oil and gas wells and mines, funded by the oil and gas industry, an activity that could employ thousands in Alberta and Saskatchewan, as the group Reclaim Alberta has documented.[68]
- New public housing development corporations focused on building low-income carbon-zero homes.
- New building retrofit corporations, staffed with an army of expert tradespeople who can advise both residential and commercial owners on the best way to maximize their

building's energy efficiency, without fear of getting fleeced, and then ensure the work gets done.

- A new crown corporation to coordinate waste reuse. Given the crisis and market failures in the global recycling market, and understanding that so much of what we call "waste" is in fact a resource that has already been semi-processed, a crown agency could oversee the deconstruction (as opposed to the demolition) of buildings and the collection of other reuseable items, and then direct these resources to various remanufactures. Where no local remanufactures exist, the corporation could seek to establish them.[69]
- New provincial and/or federal crown corporations to undertake mass reforestation and forest revitalization.
- A "Right to Repair" crown corporation that would allow consumers (including farmers) to open and repair consumer goods, combined with a national network of neighbourhood repair centres staffed with skilled technicians who can fix electronics and other household items. Such centres could even spearhead the use of 3D printing for replacement parts.
- Historian Alex Souchen has suggested a new crown corporation to develop and produce an alternative to asphalt, using recycled plastic, as some companies have already done in other countries.[70]
- The team at 350.org's Our Time initiative proposed a "No-Crown Corporation," an Indigenous-run project to end all boil-water advisories in a single year, "while upgrading every Indigenous community across the land to 100% renewable energy by year two."
- A new public steel manufacturer, using innovative electric arc furnaces. Right now, Canada exports most of the metals we recycle, as well as the metallurgic coal we mine that is used for steel production overseas. A crown corporation in this field could innovatively combine the metals we recycle with some modest amount of mined metallurgic coal to produce steel in Canada, and could, as occurred in the war,

ensure that steel resources are deployed manufacturing the goods we need for the transition — electric buses, new rail lines, bikes, wind turbines, etc.

- Dennis Bartels, the Memorial University professor who first wrote about climate lessons from the Second World War, has suggested we may need a crown corporation to coordinate the mass production of non-methane plant-based protein foods.[71]
- A labour market matchmaking corporation to assist with just transition — a public agency that would link current fossil fuel employees and other workers with the green jobs in the above corporations and building new climate and social infrastructure.

There are no doubt more possibilities worthy of consideration. We just need to approach the challenge creatively. The point is simply this: if something that needs to happen isn't happening through the market at the scale and speed that the emergency requires, then — through our governments — we can and should damn well do it ourselves.

In some cases, it will make more sense for "public" or "collective" ownership of a new enterprise to be at the municipal level, the Indigenous nation level or in some other non-profit or social enterprise form. For example, Aki Energy is an Indigenous social enterprise that installs geothermal heating in northern Manitoba communities (working in partnership with the larger Manitoba Hydro crown corporation).[72]

Governments could also use either existing or new crown corporations to support fossil-fuel-dependent communities going through transition. If a community reliant on fossil fuels, like a coal town, is facing collapse, the government has the power to move a crown corporation or agency there, as an anchor employer, so people can stay and their home values are sustained.

Sadly, thus far, the Trudeau government has seen fit to establish only two new crown corporations during its time in office — the Canada Infrastructure Bank, a vehicle for effectively privatizing the building and operating of public infrastructure, and the Trans

Mountain Corporation, the agency established in 2018 so that the government, on our collective behalf, could purchase the aging oil pipeline from Texas-based Kinder Morgan and then build the expansion project that will triple the volume of bitumen carried from the oil sands to Metro Vancouver.

We Know Enough to Launch This Fight

The ideas above give shape to a new vision of how we will live, play, get around and make a living in a post-carbon society. True, getting there will be hard work. But the picture that emerges is not one of privation and endless sacrifice. Rather, it is of a pretty nice life. These changes don't just get us off fossil fuels — they come with a host of what medical practitioners call "co-benefits": cleaner air, and therefore less asthma and a healthier future for our children; more attractive communities and more contact with our neighbours; more green space, as we shift roadway and parking space away from private vehicles; more affordability, as our transportation shifts from private/individual gas, insurance and maintenance costs to public transit and electricity; better personal health, as our mobility shifts to walking and bike and our diets shift away from meat and processed foods; more meaningful and decent work; and a more equal and fair society with greater social cohesion and less social stress and strife.

But getting there will take coordination and planning — and leadership. Maybe we need to reconstitute a new variation on the War Production Board or Wartime Industries Control Board or some other joint federal-provincial-Indigenous-municipal body tasked with coordinating the overall economic conversion, guided by a national framework for a just transition and clear targets and timelines for the phasing-out of fossil fuel use. We also need a human resources development plan to run alongside this transition — a national education and apprenticeship plan to arm people with the skills required to operationalize the economic transformation before us.

Some remain fixated on those few areas — such as air travel and some industrial processes — where complete technological solutions are not yet good-to-go, and they persist in their skepticism that we are capable of this task. It's true that we don't have all the answers. But we know enough to launch ourselves into this challenge. As former clerk of the Privy Council Alex Himelfarb reminded me: "FDR's language of a great national experiment is also useful. We will learn and adapt; we need some answers but not all the answers."

In a similar vein, when Swedish activist Greta Thunberg was invited to address U.K. parliamentarians in April 2019, she told them, "We have to start treating the crisis like a crisis — and act even if we don't have all the solutions." And then, invoking the still-fresh memory of the devastating fire that destroyed Notre-Dame Cathedral in Paris a mere one week earlier, she brilliantly told the MPs, "Avoiding climate breakdown will require cathedral thinking. We must lay the foundation, while we may not know exactly how to build the ceiling."

One frequently asked question is whether solving the climate crisis is fundamentally at-odds with capitalism. I have argued that climate action at the speed and scale now required is certainly not compatible with neoliberalism, unfettered free markets and corporate globalization. One of my core contentions is that cracking the climate emergency puzzle requires careful economic planning and government regulation. But I've always been open to a mixed economy, and I still am.

Part of what is attractive about the lessons from the Second World War is that it liberates us from an overly polarized debate on this question. The private for-profit sector had an important role to play in the war — it produced much of what we needed, and people made money doing so. But critically, the private sector — at both the corporate and household level — didn't get to decide on the allocation of scarce resources in a time of emergency. Rather, our governments directed what needed to happen to rise to the task at hand. That's the kind of thinking we need today.

WikiCommons

CHAPTER 7

Mobilizing Labour: Just Transition, Then and Now

"Those of us who had served in the armed forces during the 1939–45 bloodbath, and had reflected on the experience, could sense an even more disastrous adventure in the making, this time with the added calamity of atomic and hydrogen weaponry… If we spoke out, perhaps we could inject a note of sanity into the debate on where Canada should stand on issues fundamental to human survival."

— Kell Antoft, RCAF navigator during the Second World War, and co-founder of Veterans Against Nuclear Arms

"We need the federal government to step up and guarantee that every single worker, family and community impacted by this transition will be supported. The best way to do that is to borrow from the Green New Deal and implement a federal job guarantee that tells every single person in Canada that they don't have to choose between putting food on the table and ensuring our children inherit a liveable planet."

— Clayton Thomas-Müller, Cree activist and campaigner with 350.org Canada

The Second World War saw the wholesale mobilization of labour for both war production and military service, careful planning for the return of soldiers to civilian life and, more broadly, a fundamental transformation in the role of labour within Canadian society. A new social contract emerged from the war along with a new balance of power between workers and bosses. The war witnessed major gains for working people, men and women alike, in their wages, their rights, and their "social wage" — the programs and supports we collectively provide to one another. That history has valuable lessons for us today as we mobilize for the climate emergency and, in particular, as we think about how to offer a just transition to workers currently employed in the fossil fuel industry.

FROM BREADLINES TO FULL MOBILIZATION OF LABOUR TO POST-WAR REINTEGRATION

After ten long years of the Great Depression (1929–1939), marked by stubbornly high unemployment rates, deprivation and hardship, the Second World War saw the quick attainment of full employment and a marked jump in household incomes. And unlike past industrial revolutions that abandoned huge swaths of workers and their families to the scrap heap of history, this economic transformation saw leaders prepared to make concessions to workers whose labour was desperately needed.

Marshalling a Military

Prior to 1939, Canada's military was small and minimally equipped. The army consisted of a permanent force of just 4,000 men and two tanks (14 more tanks arrived in 1939, all imported from Britain). The air force had another 4,600 personnel sharing 270 mostly antiquated aircraft. And the navy had 13 ships and 1,500 regular personnel.[1]

Within a few weeks of the September 1939 declaration of war, 64,000 Canadians signed up to serve. Those numbers would soon swell, particularly after the fall of France in the summer of 1940. Interestingly, the first major recruitment drive for overseas enlistment only happened in May 1941. By then, the main opposition the Mackenzie King government faced was not from those demanding less mobilization, but more. Many, particularly in English Canada and among Conservatives, were demanding full conscription, seeing it as the only way to properly prosecute a major war, and the only fair means to equalize the sacrifices. Ultimately, almost 1.1 million Canadians had enlisted by the end of the war, from a total Canadian population at the time of less than 11.5 million people. It's a staggering level of participation in a common cause, worth contemplating in the face of today's emergency.

Of that total, over 50,000 were women who enlisted, serving in non-combat roles with the Canadian Women's Army Corps, the Women's Division of the Royal Canadian Air Force, and the Women's Royal Canadian Naval Service. As the short 1942 NFB film *Women Are Warriors* explained, women took on many military functions, becoming mechanics, parachute riggers, wireless operators, decoders, drivers, clerks and photographers, freeing men up for combat duty. About 4,500 served as nurses.

What motivated all these people to enlist? For many, it was a sense of purpose and duty to "King and Country," or a belief in the righteousness and necessity of fighting the Nazis and preventing a fascist "takeover of the world." For others, still coping with the aftermath of the Depression, it was a ticket to secure employment with food, shelter and a paycheque. Still others, like George Chow whom I interviewed in his late 90s, did it "for the adventure." But virtually all emerged from the experience with a heightened sense of pride in our country and a new appreciation of comradery.

The Canadians who enlisted played a pivotal role in numerous battles across many military theatres. They were a key part of retaking Italy from the south. And they were an essential component of the Normandy invasion in spring 1944, an extraordinary logistical operation. Over one million Allied troops were involved, making it the largest amphibious invasion in history, including 156,000 soldiers who landed on D-Day, of which there were 14,000 Canadians tasked with taking Juno beach. The Allies employed over 6,900 ships for the invasion, of which Canada contributed 110.

Canada also took on a special role training air forces during the war. Given the safety of Canada (compared to Britain, which was under frequent air assault), Canada assumed the task of training pilots and air crew for much of the Allied force, an expensive contribution called the British Commonwealth Air Training Plan. With planes being shot down and pilots killed at an alarming rate, the requirement to replenish these air force personnel was pressing. By war's end, over 130,000 air crew from across the British Commonwealth and the U.S. had been trained on air bases across Canada, more than were trained in the rest of the Commonwealth combined. The program employed over 100,000 Canadians and involved more than 100 training schools across the country. It was a massive undertaking by a relatively small country in the face of an urgent need. The scale of the operation has clear lessons for us today.

Those who served overseas during the Second World War came back forever changed, and some — much like this book — drew lessons from their war experience that motivated them to fight for peace and human survival for the rest of their lives. They never stopped doing what they could to secure our collective future.

When I was a teenage peace activist in the 1980s, I had the great fortune of getting to meet some of those people, including a lovely man named C.G. "Giff" Gifford, a co-founder and long-time chair of Veterans Against Nuclear Arms (VANA).

Gifford was raised in a pacifist household, but as he watched the rise of fascism as a young man in the 1930s, he became convinced that Hitler had to be stopped. He joined the Royal Canadian Air Force in 1941, became a navigator, flew 49 bombing raids, mostly in Lancaster bombers, and was awarded the Distinguished Flying Cross.

One of his last raids was the bombing of the German city of Dresden. I first heard of that fateful bombing (like many of you, I suspect) from its immortalization in Kurt Vonnegut's 1969 anti-war novel *Slaughterhouse-Five*. On the night of February 13, 1945, as the war was nearing its end, over one thousand bombers flew across the English Channel and dropped as many as 200,000 bombs on the historic city, causing a firestorm that killed 25,000 people and destroyed much of the city. The "necessity" of the operation has long been a source of controversy, but Gifford recalls that he and fellow flyers didn't feel good about the mission even before they left — they knew there were civilian refugees in the city, that Allied forces were already closing in, and they doubted the city's military value.

When I got to meet Gifford in the late 1980s, he still looked dashing. I'd seen him before in a wonderful 1986 NFB documentary called *Return to Dresden*, by director Martin Duckworth. The film follows Gifford on a trip back to Dresden in 1985, where he attended events marking the 40th anniversary of the bombing. In one scene, which I can still remember watching in awe as a teenager, Gifford is mobbed by dozens of city residents once he is identified as a veteran-turned-peace-activist who was part of the bombing mission. It is a very emotional moment, but Gifford remains remarkably calm as he listens and talks with a crush of people who lived through the destruction. "Tell him about the firestorm," one older woman asks the translator. "Tell him that three of our sons lie here under the rubble," asks another. "Tell him, 'Thank you for coming here, and that he should fight for peace, together with us,'" another requests. Later Gifford lays a memorial wreath on behalf of Canadian veterans for disarmament.

Gifford and a handful of other vets founded VANA in Halifax in 1982, when the Cold War and nuclear arms race were still in full swing, and U.S. President Ronald Reagan talked about fighting and winning a "limited nuclear war." Alarmed by such talk, fearing for the future and wanting to speak out about the dangers of modern war, it wasn't long before VANA chapters had sprung up in cities across Canada and grew to 800 members, most of them Second World War vets. They would attend peace marches and Remembrance Day ceremonies wearing white berets, to symbolize both their past in the military and their desire for peace.

"We'd best remember our dead by working to end war," recalled Gifford in a 1985 interview in *Peace Magazine*. In an essay explaining the origins of VANA, Gifford wrote, "After 1945, when a weapon powerful enough to turn a whole city into radioactive rubble came into the leaders' hands, some veterans remembered past failures [military errors they had witnessed]. When such weapons grew from a handful to tens of thousands . . . some of us felt desperation. We had a terrible sense that the men at the top were quite capable of drawing us all into a final catastrophe . . . To sit on the sidelines while nuclear weapons and international hostility spread leaves me feeling upset, hopeless and helpless. Participation in VANA's work gives me a healthy feeling of being a member of an army of good citizens who are building the necessary future."

After the war, Gifford became a social worker and then university professor, retiring in 1984 from his position as director of Dalhousie University's school of social work. He died in 1993.

Those VANA vets are no longer with us, as we now face down the climate emergency. But if they were, I've no doubt they would be with us in this fight, marching with the student climate strikers. They laid the path of taking lessons from the Second World War to confront global crises in the present.

Full Employment on the Home Front

The story of Canada's wartime labour mobilization goes far beyond those who enlisted for military service. It is also the inspiring story of how civilian labour was mobilized on the home-front to meet wartime production demands.

The government didn't produce regular unemployment data in the 1930s and early '40s, but we do know that trade unions reported in 1939, just before the declaration of war, that 17% of their members were unemployed. By 1943, the peak of war production, that had dropped to 0.3%.[2] Similarly, a special report produced by the Dominion Bureau of Statistics (the precursor to Statistics Canada) in 1957 estimates that the number of "persons without jobs and seeking jobs" fell from 529,000 in 1939 to a mere 76,000 in 1943.[3] In short, the unemployment of the Depression was banished.

Canada's civil service, charged with organizing the war effort, grew from 50,000 to 140,000. Outside government, over 600,000 Canadians became employed in munitions industries, and about 1.2 million more were involved in war-related jobs in some way. Indeed, the demand for labour was now so great that the military and war production found themselves in competition, a source of tension between Minister of National Defence Ralston and Minister of Munitions and Supply Howe.

The solution, as thousands of Canadian men went to fight overseas, was a dramatic movement of women into the paid labour force. And to facilitate that, the government introduced (albeit temporarily) public child care. The number of women in permanent paid employment doubled during the war, from 600,000 to 1.2 million, and female farmers found themselves taking on much more of the work as husbands and sons enlisted.

The Can Car company, under the direction of chief engineer Elsie MacGill, landed the contract to build Hawker Hurricane fighter planes at its plant in what is now Thunder Bay. The first one was completed in January 1940. A year later, the plant was employing 3,000 workers, 500 of whom were women. By war's end, Can Car had produced 1,400 Hawker Hurricanes. Mid-war the plant also took on production of Curtiss Helldivers for the American air force, and ultimately produced 800 of them. More workers had to be brought on for that contract, so many that they had to be recruited from western Canada and the company had to build barracks to house them all. At its peak, the plant was employing 7,500 workers on three shifts, 3,000 of whom were women. As they recounted in the 1999 NFB film *Rosies of the North*, many of the female workers found the experience liberating; they were making real money, although not equal pay with the men, and engaged in meaningful work.[4]

The end of the war brought major layoffs at the munitions plants, with women being the first let go. Only three women remained employed at the Can Car plant. The government launched a major propaganda campaign designed to persuade women that they should return to the home. Elsie MacGill later went on to help found the National Action Committee on the Status of Women and played a key role in ensuring the recommendations of the Royal Commission on the Status of Women were realized.

Recognizing the need to quickly train people in the skills needed for war production, the government in the summer of 1940 established the War Emergency Training Program. Provinces and municipalities contributed their vocational shops and technical schools' capacity, making these facilities available over the summer months and after school hours. By 1943 about 120 schools across the country were assisting in the program, many operating two to three training shifts a day on a 24-hour basis.[5] Over the course of the war, more than 300,000 people (about 50,000 of whom were women) received training under the War Emergency Training Program. That is what training looks like in an emergency.

As any worker who has lived through a strike knows, fights such as these are transformative for both the individuals and organizations involved — they embed in people a new understanding of solidarity and collective action. And arguably, Canada's Second World War experience was the most transformative event in the history of the country's labour movement. The mobilization experience led many unions to emerge with a stronger sense of purpose and with renewed demands for a better and more caring society.

The war years saw unions gain important new rights and benefits. As journalist Rod Mickleburgh writes in his history of the B.C. labour movement, "Government saw need to avoid the class turmoil sparked by World War One and end their standoffish approach of the Depression."[6] Consequently, the Second World War saw the introduction of Canada's first two major national income-support programs — federal unemployment insurance in 1940, and the universal family allowance in 1944. And, after decades of struggle, "the fight for compulsory recognition of unions and collective bargaining was finally won, first in British Columbia and then across Canada," although its application in practice remained a struggle.

The war years saw a huge upswing in unionization rates. Wages and incomes went up and new rights were secured. Mickleburgh, writing of the phenomenal growth in B.C.'s wartime shipbuilding industry, recounts: "Besides across-the-board wage increases, those big gains in the shipyards included extra pay or 'dirty money' for working in cramped, fume-filled quarters (won after a series of sit-down strikes), the first holiday pay for B.C. workers, longer rest periods, a good minimum wage, equal pay for women and a virtual union shop [near full union coverage of all workers]."[7]

The war years saw a significant increase in strikes and labour militancy. Shipbuilders went on strike in Collingwood, Kingston and Montreal in 1940, and in Quebec, North Vancouver and Halifax the following year. By 1943 a total of 401 strikes occurred across the country, resulting in over one million lost days of work.[8]

With war production needing plenty of timber, the International Woodworkers of America (IWA) grew during the war to become the third largest union in Canada. The Vancouver shipbuilders union local become the largest single local in Canada. Dockworkers in Vancouver kicked out their phony company union and joined the International Longshoremen and Warehousemen Union (ILWU).[9]

The new sense of social solidarity that emerged with the war cracked open another long-held obstacle to justice and equality. Mickleburgh writes that it was at this time that "labour's long hostility towards Asian workers was also transformed. The International Woodworkers of America showed the way by hiring Roy Mah, Joe Miyazawa and Darshan Singh Sangha to help break down the barriers of race during the union's intense organizing drives of the 1940s."[10] Mah, who would later become a well-known leader in Vancouver's Chinese Canadian community, "put out a Cantonese edition of the union newspaper, the first of its kind in North America." Miyazawa organized Japanese Canadian sawmill and forestry workers in B.C.'s interior, who had been forcibly relocated there. And Sangha became a major organizer of South Asian mill workers, until he returned to India in 1947 to join the independence movement. While Japanese Canadian fishermen and Indigenous workers had organized into their own unions before the war, these men were among the first people of colour hired as union organizers with "mainstream" labour unions. Again, one would not want to overstate the change — society and the labour movement remained home to much racism. But a door was opened during the war.

The war was, without question, a transformative time for labour. It is certainly understandable that many feel anxious today about what the post-carbon industrial revolution will mean for the future of jobs and economic security. But as workers look to the economic transformation before us, it is worth recalling that historic moments like this can bring great new gains for workers as well.

Support for Returning Soldiers: A Model for Just Transition

Western University history professor Peter Neary, in his book *On to Civvy Street: Canada's Rehabilitation Program for Veterans of the Second World War*, writes about the benefits and programs put in place for Canadian soldiers returning to civilian (or "civvy") life.[11] With government determined to do a better job reintegrating soldiers than had occurred at the end of the First World War (which had resulted in considerable civil turmoil, most famously in the form of the 1919 Winnipeg General Strike), no sooner had the Second World War commenced than a civil service team and House of Commons committee in Ottawa began to plan for the re-entry and rehabilitation of soldiers who had completed their service.[12]

Two takeaways stand out from that preparatory period. First, the government was clearly shaken by the popular unrest that had marked the pre-war period and nervous about a new militancy in the land and rising support for the CCF, particularly among soldiers. They were therefore keen to develop programs that would dampen such radicalism. And second, as with the economic production detailed earlier, they recognized the need for extensive government-led planning. Due to the war's length, planners had time to pilot various programs and make refinements. What emerged from this forward-thinking initiative was a Veterans Charter, which created the Department of Veterans Affairs along with a suite of benefits and transition supports.

The task was immense. Over a million men and women enlisted — far more than the numbers employed in the fossil fuel industry today — and thus had to be returned to civilian life, hopefully as smoothly as possible. In 1944, as the end of the war came into sight, C.D. Howe was made minister responsible for reconstruction, and his team set about the task of ensuring a million jobs for those returning from war.[13]

The benefits and supports provided to returning soldiers, and enacted in various acts and by orders-in-council during the war, played a crucial role in helping returning soldiers re-find their feet in civilian life.

The government provided grants (based on one's time in service) and transitional income support in the form of monthly allowances, a clothing allowance for new civilian clothes, and various supports to find employment. Income support was provided to those ex-service people who had not yet found work, those in a training program, those attending university, those temporarily incapacitated, or those waiting for a new business to start producing returns. The government also offered assistance setting up a business, if that is what a veteran desired, including low-interest loans based on years of service.

Pension and rehabilitation benefits were established for those with disabilities or medical needs stemming from their service. In these days before Medicare, veterans were offered free medical care to deal with post-war injuries.

Post-secondary training and education support was extensive. In particular, the government offered free training or education to any service person who wanted it, not only covering tuition but also providing a family allowance during that time. Under the Canadian Vocational Training program, which began in March 1944, the government "made use of 106 private trade schools, 200 business colleges, 48 provincial or municipal schools, and 68 special training centres."[14] By March 1951, 80,110 veterans had taken part in the program. If a soldier had interrupted a university education to enlist, for each month of service, the government would cover a month of tuition and family allowance after their service. This program fundamentally transformed the face of Canadian universities. "In 1946 about 35,000 were supported by the government to attend university," writes Neary. "By comparison, the entire full-time undergraduate enrolment in Canadian universities and colleges in 1939 had been only 35,164."[15] By 1951, 53,788 veterans had availed themselves of the university program.

The Wartime Housing program was initially developed for war production workers but was later adapted for returning veterans. "Between 1941 and 1947 Wartime Housing Limited, a Crown corporation, built 26,000 dwellings for rent, although most were soon sold off."[16]

Under the Veterans' Land Act (passed in 1942), veterans were offered small holdings, properties they could use either while engaged in paid work or from which to undertake farming or commercial fishing, and homes were built if the land did not have one already. For those who did pursue farming, the government offered grants and special loan financing rates. Those wishing to return to fishing were offered similar supportive financing to own and operate small boats. As historians Richard Harris and Tricia Shulist explain, between 1943 to 1975, the Veterans' Land Act financed the construction of almost twice as many housing units as the general Wartime Housing program — in total 44,222 units were built for veterans, and "as many households again were given assistance to acquire land on which there was already a dwelling, so that, in total, the Land Act financed 102,025 small holdings."[17]

Veterans offices were established across the country, staffed with counsellors and welfare officers to help walk veterans through their options. Local "rehabilitation boards" were set up in every major city, tasked with reviewing applications from any ex-servicemen and women for additional assistance. This might include help landing a public service job — civil service preference was guaranteed in law for every ex-service person — or family allowances while starting up a small business. The government was actively promoting all of these programs while the war was still underway, as an enticement for people to enlist, but also as an encouragement to society at large to assist in the successful transition of returning soldiers.[18]

Alice Sorby, in her study of post-war rehabilitation plans, and with resonance for oil and gas workers today, writes that it was understood "that the Second World War veteran, with his knowledge of the use of the complicated machinery of modern war, might have acquired skills with some transfer value, unlike the front line soldiers of France and Flanders [in the First World War], whose skill in trench warfare had little or no relationship to the requirements of civil life. In fact both the Navy and Air Force issued manuals in which they related service skills to peace-time occupations for the benefit of employers."[19]

It bears emphasizing: these post-war programs weren't simply the result of government largesse and goodwill. They stemmed from the demands of labour and social movements, who, after the ravages of the Depression and the sacrifices of the war, insisted on a new deal.

Not everything returning soldiers needed was present. Housing remained in short supply. As homelessness advocate Cathy Crowe writes:

> Veterans returning to Canada from the war faced a drastic shortage of homes. The vets were vocal about their right to housing and, experienced in organizing, they engaged in dramatic and effective protests with their families, supported by seniors groups and faith-based communities. Their actions included mass parades, picketing and squatting — one group occupied an empty military hospital in Montreal while another took over an army barracks in Ottawa by moving their families and furniture inside. The barrack conditions were so poor that two children came down with polio, presumably from milk not kept cool due to lack of electricity. That only caused public opinion to swell in support of the veterans.
>
> The result of their fight? Their demands for housing were answered with decisive action: the building of wartime housing followed by amendments to the National Housing Act in 1949 led to expanded federal funding for social housing — a national housing program that by 1964 was close to being a universal program for several decades.[20]

Also missing was mental health support. Many soldiers returned suffering from what we now understand to be post-traumatic stress disorder (PTSD), but these soldiers were often treated appallingly. One would not want to extend this comparison too far, but it is worth flagging that work of the oil and gas industry is often extremely intense, and frequently involves long stretches of time away from loved ones. The high rates of drug and alcohol abuse and addiction in the oil patch are well known. Sometimes just transition isn't only about jobs and training and income support; it should also include compassion and counselling, as we set off on a path to become a different society than the one we are leaving behind.

Its shortcomings notwithstanding, the Veterans Charter to emerge from the war changed the lives of hundreds of thousands of people. Neary calls it "a nation-building initiative," and concludes, "There had never before been a social welfare scheme of such scope and magnitude in Canada." It allowed people to attend university whose families likely never dreamed it possible. Kai Nagata (who we will meet again in forthcoming chapters) recalls that his maternal grandfather, Robert, was one such beneficiary. He enlisted in the Royal Canadian Air Force midway through the war, as soon as he was old enough to do so, and trained flying Avro Ansons at a base near Winnipeg. The war ended before he was shipped overseas. Even so, the military thanked him for his service by paying his tuition and costs while he attended UBC — a life-changing opportunity for a kid from a working-class immigrant family.

Indeed, Neary's view is that the programs established for Second World War veterans didn't merely change the lives of those who directly benefitted from the programs — they effectively set the stage for many of the broader social welfare programs that would benefit all Canadians. "Veterans stirred the pot in Canadian society . . . The benefits provided for them showed what was possible when Ottawa acknowledged an obligation and mobilized resources in support of a needed national social program."

A MADE-IN-CANADA GREEN NEW DEAL

As we prepare to leap into a carbon-free society and economy, it is entirely understandable that many, particularly those whose livelihoods are currently derived from the fossil fuel industry, feel anxious and economically insecure. But the good news is that, as we experienced in the Second World War, the climate mobilization means there are many jobs to be had and a lot of work to be done. The transition requires care and planning, as the jobs of the future will not be all the same jobs in the same places. Nor will most of the new jobs pay as highly as what some in the oil patch have come to expect. But overall, there is every reason to believe that there will be more jobs, and that those jobs will be meaningful, decent, well-paid and considerably more secure than the jobs in or related to the fossil fuel industry today.

We also need to ensure that those new jobs are complemented by enhanced public services and social supports that result in a better standard of living for all. When we decide as a society to collectively pay for key needs like child care, elder care, post-secondary education, pharmaceutical drugs, dental and eye care, decent pensions, public transit, and affordable housing, we are all liberated from a great deal of anxiety about how much our employment pays. In such a rebalancing of what we pay for by ourselves and what we pay for together, many will find more work-life balance, and the fulfilment that comes from more time with family, community, nature, the arts and more.

Atiya Jaffar is a young climate activist. Based in Vancouver, she is a campaigner with 350.org, and one of the organizers behind the call for a Canadian Green New Deal (GND). In 2018, she was one of hundreds who were arrested on Burnaby Mountain protesting construction of the Trans Mountain pipeline expansion project. Then in the lead up to the 2019 federal election, Jaffar helped to launch the Our Time initiative, a push to mobilize millennials, asking younger voters to get behind those candidates who were serious about the climate emergency and committed to implementing a made-in-Canada Green New Deal.

What made her a climate activist? "Every year since I graduated high school, there has been a massive climate disaster. I'm an immigrant from Pakistan, and I was visiting Pakistan in 2010 when there were the massive floods. The impact was everywhere. My entry point [to activism], when I got home to UBC, was fundraising for flood relief. But within two years, I developed a lens where I could see the connections between the fossil fuels we burn here and the impacts elsewhere."

As Jaffar explains, a great appeal of the GND is how it challenges the "jobs-versus-environment" narrative, turning this trope on its head and seizing the initiative with a bold plan to create a million-plus jobs building the future economy we want and need. In the face of a federal government that ceaselessly insists that pipeline expansion is necessary to the Canadian economy, Jaffar and her colleagues realized that "winning this broader, populist narrative meant moving beyond fighting for climate action one pipeline at a time or one fossil fuel project at a time."

Just as Jaffer and her 350.org team were re-evaluating their strategy, they started to witness the rise of the youth-led Sunrise Movement and Alexandria Ocasio-Cortez in the U.S., who were taking on the Democratic Party establishment and demanding real climate action, and whose campaign was focused around the call for a Green New Deal, a federal jobs guarantee, and a 100% renewable energy economy. They could see the appeal taking off, not only in the U.S., but in other countries as well. "We were really inspired by that, and of course we were planning for what our campaigning would look like in an election year and recognizing that youth and millennials will be the largest voting bloc in this election. And so we turned our attention to this more populist narrative around the Green New Deal and toward what it would take to have a massive shift in the structure of Canadian society."

"The Green New Deal packages everything really nicely," notes Jaffar. "It invokes historical examples of massive state spending for a massive transition, social safety nets for people that are impacted, and it is really what we need in order to address the scale of the climate crises. If we were to win something like the Green New Deal, things like the Trans Mountain pipeline would not even be on the table."

The coalition of Canadian organizations calling for a GND has now pulled in over 60 environmental organizations and civil society groups, including Leadnow, Greenpeace, The Leap, Dogwood, the Canadian Union of Postal Workers, the Union of BC Indian Chiefs, Indigenous Climate Action, Confédération des syndicats nationaux (CSN), the Council of Canadians and CUPE Ontario.

Those behind the GND have sought to link their call for real climate action with migrant and Indigenous rights. As Jaffar explains, "We need to codify the resistance to white supremacy in our climate change fight. The rallying populist message for the right-wing right now is racism, it is anti–climate change action, and we need to present an alternative to that that is inclusive, that refuses this divisive narrative that the right is putting forward. And we want to lead with hope. They are leading with fear. I think our greatest tool, our greatest counter to the fear-based message they are bringing, is the message of hope. That is really what we are rallying behind."

The Green New Deal, which exploded into the popular debate just as the IPCC issued its urgent timeline in late 2018, is indeed that positive and hopeful vision. Naomi Klein, whom Bill McKibben calls "the intellectual godmother of the Green New Deal," says this of a truly sweeping GND:

> The idea is a simple one: in the process of transforming the infrastructure of our societies at the speed and scale that scientists have called for, humanity has a once-in-a-century chance to fix an economic model that is failing the majority of people on multiple fronts. Because the factors that are destroying our planet are also destroying people's quality of life in many ways, from wage stagnation to gaping inequalities to crumbling services. Challenging these underlying forces is an opportunity to solve several interlocking crises as once.
>
> In tackling the climate crisis, we can create hundreds of millions of good jobs around the world, invest in the most systematically excluded communities and nations, guarantee health care and child care, and much more.[21]

Among the first to call for a Canadian version of the GND was Cree climate activist Clayton Thomas-Müller of 350.org's Canadian team.[22] Inspired by what he saw unfolding south of the border, Thomas-Müller wrote in late 2018, "That's the same thing we need here in Canada — a climate plan that stops fossil fuel expansion, gets us to 100% renewables and guarantees a good job for impacted workers. And, it has to do it all in full partnership with Indigenous peoples."

Avi Lewis, in a cross-Canada tour promoting a Green New Deal ahead of the 2019 federal election, called it "the greatest anti-austerity program of our time." In that sense, it is also, poetically, the surest path to preventing a rise in neo-fascism in the present, given the connections between austerity and the far-right revival.

Like Jaffer, Lewis emphasizes that a great benefit of the GND (like the earlier *The Leap Manifesto*) is that it turns the "jobs-versus-environment" barrier to action on its head, and instead seizes the mantle of job creation. "The Green New Deal will be a massive job creator, swell the ranks of unions, and increase workers' rights for all, especially the most vulnerable," declares Lewis. Just as occurred in the Second World War.

Clean Energy Canada has calculated that "Canada's clean energy sector will employ 559,400 Canadians by 2030 — in jobs like insulating homes, manufacturing electric buses, or maintaining wind farms. And while 50,000 jobs are likely to be lost in fossil fuels over the next decade, just over 160,000 will be created in clean energy — a net increase of 110,000 new energy jobs in Canada."[23]

But the job creation potential is far greater than that, if and when we marshal green and social infrastructure investments at the scale called for by a GND. The number of jobs created per dollar invested in these sectors far outstrips the fossil fuel industry. That's mainly because the latter is so capital-intensive, whereas the alternatives are more labour-intensive. This is especially true for public services and the caring economy — to state the obvious, services are delivered by people. As Marjorie Griffin Cohen notes, these service and caring jobs — more often done by women — are just as much a part of the "real" economy as goods-producing ones are, and "the huge growth in the services sectors of advanced industrial nations indicates that real prosperity may not be tied, ultimately, to manufacturing and resource extraction."[24] Similarly, zero-waste management, agriculture, value-added forestry, energy efficiency work such as building retrofits, and construction work building new low-income housing are all more labour-intensive than fossil fuel extraction and processing. The comparative job-creation can be seen in the accompanying chart.

DIRECT JOBS PER $1 MILLION OF OUTPUT (2016)

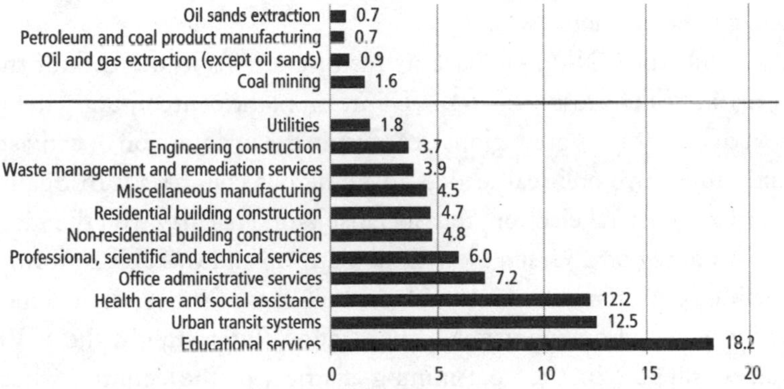

Source: Statistics Canada: Tables 36-10-0594-01 and 36-10-0013-01.

"So, when people tell you that the GND will hurt workers, set them straight," insists Lewis. "Tell them that the Green New Deal is a job program of epic proportions."

If Naomi Klein is the GND's "intellectual godmother," then rookie U.S. congresswomen Alexandria Ocasio-Cortez (AOC) is surely its global popular champion. Her Green New Deal congressional resolution, introduced with Senator Ed Markey in February 2019, outlines the goals and principles that should guide a U.S. GND. Importantly, it grounds the call for an ambitious federal action plan as much in justice, anti-racism, job-creation and inequality as in confronting the climate crisis, calling for massive investments in a host of initiatives to meet the IPCC targets while leaving no one behind.

Interestingly, the Ocasio-Cortez resolution also specifically casts the climate crisis as "a direct threat to the national security of the United States," and also speaks of a GND as a "new national, social, industrial, and economic mobilization on a scale not seen since World War II and the New Deal era." Indeed, the language of "mobilization" is embedded throughout the document.[25] AOC's work has been informed and supported by a network of civil society groups, including the Sunrise Movement, the Justice Democrats and a new Washington-based think tank called New Consensus. Notably, the leadership of all of these groups is primarily made up of young people and people of colour.

While the GND may have its historic origins in the U.S. of the 1930s and FDR's response to the Depression, the contemporary iteration of the idea has gone global, capturing the imagination of activists and progressive political leaders in numerous countries.[26] In Spain's April 2019 federal election, the Socialist Workers Party won the most seats running on a Green New Deal platform and was able to form a coalition government, mainly with the help of the further left Podemos party, which also supports decarbonization. Interestingly, the SWP gained support in the coal-mining districts of the country, where they have negotiated a just transition package as part of phasing-out the industry.[27]

The GND idea first found traction in the U.K. over a dozen years ago, and in September 2019 the Labour Party formally adopted a Green New Deal program. The civil society think tank Common Wealth has developed a compelling, radical and original series of papers outlining what a U.K. Green New Deal should consist of, including innovative ideas for democratizing the economy and public institutions, public ownership, industrial policy, monetary and fiscal policy needed to finance the transition, and transforming communities, homes and transportation.[28]

Real Just Transition

Various components of what should be in a full Green New Deal are found throughout this book. But let us focus here specifically on the *just transition* elements that are needed.

The idea of "just transition" was first coined in the early 1990s by U.S. labour leader Tony Mazzocchi, who developed the concept for defence industry workers. Mazzocchi, a leader with what was the

Oil, Chemical and Atomic Workers International Union (later amalgamated into the United Steelworkers), was a pioneer in worker health and safety. He was also a disarmament and environmental activist who wanted to see a conversion of the U.S. economy away from war production, but wanted to ensure that the workers and communities who relied on those industries were not abandoned. He envisioned a Superfund for Workers displaced by needed economic restructuring, which he modelled on America's GI Bill for returning Second World War veterans. As Mazzocchi explained years ago, "The GI Bill helped more than 13 million ex-servicemen and women between 1945 and 1972 make the transition from military service to skilled employment in the private sector. This program had a formidable price tag, but the country overwhelmingly approved it as an investment in the future. Education became the key to national economic recovery."[29]

The notion of just transition, however, should be understood as bigger than the issue of support for individual workers in industries we seek to wind down. The CCPA's Shannon Daub proposes three levels of just transition: first, at an individual worker level, it captures issues such as income support, early retirement, training and education, relocation costs, counselling and, ideally, help finding new "green jobs" in the renewable economy or other industries; second, at a community level, it can address issues such as economic diversification for communities that have been heavily reliant on one industry such as fossil fuels, the recruitment of new industries and employers, and green infrastructure and renewable energy projects; and third, at the macro level, just transition captures broader issues of climate justice and the transition to a post-fossil fuel economy, decolonization and Indigenous rights, particularly for Indigenous communities that have been harmed by fossil fuel extraction, tackling inequalities in the labour market, particularly between traditionally male-oriented versus female-oriented jobs, and energy democracy, meaning who owns and controls energy resources and infrastructure.

A truly just transition plan would see us address all these dimensions. And for some guidance, there are useful lessons to be learned from past just transition experiences.

Lee Loftus is a labour leader who has been working on just transition issues for years. Recently retired as business manager of the B.C. Insulators Union (formally, Local 118 of the International Association of the Heat and Frost Insulators and Allied Workers) and past president of the B.C. Building Trades, Loftus has long been a labour voice in support of climate action, but one infused with an understanding of workers' needs and experiences.

Confronting climate change has been a fraught issue for the B.C. building trades. The unions affiliated with the sector have frequently been among those who support new fossil fuel projects such as pipelines and LNG, given the construction jobs these projects promise. Loftus, however, has long been among those seeking to bridge these divides between the status quo and necessary change. As an insulator by trade, he sees the huge employment potential that would come with the building retrofits that climate action requires. Loftus was among the founders of Green Jobs BC, an alliance of labour and environmental organizations "with a shared vision of an inclusive, sustainable economy that provides good jobs that are socially just, protect the environment and reduce carbon emissions."

Much of the necessary transition before us is something Loftus has lived through before, and it hasn't often been pretty. He's experienced a history of transition failures — promises that were made to workers that never materialized, and economic shifts that left working people behind. "Transition in our Canadian environment has been an absolute failure," Loftus told me. "Politicians are not interested in developing real transitional plans, they are interested in silencing those that are speaking. You know, if you take a look at fisheries, you take a look at forestry, you take a look at shipbuilding, you take a look at asbestos. The list goes on and on. There has not been a successful and just transition. I have to tell you that saying that we are going to give you ten weeks of unemployment insurance and teach you how to write a résumé, if you are under the age of forty-five, it is not a fucking transition plan. Excuse the phrase. It is a scam." Consequently, like many labour activists, Loftus dislikes the language of "just transition." He prefers to talk about "pathways to tomorrow's employment."

Before climate was on the agenda, Loftus was spearheading

another industry transition — the need to eliminate the use of asbestos, which for generations has caused severe health problems and death for those in the building trades. As with the fossil fuel industry, the asbestos industry sought to deny and obfuscate. "There is documented evidence back in 1947 that asbestos was killing people," recounts Loftus. "That was suppressed by epidemiologists that were employed by companies that were manufacturing and distributing it. But the knowledge was there." And like today, the need for change divided labour, particularly in Quebec, where asbestos was mined. "The union there [in Quebec] that did the work that I do, meaning finished products and installing them, was driving change," says Loftus. "But the miners, the people that were actually mining it and producing it, were blocking change because they did not have certainty. And there was no offer to them of what was going to happen. There was no plan, there was no transition."

That experience informs Loftus's climate work today, and how he seeks to engage workers. "Awareness and education and meaningful discussion is everything. If you do not offer somebody hope, they are not going to transition. If you do not have a plan in place for them to move to that next level, they are not going to participate."

Loftus was one of the few voices within the B.C. building trades expressing a dissenting view on LNG development, warning his fellow workers "not to drink the Kool-Aid." Quite literally. When the B.C. Liberal government of Christy Clark sought to make the building trades partners in her quest for LNG, Loftus recounts, "I actually went out and bought an old Kool-Aid cooler and put it on the boardroom table." But Loftus also understands the predicament facing workers in his sector. As he explains it, when the industry asks the building trades unions to politically support a fossil fuel project, looming behind that request is a threat — that if their support is lacking, the jobs will go to a non-union or fake-union contractor. And those threats are real. Industry has proven itself willing to do just that.

According to Loftus, most people in the building trades do not "get" the IPCC emergency. "People in construction think about tomorrow, they do not think about next year, they do not think five years, and they do not think ten years. We are trained to think about the next job." Many in the building trades face a constant need for work and a perpetual sense of uncertainty means many are prepared to say yes to whatever construction project is on offer, regardless of its climate impact.

Loftus wanted to throw a wrench in the works: "I am tired of us defending pipelines. I am tired of us defending oil refineries. I am tired of us talking about the oil sands. Let us talk instead about tomorrow's employment." So Loftus, in his then capacity as president of the B.C. Building Trades, initiated a research project entitled *Jobs for Tomorrow*. The report offers estimates of the number of construction jobs that will be needed in Canada to supply clean energy, retrofit buildings, build district energy and build new public transit infrastructure, and arrives at a grand total of 3.3 million direct construction jobs (each representing one person-year of employment) between now and 2050.[30]

These days, Loftus has a terrific slideshow based on *Jobs for Tomorrow* and he spends his retirement taking it around to various Building Trades locals. Loftus is trying to drive home the reality of the climate crisis and that change is inevitable. But he's also trying to communicate, as one trades worker to others, a sense of excitement about all the future work that will result from bold climate action. Loftus is inviting his fellow building trades workers to embrace a special role in this historic moment, and he's trying to convince his fellow workers that they need not fear the change that must come. "And the last slide on every one of my decks is that we need to lobby the federal, provincial, and municipal governments to put in substantive climate actions plans to get there."

"The only way that I can move the construction sector is to talk about hope," says Loftus. "But if there is hope attached to employment, I think that there is a campaign to launch that says, 'Let us go for that ride.' Because I think it is a good ride, and if they climb on this bus and go for this ride, they are part of the solution, rather than being an obstruction. And I do not want to find ourselves, like steel workers in Quebec [the asbestos miners], being an obstruction."

Loftus also sees a generational change. "I think the construction sector will always support someone who offers them employment. I do not think that will ever change. But I also know that the demographics within the construction sector is in a significant shift. The new people coming into the sector are from an age that will not support the traditional thoughts in regards to that. They are coming to the table educated very differently and certainly aware of climate issues. So, I think that the next generation of leaders, I think the next generation of workers, I think the next generations of apprentices will all be on the page that they need to be mindful of climate."

Stephen Buhler is one of that new generation of tradespeople hungry for change.

A journeyman machinist in his late 20s, born and raised and living in Alberta, Buhler figures about 98% of the contracts he works on are related to the oil and gas industry. Buhler's been engaged in this work for over ten years. His dad is a journeyman pipefitter and steamfitter who has likewise derived much of his employment income from the oil sands. But that's not the future Buhler wants.

While by day Buhler is earning an income machining parts for the oil patch, in his off-time he's become an activist with Climate Justice Edmonton. In the lead-up to the 2019 federal election, when I spoke with him, he was working with the Our Time millennial initiative pushing federal parties to offer up a made-in-Canada Green New Deal.

He credits his dad for raising him to keep a critical mind, and in particular remembers his father introducing him to the music of British punk legends The Clash at a young age. He believes that music helped to shape a worldview that sees him second-guessing the claims of the oil industry and those in authority.

But it was the recent Alberta wildfires — including the 2016 blaze that forced the evacuation of Fort McMurray in the heart of the oil sands — that served as Buhler's revelatory moment and led him to climate activism. The smoke that now characterizes the summer months caused a pain in his lungs that "got me pretty pissed, that it is getting hotter and we are having drier and drier summers, and nobody is making the connections, that this is because we have a high carbon economy, that this fundamentally needs to change."

When asked how "typical" he is as a young oil and gas worker with respect to his climate change concerns, he replied, "All oil workers are on a spectrum. I am definitely on the further end, on the more climate activist end of that spectrum. And honestly, there is probably a very small percent of us that are down there. There is one guy I know that is a big proponent of carbon taxing; he is right now looking to buy an electric van for his family. So there are definitely people working in oil and gas who do believe that climate change is coming and that something needs to be done about it. Where a lot of people differ is how much time they are willing to give, both emotionally and just hours in a day of pushing to make that change. Honestly, I think most people are kind of in the middle. There are a lot of people who do not really want to engage with it either way. I do not think that most oil workers are buying 'I Heart Canadian Oil and Gas' shirts. Most oil workers are not going to go to a rally for pro–oil and gas groups. I think the vast majority of oil workers are trying to get through the day. They are trying to make sure they can pay their bills and that requires putting in long hours."

When asked his thoughts on just transition, and if he thinks most oil and gas workers would want their own children to work in the industry, Buhler again sees a spectrum. Buhler figures older workers like his dad just want to make it through a few more years until they qualify for their pensions. This is, of course, entirely reasonable and, as we wind down the oil sands over the coming two to three decades, ought to be perfectly possible.

Buhler has some friends who don't want their children to follow them into work in this industry and who don't see a future in it. And he knows others who would encourage their kids to follow them into the industry, where they have made good money, "and they are afraid for their kids too, which I can totally understand."

Buhler is among those who are tired of being told that his interests are the same as the oil and gas companies'. As he told an audience at a GND event in Edmonton in 2019: "I have watched as workers like myself have been exploited time and time again by an industry that is leaving us behind. The oil and gas industry likes to tell us that their fight is our fight, that their interests must be our interests, all while they line their pockets with billions of dollars as workers like me are subject to the precarious boom-and-bust cycle."

"My vision for myself and other machinists," he says, "is that we need machine parts for wind turbines or for any kind of power generation. More broadly, there is still tons of work that we can do as tradespeople in renewable resources." It frustrates him that people often think oil workers have to be "retrained." His view is that the skills needed for the new economy are largely the same, just working on different kinds of projects. "You know, a welder can burn metal whether it is for pipeline or whether it is for wind turbines. It is all the same." Buhler sees the Green New Deal as key to winning over Albertans and oil workers.

And he's not alone. Local 424 of the International Brotherhood of Electrical Workers (IBEW) operates its own training facility in Edmonton where they have installed a large solar array and have already trained over 1,800 of their members in solar PV systems. A majority of the local's members currently work in the oil and gas sector but the union is committed to working on the green jobs agenda, and many of their members are keen to move into solar installation, wind, geothermal and electric vehicle maintenance.

And of course, planning for a just transition isn't just about the "guy jobs." Edmonton-based climate justice campaigner Emma Jackson points out, "We get caught in this trap where we talk about a 'just transition' and we always tend to frame it around oil workers." Her MA research, in contrast, focused on live-in caregivers in Fort McMurray. "When we talk about just transition, we really need to broaden the scope to think about how we build an economy that is less dependent on extractive industries . . . and that is massively expanding our public sector. It should be much more deeply rooted in caring economies."

Jackson notes that what the workers she interviewed need, first and foremost, is permanent landed immigrant status upon arrival, freeing them to leave crappy or abusive employers and to find other work — the basic right to quit and be mobile that all Canadian workers expect. Reflecting on the workers she met, Jackson emphasizes that permanent status "would open up all kinds of opportunities for work. A lot of the time they really wanted to stay in Fort McMurray, which I found really surprising, but they did not want to do live-in care work anymore . . . They wanted to work in administration, or they wanted to continue working in a daycare instead of being tied to one employer and having to live in their home because housing is so unaffordable."

Alberta's Coal Phase-Out: Just Transition Lessons

While many lament that the NDP government in Alberta did not do more to move that province away from oil and gas, Notley did receive kudos for her move to accelerate the phase-out of coal-generated electricity in Alberta. How that was done is now often held up as a global model for just transition, albeit an inadequate one. Its design offers both positive and negative lessons.

The earlier phase-out of coal-generated electricity in Ontario was, in all likelihood, easier. The coal burned in Ontario wasn't mined there (whereas Alberta's phase-out impacted both coal miners and those working in power generating plants), and most of Ontario's coal-based power plants were near major metropolitan areas, so a diversified economy and other jobs were close by.

In Alberta, phase-out meant closing five thermal coal mines and shutting down or converting six coal-powered plants with 18 generators, affecting about 3,000 workers directly, and many more indirectly. Numerous communities were almost entirely reliant on coal power. At the start of the phase-out, Alberta accounted for roughly half of the coal-fired electricity in Canada.

Alberta Federation of Labour President Gil McGowan had a front-row seat during the planning of Alberta's coal phase-out and the associated development of a just transition plan for impacted workers and communities. He chaired the Alberta Coal Transition Coalition and served as well on the federal government's 2018–2019 Just Transition Task Force on phasing out coal. While McGowan believes the just transition plan developed for coal workers in Alberta was trendsetting, he cautions that what unfolded in Alberta is not a victory to be celebrated. Rather, he contends, "it was a necessary tragedy," from which some people and communities will never recover. He urges that just transition be front-and-centre in any plans to wind down fossil fuels, if we wish to stave off an ugly backlash. "The path to economic transition runs through real people's lives," he warns.

McGowan offers a number of insights from Alberta's coal phase-out experience. He stresses that planning needs to include labour representation and strong just transition commitments, and it must bring impacted workers into the process. To that end, the labour movement organized numerous town halls with workers affected by the phase-out in their own communities. Not surprisingly, those workers weren't worried about climate change. Rather, they were worried about the loss of good jobs with few prospects for well-paying alternatives, the collapse of home values and being forced to leave communities that in some cases have been home for generations. Those were "hard conversations."

McGowan also contends that impacted workers need to understand that it isn't just climate change that is necessitating an end to their jobs — it's market forces as well, such as the explosion in fracked gas and oil from the U.S. oversupplying the market. Once people appreciate these factors at play, even if reluctantly, they often understand that it is better to negotiate the terms of a just transition, particularly with a friendly government, than to be tossed aside by companies and market forces that don't actually care about local workers. But again, getting people to that place is hard, and it's vital to have trusted leaders helping to have that conversation.

The coal phase-out negotiations in Alberta resulted in a deal that saw the government dedicate $40 million for affected workers (funded by revenues collected from Alberta's former carbon tax). That money was used to provide $12,000 per worker for retraining or education; pension-bridging income support for those workers close to retirement; income supplements of up to 75% of previous earnings (to top up meager EI payments) as workers went through the transition process; relocation allowances for workers who chose to move; and $5 million for community economic development for affected communities, with the goal of allowing as many workers as possible to remain in their home communities.[31]

Unfortunately, while the issue of falling home values was discussed, dealing with this issue was not part of the package. The government deemed the matter "too expensive," although given the limited number of people involved, having the government simply purchase many of the impacted homes should not have been seen as impossible.

A much larger claim on transition money went to the coal companies themselves. The three main companies affected were TransAlta, ATCO and Capital Power, who combined will receive "compensation" from the provincial government of $1.36 billion over 14 years. Emma Jackson worked on a research project that studied the coal phase-out on behalf of the Parkland Institute and Corporate Mapping Project. Tasked with investigating the financial position of the coal companies, she concluded that they likely didn't need the money,[32] but that the government disbursed it to them in the interests of avoiding legal action. Given the wealth these companies had extracted from these communities and public assets over many decades, those payouts would have been more justly directed towards those workers whose home values needed bolstering or replacement.

The government took many ideas from the coalition McGowan chaired, but not all.

He had recommended that the province establish an Alberta Economic Adjustment Agency to manage the transition for workers. That didn't happen and represents a key missing piece from Alberta's just transition plan because, as McGowan emphasizes, what people most want in a just transition isn't training or income support — it's another good job. That's vital if we are going to offer people hope during the great transition to come. But the current system simply isn't well set up to connect workers losing their jobs with the new jobs being created in the green economy. Navigating that is left to individuals themselves.

And that's a curious omission when you think about our wartime experience, because Canada didn't just offer training and relocation allowances to people then — we actively connected people to the jobs that the wartime mobilization needed filled. We must do the same today — create a government "matchmaking" agency or crown corporation that links workers in sunset industries to new jobs in the emerging economy. Better yet, if our governments truly embraced the logic of the Green New Deal, they would be actively creating new climate-action crown corporations like those proposed earlier and massively spending on green infrastructure. This would result in hundreds of thousands of well-paying jobs for transitioning workers, without having to rely solely on policy incentives that hope to attract new private businesses to impacted communities.

Scott Lunny,[33] assistant to the president of the United Steelworkers District 3 in western Canada, proposes a few key guiding principles for successful just transition. First, he emphasizes that people need to see a clear "pathway" from their current job to a new one — what is the new job, where is it, what if any new skills will be needed to secure it, and what will it pay? Second, workers need to have confidence that the new jobs on offer will be "quality, family-supporting unionized jobs." Third, we need to be thinking about the indirect jobs in the supply chain of the current fossil fuel sector, many of which are also well-paying jobs and will also require just transition plans. Finally, just transition needs to be truly funded — not just an empty promise. And the promise needs to be in writing and binding, "just like a collective agreement."

Based on all the insights above, here are what I see as core government commitments that should be at the heart of a robust just transition plan:[34]

- Massive public investment in new green and social infrastructure, as outlined in the previous chapter.
- A "Good Jobs Guarantee," not only for workers and communities that currently rely on the fossil fuel sector and for those whose communities have been most harmed by the fossil fuel industry, but for anyone who wants

employment building the green and social infrastructure we will need. The guarantee should be for well-paid and unionized jobs. This may sound too radical or ambitious. But why? We did it during the war and we can do it again. Moreover, as in the war, our most likely challenge in the coming decades is not going to be finding enough jobs for workers, but rather the reverse — finding enough workers for all the jobs that need doing.

- Given that some will need to retrain and/or relocate for some of those new jobs, the Good Jobs Guarantee should include training commitments (with income support during that time) and relocation allowances. The training should include a renewed focus on apprenticeship programs, with minimum quotas for women, Indigenous people and other underrepresented groups in the building trades that will be most in demand.[35]
- Early retirement bridge financing for older workers whose jobs in the fossil fuel sector are lost.
- Community transition support will also be needed and, arguably, is often more necessary that individual transition supports. Most current fossil fuel workers will retire over the next 20–30 years, in line with the wind-down period, and those still too young to retire have transferable skills. Given this, just transition requires specific supports and new opportunities for small and rural communities that are currently highly reliant on the fossil fuel sector. That could mean the government buying out homes that lose their value, but preferably it would mean locating major anchor employers in these communities (either crown corporations or enticing other employers). Our governments should seek to foster renewable energy projects, value-added forestry ventures and sustainable agriculture growth in smaller and rural communities so that people see local options beyond the fossil fuel sector.
- Generous income transfers — a "Climate Action Dividend" — for all low- and middle-income households in Canada

to offset transition costs, an escalating carbon tax and rising energy prices. These should come in the form of a refundable tax credit like the Canada Child Benefit, so that cheques go to all who qualify regardless of how much tax they pay. And like the CCB, the Climate Action Dividend should have a long phase-out tail, so that most in the middle class receive at least some benefit.

- In addition to the Climate Action Dividend (which will likely need to be provided for at least a couple decades), low- and middle-income homeowners should be provided with one-time rebates for home energy retrofits and fuel-switching their homes and appliances (such as furnaces and stoves). A separate program will be needed to offset these costs in rental properties.

We would be well-served by a public agency or crown corporation to oversee all the elements of this just transition plan. The U.S.-based The Climate Mobilization's *Victory Plan* proposed establishing a Mobilization Labor Board, a tri-partite body composed equally of labour, business and federal government representatives to "monitor and manage industrial labor relations and America's federal job guarantee program for the duration of the emergency transition."

A Change Is Gonna Come

There will be a transformation — a response to the climate crisis — and whether it occurs in a manner that is just and fair or unjust and repressive remains an open question. Past industrial revolutions have cared little for those whose lives were turned upside down by change. Another is coming. Our challenge is to try as best we can to ensure this one unfolds differently.

The climate reality requires that our governments talk honestly about the future of Canada's fossil fuel industries. We need our leaders to acknowledge that we cannot continue to expand fossil fuel sectors, but rather, these industries will need to be managed for wind down. This transition, however, need not be as jarring and anxiety-producing as what some communities are already facing given low oil prices.

Instead, as in the Second World War, a planned adjustment with just transition programs in place can make a huge difference.

In the face of the historic changes before us, the leadership of each labour union also has a decision to make — a moment of reckoning as we confront this emergency: do you want to resist this call for climate mobilization, fight a rearguard action against the grand transition that needs to happen and make common cause with the fossil fuel corporations who seek to block progress? Or, do you want to boldly engage your members, invite them to join you in leading change and ensure it unfolds in a fair manner? Which path will your grandkids take pride in?

What might it look like for unions and labour leaders to seize the initiative? Rather than merely banging on the doors of government demanding that workers have a voice at the table in climate deliberations, what if unions instead took charge in retooling our economy and forced our governments and employers to run to catch up? While the core duty of any union is to secure better pay and benefits and working conditions, surely, at this critical juncture, equally important is the task of securing a safe future for our children and grandchildren. Given that, for every workplace a union represents, why not have the union drive the development of a carbon-zero plan? Take that plan to the bargaining table. If it is rejected, occupy the workplace — think of it as the next generation of sit-down strikes. Invite local teenage climate strikers to join you in common cause. As they do so, they will learn — for the rest of their lives — the power that comes from young people and workers joining forces. They will never look at the labour movement the same way again, and you and your own children may share in an experience of solidarity that you will cherish for the rest of your lives — the time you acted together when called upon to confront this emergency.

A Canadian Green New Deal is how we can truly make "common cause," as we link climate action with tackling inequality and economic insecurity, such that those who currently feel tied to the fossil fuel economy may be prepared to leap into the next one.

Any way you cut it, a great transition is coming. One way or another, on our terms or not, driven by well-planned policy or by the convulsions of the market — it's coming. A few short decades from now, we will not be extracting and using fossil fuels. Far better that the inevitable transition be well-managed and labour-led than subject to the tumult of an unfettered free market.

I'm making bombs
and buying bonds!
Buy VICTORY BONDS

Archives Canada

CHAPTER 8

Paying for Mobilization, Then and Now

"War is the strong life; it is life in extremis; war-taxes are the only ones men never hesitate to pay, as the budgets of all nations show us."

— philosopher William James, in his 1910 essay "The Moral Equivalent of War"

"The war on the climate emergency, if correctly waged, would actually be good for the economy — just as the second world war set the stage for America's golden economic era."

— Nobel economist Joseph Stiglitz

"One thing is clear: public scarcity in times of unprecedented private wealth is a manufactured crisis, designed to extinguish our dreams before they have a chance to be born."

— 2015 *The Leap Manifesto*

The ubiquitous question that invariably comes when we propose saving civilization: but how will we pay for it? It's a fair enough question. So we now explore how the needed finances were mobilized in the Second World War and how they might be marshalled once again to fund a Green New Deal–type mobilization in Canada today.

MOBILIZING MONEY: WAR LIBERATES THE PURSE STRINGS

A great benefit of an emergency or wartime mentality is that it forces governments out of an austerity mindset and liberates public spending. Tommy Douglas, in his farewell speech to the NDP in 1971, recounted the following from his time as a young MP in the Depression years:

> My message to you is that we have in Canada the resources, the technical know-how and the industrious people who could make this a great land if we were prepared to bring these various factors together in building a planned economy dedicated to meeting human needs and responding to human wants.
>
> Mr. Coldwell [leader of the CCF from 1942 to 1960] and I have seen it happen. In 1937 when the CCF proposed in the House of Commons a $500 million program to put single unemployed to work, the Minister of Finance said, "Where will we get the money?". . . My reply at that time was that if we were to go to war, the minister would find the money. And it turned out to be true.
>
> In 1939, when we declared war against Nazi Germany, for the first time we used the Bank of Canada to make financially possible what was physically possible. We took a million men and women and put them in uniform. We fed and clothed and armed them. The rest of the people of Canada went to work . . . We manufactured things that had never been manufactured before. We gave our farmers and fishermen guaranteed prices and they produced more food than we had ever produced in peace time. We built the third largest merchant navy in the world and we manned it. In order to prevent profiteering and inflation, we fixed

> prices, and we did it all without borrowing a single dollar from outside of Canada.
>
> And my message to the people of Canada is this: that if we could mobilize the financial and the material and the human resources of this country to fight a successful war against Nazi tyranny, we can if we want to mobilize the same resources to fight a continual war against poverty, unemployment and social injustice.[1]

Douglas was right. Once the need to ramp-up war production became apparent in 1940, C.D. Howe brushed aside concerns about cost, declaring, "If we lose the war nothing will matter."[2] Similarly, early on in the war, when then Finance Minister Ralston (soon to be shifted to the National Defence portfolio) was made aware of how dire the situation was in Europe, he wrote to the Bank of Canada's first governor, Graham Towers, "We have to, and should, take chances."[3]

And so they did. The senior civil servants at the Department of Finance and the Bank of Canada, along with the relevant ministers, were prepared to push the outer limits of how much they believed it possible to increase spending (and loans and gifts to Britain) relative to the size of Canada's overall economy.[4] Consequently, federal government spending increased dramatically during the war. By 1943–1944, it reached $5.3 billion, "eight times what it was a mere four years earlier when the war began."[5] The wartime government issued public Victory Bonds to help finance the war — meaning it borrowed from both Canadian households and corporations, taking on historic levels of national public debt. The government also did some of that borrowing directly from the newly created Bank of Canada, in effect borrowing from itself (as the Bank of Canada, our central bank, is a federally owned crown corporation established in 1934).[6] And new forms of progressive taxation were also instituted to raise additional revenues.

Other Allied countries were also ripping open the public purse strings. In the U.K. defence spending exploded from 9% of GDP the year before the war to a peak of 52% of GDP in 1945. U.S. government spending during the war, at over $4 trillion in contemporary dollars, was astronomical. This war was the most expensive in U.S.

history, consuming 40% of the nation's GDP in the final year. To give one example of what the U.S. government was prepared to expend for a single notable initiative: the Manhattan Project that created the atomic bomb cost $24 billion (in 2015 U.S. dollars).[7]

In early 1940, the Canadian government revived the Victory Bonds model it had created during the First World War. Hundreds of volunteers and major promotional campaigns were mobilized to sell the bonds to everyday Canadians. Bond purchases were encouraged via posters, movie trailers, radio commercials and full-page ads in all the major newspapers and magazines of the era. They could be purchased from financial institutions, post offices, even from door-to-door salespeople. The bonds offered interest rate returns of 1.5% to 3% with maturity dates of between six and fourteen years.

Over the course of the war, the bonds were issued nine times, raising a total of more than $12 billion for the war effort (equivalent to about $175 billion in today's dollars), an amount that covered about half the government's wartime expenditures. It wasn't just individuals buying Victory Bonds. Businesses also purchased billions in war bonds, about half the total.[8] Given that about half the Victory Bonds were sold to individuals, this represented over $520 — or about $7,500 in today's dollars — per capita for every man, woman and child in the country. A remarkable level of participation.

Similarly, much of the U.S. spending was financed by the issuing of war bonds. Over the course of the war, the U.S. raised over half its war spending — $2.4 trillion in today's dollars — through the sale of war bonds. And there too the promotional campaigns were a major undertaking, with Hollywood stars helping to sell the bonds to the public.

Effectively, the bond sales represent the use of expansionary monetary policy to raise revenues. There is always the risk that all that new borrowing (and hence new money creation and spending flowing through the economy) could spark a surge in inflation. But, as we saw earlier, the government used other tools to keep a lid on that, such as rationing, price controls and by convincing so many Canadians to lend money back to the government through war bonds rather than spending their newfound earnings. Interest rates were also kept low and this made it easier for government to borrow such large amounts.

But it wasn't just borrowing and monetary policy that financed the war. Taxes were also raised. A lot.

"When Canadians went to war in 1939, the full force of the Canadian State went with them," write Maude Barlow and Bruce Campbell. In contrast to the government's parsimonious response to the Depression, "much of the Liberal government's change in outlook resulted from pressure from the CCF. Canadian support for the young party . . . peaked in a Gallup poll in 1943 . . . The mainstream of Canadian political thought was moving closer to the CCF's call to 'conscript wealth as well as men.'"[9]

In the interest of financing the war effort in a coordinated manner, the provinces vacated certain areas of individual and corporate taxation, leaving these to the federal government in exchange for fixed transfers from the feds. Hard to imagine in view of provincial temper tantrums about a federal carbon tax today.

The federal government then substantially increased sales taxes (including a special tax on gasoline), import tariffs and personal and corporate tax rates. In 1941, the federal government introduced "succession duties" — an inheritance tax on the transfer of assets after death.

Individual income taxes, which had been introduced but at quite a modest level in the First World War, were significantly increased in the Second World War with the Income War Tax Act, and Canada saw the dramatic expansion of a multibracket progressive income tax regime.[10] (After the war, a ten-bracket income tax system was maintained under the new Income Tax Act.) But from the outset, the basic exemption was set high enough that the majority of Canadians were not subject to the income tax.

The corporate income tax rate went from 18% to 40% during the war. And, as noted previously, to deter wartime profiteering, an "excess profits tax" — a cap on corporate profits set at the average profit made during the four years prior to the war — raised even more revenues.[11]

Overall, during the six years of the war, personal income taxes paid to the federal government went from $112 million in 1939 to $809 million in 1945, while corporate taxes paid rose from $115 million to $599 million over the same period, increases of 662% and 421% respectively.[12]

This is what raising revenues looked like in the face of an existential threat.

And did all that taxing and spending and debt-financing wreck the economy? Hardly.

Public debt did indeed soar during the war. By war's end, Canada's federal debt-to-GDP ratio stood at an historic high of 108% (well higher than the 67% level reached in the mid-1990s, at the peak of the "debt-scare" era, when the Chrétien/Martin government claimed the debt left them "no choice" but to dramatically slash spending). But the wartime debt was owed largely to ourselves in public bonds; less than 5% of Canadian debt was held outside the country in the three decades that followed the war, whereas in recent decades, 15–30% of our debt has been held by investors outside Canada.

Rather than harming the economy, Canada's wartime spending and debt levels presaged a 30-year period of unparalleled economic performance. After the war, the government repaid a small portion of its accumulated debt — but most of it was simply refinanced or "rolled over." Since under Keynesian full-employment policies, the economy then experienced a sustained period of strong growth and rising incomes, which in turn boosted government revenues, the remaining debt shrank rapidly relative to GDP and relative to the government's total revenue base. This experience proves that through a combination of strong fiscal expansion, complementary monetary policy, and forms of economic planning, it is not just feasible for government to take on a huge expenditure — it actually can make the economy stronger. As economist Jim Stanford told me, "The greatest irony of Canada's war effort in the 1940s is that average living standards for those on the home front (not counting soldiers, of course) actually improved compared to the Depression: incomes rose, nutrition improved and life expectancy was extended. The net economic and social benefits of this type of intervention would be even larger if we undertook a similarly ambitious expansion, but this time to do something creative and helpful — rather than financing war and destruction."

By the early 1970s, that strong economic run had led to the debt-to-GDP ratio shrinking to less than 20%. And, as we saw earlier, the taxation and spending choices of that era also resulted in the lowest level of income inequality we have known.

CANADIAN FEDERAL DEBT AS PERCENTAGE OF GDP

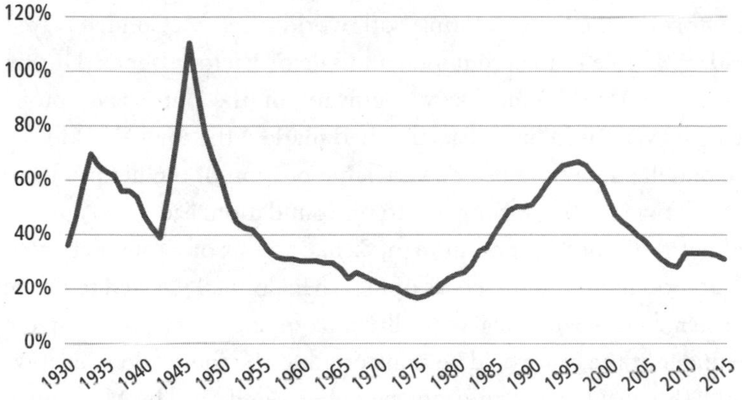

Source: Statistics Canada: Tables 10-10-0048-01, 36-10-0202-01, 36-10-0222-01 , Fiscal Reference Tables 2015 and 2019.

Once the war was on, it is notable the degree to which various business leaders threw themselves into the effort, in a way that is hard to imagine with respect to the climate fight today, at least so far. Many dollar-a-year men left their own businesses, sometimes for years, to head up various wartime crown corporations or serve as sectoral controllers. Numerous business leaders dove into the task of raising funds for the war and selling Victory Bonds. And of course, many businessmen and professionals left lucrative careers and enlisted, offering themselves up for, quite possibly, the ultimate sacrifice.

Much of my funding for researching and writing this book came from the Montreal-based J.W. McConnell Family Foundation. John Wilson McConnell was born into a modest farming family of Irish immigrants in rural Ontario. As a young man, he found his way to

Montreal, where he worked his way up the ranks into senior management at the Standard Chemical Company, before going off on his own and becoming the majority owner of the St. Lawrence Sugar company and later the *Montreal Star* newspaper. By 1920, he was one of the wealthiest people in Canada. Along the way, he became a major philanthropist, and in 1937 established the foundation in his name.

During the Second World War, C.D. Howe approached McConnell, asking him to serve as one of his dollar-a-year men. McConnell turned him down. But he vigorously contributed to the war effort in other ways. During both world wars, McConnell played a leadership role in the promotion and sale of Victory Bonds. During the Second World War, likely cognizant of the outrageous profiteering by various industrialists that had marked the First World War, McConnell chose to transfer a very large portion of the net profits of the St. Lawrence Sugar company to his foundation: $300,000 out of a total net profit of $633,000 in 1940, $479,000 out of a total net profit of $540,000 in 1941, and then, as of 1942, McConnell pledged to Prime Minister Mackenzie King that all the company's net profit for the remainder of the war would be transferred to the foundation's endowment.[13] (So, in a twist of fate, money that was endowed by McConnell during the war, likely to inoculate against any perception of profiteering, has been the main source of funding for this book project.)

McConnell also personally made major donations to the war effort. In particular, he donated $1 million (a very considerable sum back then) to the Canadian program training Commonwealth fighter plane pilots and crew, which resulted in a squadron being named for him.

While I would never suggest that philanthropy will solve the climate crisis, it is nevertheless notable that we have yet to see the private sector and businesses in Canada today rally to the cause of fighting the climate emergency in anything close to this manner. To offer one case in point, environmental philanthropy currently represents only about 2% of overall charitable giving in Canada, and of that most is for conservation efforts. Not exactly indicative of a perceived crisis.[14]

PAYING FOR FULL-SCALE TRANSITION TODAY

Each year, our provincial governments allot a certain amount in their annual budgets for fighting forest fires. In this era of climate change, it is usually insufficient. Yet can any of us imagine a scenario in which summer forest fires are raging, and then, as communities are facing evacuations and thousands of hectares of our forests are going up in smoke, a provincial government abruptly announces that it is ceasing all efforts to combat the wildfires because it has exhausted the budgetary allotment for such efforts?

Of course not. When an emergency is clear and present, we spend what needs to be spent.

And yet, in the face of the climate emergency, our governments oddly pretend that all is normal and that the necessary level of spending is not "realistic."

In a society as wealthy as Canada's we can absolutely afford to dramatically boost our public spending. We have the capacity to spend billions more each year on both operating programs and capital/infrastructure investments to meet the climate emergency head-on and make real the vision of a made-in-Canada Green New Deal. The 2008–2009 financial crisis laid bare just how much money the national governments of Canada, the U.S., the U.K. and others were quickly able to muster when hundreds of billions were needed to bail out the banks and auto companies, and to pump liquidity into the economy. But the wealth of our country is currently very unevenly shared. According to 2018 analysis by CCPA senior economist David Macdonald, the richest 87 families in Canada have the same combined wealth as everyone living in Newfoundland and Labrador, New Brunswick and PEI combined.[15]

How much per year should Canada be spending on tackling the climate emergency?

The path-breaking 2006 *Stern Review on the Economics of Climate Change* commissioned by the then U.K. government and headed by former World Bank chief economist Nicholas Stern, chair of the Grantham Research Institute on Climate Change and the Environment at the London School of Economics, recommended that governments spend 1% of their respective GDP on climate mitigation investments, although two years later Stern boosted that recommended target to 2% given evidence of accelerating climate change.

In either scenario, this is certainly a much lower share of GDP than Canada or other Allied nations spent on the war effort. For Canada, with an annual GDP currently of about $2.4 trillion, a target of 2% of GDP would mean our governments joining forces to spend approximately $48 billion a year on green infrastructure and climate action. In fact, we should spend more, given Canada's comparative wealth and disproportionate role in historic GHG emissions, with an additional sum directed towards financial assistance for climate mitigation and adaptation in poorer countries.

Achieving such investment and spending levels is well within our means, given the political will. Yet so far, no political party in Canada has proposed spending anything close to this. Prior to the 2019 federal election, the Trudeau government had committed to spend about $64 billion on various climate-related initiatives (mostly on public transit and green infrastructure investments),[16] but that amount is to be spread over 10 to 12 years. Meaning, the annual pledge was for about $6 billion a year.

In their 2019 federal election platform, the Liberals committed to further climate-related spending of about $900 million a year. The federal NDP, in their platform, proposed new climate-related investments of about $15 billion over four years, or about $3.75 billion per year. Even the federal Green Party, which proposes more ambitious climate spending than the others, promised climate-related investments that came in at about $11 billion per year, still a far cry from the Stern target.

The above amounts speak only to spending specific to the climate emergency. If we add into the mix investments to tackle the urgent needs of Indigenous communities, the housing crisis, instituting public child care, transforming public health care, ending post-secondary tuition and other spending related to a Green New Deal, the target should be well higher than the Stern recommendation.

Let us imagine, then, an ambitious climate and social infrastructure program that sought to boost federal government spending by $100 billion a year, an amount that would jolt a transformative leap into a new and more caring post-carbon economy. That sounds like a lot of money, and it is. It represents an increase in overall federal spending of about 25%.

But to put that into perspective, in an economy such as Canada's, with an annual GDP of $2.4 trillion, a bold spending and investment plan of $100 billion represents a little more than 4.1% of GDP. Put another way, it would mean increasing federal spending from about 14% of GDP today to about 18%. That's significant, but far from radical, and still pales in comparison to what we did in the war years.

As we did then, Canada should raise the needed revenues to confront the climate emergency with a combination of borrowing and new taxes and other revenues. Let us explore each of these in turn.

Green Victory Bonds

While a government's operating expenses (the funds needed to annually run public programs) should generally be covered by tax revenues (or some other form of annual income), it has long been common practice — in both the public and private sectors — to cover one-time capital or infrastructure expenses through borrowing. Infrastructure is rightly understood as an investment, and thus it makes sense to amortize the cost of these expenditures plus the interest charges over the working life of what they purchase. And traditionally, governments do such borrowing through the sale of some form of bonds or government securities.

In the Second World War we sold Victory Bonds. So, to finance Green New Deal capital expenditures, we should have our governments (federal, provincial and municipal) issue *Green* Victory Bonds.

Many, although certainly not all, of the expenditures we urgently need are indeed just this kind of one-time capital/infrastructure investments. High-speed rail, building retrofits, renewable energy projects, new net-zero social housing, new child care spaces, new zero-emission buses, charging stations, etc. — all of these are capital, not operating, expenditures. So let us then pay for these via new Green Victory Bonds.

Moreover, today's low interest-rate context (the government can currently borrow at 1% interest) makes this an ideal time to borrow. Indeed, interest rates are so low today that, in many cases, the returns on these public investments over time — in rail fares, energy savings, utility fees, charging fees, rent, etc. — make our failure to undertake these investments foolhardy.

All the stars — and prices — are aligned. The climate and inequality imperative, the weak price of oil and gas, the plummeting cost of renewable energy such as solar, the price of borrowing and public debt service costs at an historic low — all are telling us that now is the time to "go big" on Green New Deal investments.

I believe the Canadian public would be very keen to purchase Green Victory Bonds. The Abacus poll I commissioned in summer 2019 included a question on this topic. The online poll asked:

> One idea for how to help pay the cost of transitioning the economy away from fossil fuels would be to sell "Green Victory Bonds" (which would be like "Victory Bonds" that were sold during World War II to help pay for the costs of the war). The money raised from these Green Victory Bonds would be used to pay for public infrastructure needed to respond to and tackle climate change (such as public transit, high-speed rail, building retrofits and renewable energy projects). They would also offer a reasonable rate of return for those who buy them. Given this, how likely are you to purchase a "Green Victory Bond"?

11% of respondents said they would be "certain to," 19% replied they would be "likely to," and a further 35% said they "might consider it," for a total of 65%. And that's without any promotional campaigns, which were key to the Second World War sales and would be again.

Conversely, loans to sectors that are high-GHG emitters should be made more costly. The U.K.-based think tank Common Wealth has proposed that "private banks should be instructed [by the central bank] to hold back more reserves for loans to carbon intensive industry," reducing the overall attractiveness of such investments.[17] There is a logic to a requirement such as this, as it is increasingly understood that fossil fuel investments are at high risk of becoming stranded assets once we start to take needed action on the climate emergency. A policy such as this would free up more investment for Green Victory Bonds or other more sustainable activities in the private sector.

As occurred in the war, there is no reason the government shouldn't do some of the needed borrowing directly from the Bank of Canada by selling a large share of its Green Victory Bonds directly to our own central bank (meaning, lending to ourselves). Similarly, The Climate Mobilization's *Victory Plan* calls for "Mobilizing the Federal Reserve," the U.S. central bank, just as occurred in America during the war.

As Jim Stanford explains, "By borrowing money, the government (like any other borrower) creates new spending power in the economy, and that stimulates more spending and job creation in all industries." If that puts upward pressure on prices, then as happened in the war, we may need to employ measures other than raising interest rates to stem inflation. But it is worth noting that, over the last two decades, the U.S. has recorded huge annual deficits (mostly to finance tax cuts, corporate bailouts and military spending), and this has not resulted in any inflationary pressure.

The federal government can afford to take on more debt, and in doing so it would take pressure off private household debt. Indeed, these forms of debt are often inversely related. Household debt, at nearly 100% of GDP, is currently at record levels in Canada, as families struggle with the costs of housing, child care, elder care, transportation and post-secondary education. But Green New Deal policies could go a long way towards reducing the household debts associated with these pressures, as we choose to pay for more of these items collectively. In contrast, federal government debt is currently at about 30% of GDP, a very manageable level. Some rebalancing of public versus private debt makes good sense.

New Tax Revenues

The money needed to substantially increase annual program spending should come from a combination of raising existing taxes, closing various loopholes and deductions that disproportionately benefit the wealthy and corporations, instituting new taxes, and from other revenue sources. And the options for doing so abound.

The CCPA's annual *Alternative Federal Budget*[18] has long proposed many possibilities for raising new tax revenues and restoring fairness to the system. Some of the most rewarding ones include:

- Treating capital gains income the same as employment income would generate approximately $11 billion a year.
- Instituting a withholding tax on money in tax havens could generate $2 billion a year (and others have suggested that more actively cracking down on tax havens could raise considerably more).
- Increasing the federal corporate income tax rate from 15% to 21% (still well below what it was in the Second World War, even factoring in provincial corporate income tax rates, and also well below what it was as recently as the year 2000) would raise about $9 billion a year.
- Canada is currently one of the few industrialized countries without an inheritance tax. An inheritance tax of 45% on the transfer of estates worth more than $5 million would raise $2 billion a year.

The NDP proposed a wealth tax in the 2019 federal election — an annual 1% tax on households with net wealth over $20 million, which it estimated would raise over $6 billion a year. The federal Green Party, in its 2019 election platform, proposed a financial transaction tax of 0.5% (on the purchase of stocks), which they estimated would raise about $15 billion a year. We should also make the income tax system more progressive, by increasing the top marginal rate and/or by adding new tax brackets. There are countless permutations for doing so, all with different revenue implications. For example, in the 2019 federal election the NDP proposed raising the current top marginal rate (on incomes over $210,000 a year) to 35%, which was estimated to raise about $900 million a year.

While I do not believe that carbon pricing alone can achieve the

GHG reductions we urgently need, I remain supportive of carbon pricing/taxes as a revenue-raising tool. Carbon pricing is in keeping with a basic "polluter pay" principle. It makes sense as a means of capturing what economists call "externalities." To put it bluntly, neither households nor industry should be able to use our shared atmosphere as a free toilet. There is a logic to raising revenues from carbon pollution and then using that money to tackle the climate emergency. It is quite possible to design a carbon tax that is progressive rather than regressive. Marc Lee, in his CCPA Climate Justice Project report *Fair and Effective Carbon Pricing*,[19] has modelled a British Columbia carbon tax that rises incrementally to $200 a tonne (much higher than the $50 it is scheduled to hit in 2022). At this level, the tax would truly have an impact on the consumption and investment choices of households and businesses, helping to reduce GHG emissions. But it would also raise about $8 billion a year in B.C. alone. Lee proposes that half this income be used to fund climate action and green infrastructure, and half be used for a carbon tax credit for low- and middle-income households. He models a credit whereby the bottom half of households would receive more in the credit than they pay in the higher carbon tax, thus improving the progressivity and fairness of the overall tax system. At a national level, a carbon tax of $200/tonne would raise approximately $80 billion a year.

The overall point is this — the options for raising new tax revenues from those who can well afford it are myriad and boosting revenues by $80 to $100 billion is entirely possible. Today, federal spending and revenues clock in at about 14% of GDP, whereas during the war these reached levels of 45% and 25% respectively.[20] Nearly four decades of federal and provincial tax and spending cuts have resulted in Canada having a lower level of overall taxes and spending than most OECD countries — leaving us plenty of room to raise both (the chart on the next page shows total government revenues in Canada, including all levels of government). And as we saw in the war years, doing so would not harm the economy if those revenues are redeployed making GND investments. The net impact on the economy and employment would surely be positive.

TOTAL TAX REVENUES AS A PERCENTAGE OF GDP

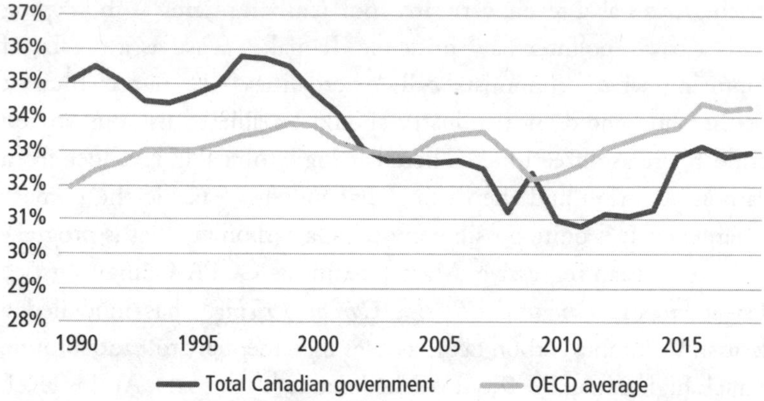

Source: OECD (2020), Tax revenue (indicator). doi: 10.1787/d98b8cf5-en (Accessed on March 24, 2020).

The above taxation measures not only raise much needed revenue, they also achieve two other core and interrelated objectives.

First, they redistribute income from the wealthier among us to lower- and middle-income households and to the public programs upon which we all rely, but particularly the poorer among us. Thus, the tax policies outlined above help us to reduce harmful income and wealth inequality, which currently undermines common action on climate. That said, while the bulk of new taxes should be paid by the wealthy and corporations, I have long believed that almost all of us (other than low-income people) should be paying more in taxes. There are not enough of the super-rich to fully fund a Green New Deal from their taxes, and more importantly, the point of a GND is that if we all consume less and contribute a little more, we can then afford to collectively provide for more of the things we really need.

Second, understanding that runaway wealth is associated with runaway emissions, a benefit of raising taxes on the wealthy isn't merely that it generates needed revenues for the transition before us — it is also of ecological benefit to lessen the disposable income and luxury consumption of the wealthy. Keynes understood this too as part of his wartime economic plan — namely, that we needed to defer and redirect consumption.[21]

Other Revenue Options

Some of what is needed for GND infrastructure investments doesn't need to come from new sources, but rather can come from a focused commitment to redirect what we currently spend on capital projects. Our governments already spend billions of dollars a year on infrastructure, but far too many of these dollars go towards traditional projects that seek to accommodate our car-centric economy or facilitate the extraction and expansion of fossil fuels. Even in the face of the climate emergency, we continue to pour billions into *expanding* roads and bridges and airports, upgrading port facilities that manifest all that is wrong with our current carbon-intensive globalized economy, and in electricity infrastructure whose sole purpose is to carry energy to the oil and gas fields. The Trans Mountain pipeline expansion project alone will (if it proceeds) clock in at over $12 billion, above and beyond the $4.5 billion the feds paid to purchase the existing pipeline.[22] All of this needs to stop. And when we do, billions can be made available for new green infrastructure with no impact on our spending relative to GDP.

Another source of revenues is higher fossil fuel royalties. Even as we wind down the oil and gas sector, we need to be extracting more returns to the public in royalties. This revenue source should not be seen as a tax but as a "rent" we require of the fossil fuel companies in exchange for the privilege of accessing resources that are collectively owned by Indigenous communities and the public at large. Resource royalties across Canada are in urgent need of review, and many should be raised considerably. What our provincial governments currently charge in forestry stumpage fees, natural gas and oil royalties, and for industrial water usage and extraction is deplorably low. Setting appropriate royalty rates could raise much needed new revenues for provincial governments and Indigenous nations (on whose territories most of this extraction occurs). In particular, higher oil and gas royalties, given that we will collect them for only two to three more decades, should be used to endow permanent public trust funds for the benefit of current and future generations, as Norway has done with great success.

The fact that Alberta's public return from oil and gas has been shrinking isn't solely because of a slow-down in the sector; it is also because Alberta charges far too little for this public resource. The Notley government did initiate a royalty review upon its election, but it was, according to many, a sham — a conscious choice not to confront the industry on this issue, in exchange for gaining industry support for the province's climate plan. The Alberta government of Peter Lougheed (1971–1985) was really the only time the province bargained hard and charged an appropriate rent for access to the province's collectively owned resources.

A related source of revenue can and should come from ending public subsidies to the fossil fuel sector. The *Alternative Federal Budget* estimates this would save between $300 to $400 million a year at the federal level, but others take a broader view of what constitutes a subsidy and put the savings considerably higher for federal and provincial governments combined.[23]

A further related revenue source: suing the fossil fuel companies for damages. A growing list of municipalities and U.S. states are pursuing this option, arguing that the oil and gas companies misled the public for years, delaying climate action, and should thus be held liable to help cover the costs of climate adaptation infrastructure. And let us be clear, those major fossil fuel companies continue to do very well. In 2017, the five biggest oil companies in Alberta made $46.6 billion in profit. They can afford to pay up.

It is important to stress that much of what we need to do to confront the climate emergency does not require government spending. A great deal of the transformation before us will be driven by law and policy — meaning by regulatory mandate. And much of the spending needed will be private — undertaken by households and businesses — as they invest in new equipment and technology that gets us off fossil fuels. It is appropriate that much of this spending should be private, as those investments will save both families and industry money down the road in reduced energy costs.

For example, I recently installed solar panels on my home. The cost was considerable but there is no question that I was rightly required to pay for this investment myself. That's because the panels will now dramatically lower my monthly electricity bills, such that the investment will pay for itself three to four times over the next

30 years. The same logic holds with electric vehicles, which dramatically lower fuel and maintenance costs over time. If a public subsidy is to be provided, it should be focused on low- and modest-income households who most need the help.

For homeowners wanting to transition their homes off fossil fuels, what matters more is offering favourable financing, such as low-interest loans. Ideally, financing for items such as heat pumps or solar panels would be offered by public utility companies or city governments, and amortized on utility or property tax bills. That way, the payback over a number of years stays with the home that will benefit from the purchase and not with the individual purchaser who may want to sell the property before the capital expense has been paid off. Of course, if households could purchase their solar panels and heat pumps from a new crown corporation, lowering the price through economies of scale and eliminating the profit margins, so much the better.

On the business side, we could be using new public banks (such as postal banks) to offer low-interest and/or higher-risk loans to businesses wanting to invest in carbon-zero technology, or to new ventures that seek to help us through this transition. Ironically, Alberta itself is home to the largest public bank in North America — ATB Financial, or the Alberta Treasury Branches, founded 80 years ago. As Bob Ascah and Mark Anielski explain in a 2018 report for the Parkland Institute, with $50 billion in assets, ATB's investments and lending could be reimagined. Alberta's public bank is well situated to provide low-interest loans for social housing, renewable energy projects, energy efficiency programs, and agricultural investments. And like the Bank of Canada at a federal level, the ATB could purchase provincial and municipal Green bonds. All of this could help to expedite the transition off fossil fuels in Canada's most fossil-fuel-dependent province.[24]

There is a role for the financial sector to play in helping shift the economy off fossil fuels. We hear that articulated by Mark Carney, the former governor of both the Bank of Canada and the Bank of England, who, in a modern reincarnation of the dollar-a-year men, has taken on the role of United Nations special envoy on climate action and climate finance. Carney has been warning the business sector of the financial risks of failing to recognize the climate crisis and of investments in fossil fuel companies becoming "stranded assets." Insurance companies, for obvious reasons, have become attuned to climate risk, and are starting to press their business clients to incorporate the climate crisis into their plans. Canada's Expert Panel on Sustainable Finance (chaired by new Bank of Canada governor Tiff Macklem) has made recommendations for how to mobilize the financial sector to transition our economy.

We are now seeing a few leading corporations starting to wake up to the climate emergency. Among Canadian companies, McCain Foods announced in 2018 that it was investing $75 million over three years to reduce its GHG emissions by 10% in that time. In 2019, Maple Leaf Foods announced, in light of the crisis, it would "slash its greenhouse gas emissions by nearly a third over the next decade and will offset what it can't eliminate, while leaning on its suppliers" to do the same.[25] The company is making multimillion dollar investments into lowering its emissions, it is seeking to capture methane produced at its pig farms, and it is making major investments in expanding its plant-based protein division. There are clusters of companies, particularly in the tech and clean energy sectors, that are seeking to expedite the transition.

That's all good and welcome. And we should give kudos to the businesses that are now acting on the crisis. But with two caveats. First, these examples are, thus far, the exception. Most businesses are not there yet, and the new climate denialism remains the norm within Canada's corporate sector. Second, what is so remarkable about the Canadian private sector leaders I have recalled from our Second World War experience — the dollar-a-year men and the likes of J.W. McConnell — is that, even as they took on huge tasks themselves, made large personal and business contributions to the war effort and rallied their fellow business leaders to do likewise, they were simultaneously disabusing their colleagues of any notion that this mobilization could be anything other than government-led. Remember how H.R. McMillan told his fellow B.C. businessmen that they had to accept high taxes during the war? Appreciate that all those dollar-a-year men took leave from their private businesses to help lead state-run enterprises.

So yes, we need the corporate sector to join in the climate mobilization. But if we expect this transition to be led by the private sector, or achieved through moral suasion, or we hope to wait for market signals and the science to convince private businesses that they need to voluntarily become carbon-zero, we're fried.

A final message on this subject to today's politicians, particularly the prime minister, premiers and finance ministers: if we fail to rise to this challenge, and the world does indeed go to hell in a handbasket, posterity will not recall of you, "Well, at least they managed to stay within their budget."

We must also never lose sight of the fact that this crisis is, of course, global. And so we can't just finance our own climate mobilization. As one of the wealthiest countries on Earth, and a country that has historically contributed more to the problem than our share (based on population), we also have to help other countries finance this transition. In addition to a Canadian Green New Deal wartime-scale mobilization plan, we need to contribute our fair share toward a new global Climate "Marshall Plan," to reference another wartime legacy, the post-war fund for reconstruction in war-ravaged Europe.

Finally, as we think about the cost of the transition, remember that the status quo is far from free. In 2019, the Expert Panel on Sustainable Finance reported that "extreme weather caused $1.9 billion of insured damage across Canada in 2018."[26] As numerous reports on the economic costs of climate change have shown, failure to act — meaning, failing to keep global temperature rise under 1.5 degrees — will be far more costly than paying for the transition.[27]

Nobel Prize–winning economist Joseph Stiglitz has written with respect to the U.S.:

> We are already experiencing the direct costs of ignoring the issue — in recent years the country has lost almost 2%

> of GDP in weather-related disasters, which include floods, hurricanes, and forest fires. The cost to our health from climate-related diseases is just being tabulated, but it, too, will run into the tens of billions of dollars — not to mention the as-yet-uncounted number of lives lost. We will pay for climate breakdown one way or another, so it makes sense to spend money now to reduce emissions rather than wait until later to pay a lot more for the consequences — not just from weather but also from rising sea levels. It's a cliché, but it's true: an ounce of prevention is worth a pound of cure.[28]

Given this, Stiglitz concludes, "Yes, we can afford it [a Green New Deal], with the right fiscal policies and collective will. But more importantly, we must afford it. The climate emergency is our Third World War. Our lives and civilization as we know it are at stake, just as they were in the Second World War."

PART FOUR

BOLD LEADERSHIP — FROM THE GRASSROOTS AND IN OUR POLITICS

Photo of Indigenous Second World War hero Tommy Prince, courtesy of Princess Patricia's Canadian Light Infantry (PPCLI) Museum and Archives

CHAPTER 9

Indigenous Leadership

"For the sake of our grandchildren, our children and those generations yet to come, all of us — as grandparents, as parents, as aunts and uncles — we must take our power back. We can no longer afford to delegate that power to governments that do not listen, that continue to cater to the corporations, continue to devastate and destroy Mother Earth. That must stop."

— Grand Chief Stewart Phillip, president of the Union of B.C. Indian Chiefs, speaking October 25, 2019, at a Vancouver climate strike with Greta Thunberg

"We know that one condition will render this pipeline — all pipelines — impossible: respect for the free, prior and informed consent of Indigenous peoples. We do not consent to energy projects poisoning the drinking water, polluting the land and threatening our oceans."[1]

— Melina Laboucan-Massimo, member of the Lubicon Cree First Nation in Alberta, and climate and energy campaigner

On March 8, 2019, Prime Minister Justin Trudeau travelled to Iqaluit to issue a formal apology on behalf of Canada to Inuit communities. For about two decades (in the 1950s and '60s), Inuit suffering from tuberculosis were removed from their homes and communities, generally without their consent or their families' knowledge, and taken to sanitoriums in the south, often for years. Their families were not told where they were and were often not notified if their loved ones had died or where they were buried. It is a source of generational trauma. Trudeau called it "a shameful chapter" in our history. And the name of the coast guard ship that for those two decades took most of these people away from their homes — *The C.D. Howe.*

It was a colonial response to a health emergency. The need for help and action was real. But rather than responding with respect and working in partnership with Indigenous communities, Inuit experienced, once again, the racist and oppressive heavy hand of the federal government, with the additional historic twist that the vessel that ripped people from their families was named for the man who planned Canada's wartime economy "his way."

As we confront the urgency of the climate emergency, there is a certain understandable attraction for a "strong man" leader who can see us through this crisis. But we are, hopefully, not the people we were in 1939. It will take a new kind of leadership, and a different kind of politics, to get people on board for this challenge. And this time, much of that leadership is coming from Indigenous communities. The path forward has to be one of true cooperation and partnership, and one that honours and respects Indigenous title and rights.

INDIGENOUS PEOPLE AND THE SECOND WORLD WAR

The legacy of Indigenous experiences in the Second World War is very mixed.

Thousands of Indigenous people in Canada joined the fight against global fascism.

The Iroquois Confederacy — the Haudenosaunee or "People of the Longhouse," made up of six Indigenous nations whose territory spans the Canada-U.S. border — in an assertion of sovereignty, issued its own declaration of war upon the Axis powers, just as Canada chose to declare war independently from Britain. In so doing, these nations formally signalled to their people that they were encouraged to enlist with the Allied armies.

The confederacy's declaration of war was passed at a conference of the Six Nations on June 13, 1942, and the next day was read on the steps of the U.S. Congress. It stated:

> We represent the oldest, though smallest, democracy in the world today. It is the unanimous sentiment among Indian people that the atrocities of the Axis nations are violently repulsive to all sense of righteousness of our people, and that this merciless slaughter of mankind can no longer be tolerated. Now we do resolve that it is the sentiment of this council that the Six Nations of Indians declare that a state of war exists between our Confederacy of Six Nations on the one part and Germany, Italy, Japan and their allies against whom the United States has declared war, on the other part.[2]

Exact numbers are hard to come by, but approximately 4,300 "status Indians" in Canada enlisted during the Second World War. That figure rises considerably if one adds an unknown number of Métis, Inuit and non-status Indigenous people, both men and women.

Why did they sign up, particularly given their historic treatment by the Canadian state, and before "status Indian" people even had the right to vote in Canada?

University of the Fraser Valley history professor Scott Sheffield has spent many years studying Indigenous people's participation in the military.[3] "There were a lot of different reasons why people might choose to enlist," Sheffield said in an interview with CBC Radio's Falen Johnson. "And just the fact that they were enlisting didn't necessarily signify that they were supportive of the governance system that oppressed them." According to Sheffield, for some, enlisting was seen as a fulfillment of treaty obligations, given historic military alliances between First Nations and Britain. For others, it was about patriotism, but not a conventional Canadian patriotism. "They wanted to defend Canada, yes. But they also wanted to defend their own territories, their own communities, their own people." For some it was about following in the footsteps of fathers or uncles or ancestors, or to seek adventure, or to escape economic hardship, or a political desire to demand respect for Indigenous people.[4]

Almost all Indigenous people who enlisted joined the army, as racist rules barred them from other branches of the military. Once in the military, however, Sheffield's research found most felt their army experience to be an equalizing force. Upon their return to Canada, Indigenous soldiers expected to be more accepted and equally treated. That, of course, did not happen.

Prior to 1960, Indigenous people in Canada were offered citizenship and voting rights only if they forfeited their Indian status and treaty rights and left their communities. Understandably, few returning Indigenous soldiers pursued this option, although in some cases they were forced to "enfranchise" (and give up their status) when they signed up. Also, under the Indian Act, anyone who was away from their reserve for four consecutive years lost their status, a policy that robbed some veterans of their official status and treaty rights. For those who retained their status, they found this disqualified them from some of the veterans' benefits outlined in Chapter 7. Even worse, some returned home to find large sections of their reserve lands had been expropriated by the government, including lands of the Stony Point Ojibway people that years later became the site of the Ipperwash standoff in Ontario in 1995, when local Indigenous protestors sought to reclaim their land. One of the land defenders, a local unarmed man named Dudley George, was shot and killed by the Ontario Provincial Police. The Stony Point case was but one in a long history of the Canadian military expropriating Indigenous lands for use as training and testing sites or for waste disposal.

The war also left behind a deadly legacy at home for some Indigenous communities. Foreshadowing what Naomi Klein calls the fossil fuel "sacrifice zones" on Indigenous lands today, wartime production left a toxic imprint on several First Nations territories.

Uranium for the development of the first atomic bombs was mined at Port Radium on the eastern shore of Great Bear Lake in the Northwest Territories. The mine was owned by the Eldorado mining company, which Howe nationalized during the war and made into a crown corporation, given its role in the Manhattan Project. The radioactive tailings from that mine were dumped into Great Bear Lake and contaminated the lands of the Déline Dene First Nation, becoming a source of protracted cleanup and settlement negotiations with the Canadian government. Many Dene people were exposed to dangerous levels of radiation. Men from the village were hired to carry gunny sacks of radioactive uranium ore from the mine to the train depot, where it was then transported to Port Hope, Ontario, for refinement. The community holds this exposure responsible for high rates of cancer and premature deaths. In 1998, a delegation of Déline Dene went to the atomic bombing commemorations in Hiroshima, Japan, to share their story and "to express their sadness and compassion for the suffering that the uranium from Great Bear Lake caused elsewhere."[5]

Another of the wartime crown corporations, Polymer (the Howe-created public enterprise that made synthetic rubber for all the defence contractors), also left an environmental mess in its wake. Its chemical plant was built in Sarnia, Ontario, on the land of the Aamjiwnaang First Nation (previously known as the Chippewas of Sarnia and part of the Ojibwe people). It is hard to isolate the toxic impact of the Polymer factory alone, given that Sarnia was home over many decades to numerous oil and chemical plants, and came to be known as Chemical Valley. What is without dispute is that the Aamjiwnaang people have suffered profound long-term health consequences from this activity.[6]

The war produced positive legacies as well. It gave birth to the United Nations and led in 1948 to the UN's adoption of the Universal Declaration of Human Rights. That, in turn, sowed the seeds for recognition of further human rights, not least the United Nations Declaration on the Rights of Indigenous Peoples.

Like I said, it's a mixed history.

On May 28, 2019, Louis Levi Oakes, the last of the Mohawk "Code Talkers," died at the age of 94. The Code Talkers were Indigenous soldiers tasked with using their own languages to communicate secret military information among Allied forces.

Oakes was from Akwesasne, which spans the Canada-U.S. border. His older brother had been roughed up at the hands of the RCMP,[7] and so, at age 18, Oakes chose to enlist with the U.S. military rather than with Canada. He served mainly in the South Pacific, in the jungles of New Guinea and the Philippines, where he carried a heavy field pack with telephone lines to communicate with other Mohawks in the signal corps. After the war, like many Mohawk men, he became an ironworker and later worked in highway maintenance until his retirement.

In news reports after Oakes's death, his daughter Dora revealed that, astonishingly, Oakes hadn't told his family what he did during the war for seven decades. Having been sworn to secrecy, only in his late 80s, when stories of the Code Talkers were made more public, "he finally started talking about it. He said he was threatened not to say anything."[8] Once his role was finally revealed, Oakes was awarded a Congressional Silver Medal in 2016, the third-highest military decoration given in the U.S., followed by special honours in 2018 from the Assembly of First Nations and the Canadian House of Commons.

Oakes was one of 17 Code Talkers from the Mohawk community of Akwesasne who served in the Second World War. But there were hundreds of others from Indigenous nations across North America. As the war was unfolding, the secret codes employed by the Allies to communicate military plans kept getting broken by Nazi and Japanese forces. The U.S. Marines, however, discovered that enemy forces were unable to crack Navajo, and so, once the U.S. joined the war they began recruiting hundreds of Navajo men into military service, where they developed Indigenous code words for the alphabet and various military-related terms. Ultimately, 33 Indigenous languages were used by various branches of the Allied forces in both the Pacific and European theatres, including a number from Indigenous nations in Canada such as the Mohawk language Kanien'kéha, Cree, Tlingit and Ojibwe.

There is in this piece of wartime history a deep and tragic irony. The Canadian and U.S. states spent generations trying to erase Indigenous languages from the Earth, literally beating these languages out of children in residential schools, only to uncover that these languages were "the unbreakable code" (as they were dubbed in the war), credited as having been vital to victory in certain battles.

Similarly, after our governments have spent decades violating and refusing to recognize Indigenous title and rights, in the face of the climate emergency, the assertion of those rights today has become key to saving our collective bacon.

HOW THE ASSERTION OF RIGHTS AND TITLE TODAY IS BUYING US ALL TIME

Not only do we suffer in Canada from the new climate denialism; we are also plagued by a related affliction that functions hand-in-glove — Indigenous title and rights denialism. Governments in Canada now claim to support the United Nations Declaration on the Rights of Indigenous Peoples, and the B.C. government has even taken the important step of embedding the UN Declaration in law.[9] The Truth and Reconciliation Commission made clear that adoption of the UN Declaration was foundational to its calls to action. Yet our governments continue to act contrary to the declaration's articles, particularly when Indigenous rights and the UN Declaration's core principle of free, prior and informed consent are in conflict with the interests and desires of the fossil fuel industry.

A vital element of climate mobilization is the defence of Indigenous title and rights. We need a path forward that seeks to make right the injustices of the past and fundamentally breaks with the colonial practices that have marked so much of our politics and economic development for centuries. The new politics we now need must meaningfully put into practice the principles and articles of the UN Declaration.[10]

Over and over again, as our politics has dithered and dodged on the climate question, implementing contradictory and inadequate policies, it is the assertion of Indigenous title and rights that has come to our collective aid — blocking new fossil fuel development and buying us time, until such time as our politics comes into alignment with the reality of the climate emergency. This has been particularly true in B.C., where most of the land is "unceded," meaning land was never relinquished by treaty or conquest.

Much of the opposition to the Trans Mountain pipeline expansion (TMX) has been Indigenous-led. As I write, the Tsleil-Waututh, Squamish and Coldwater First Nations await news of whether their opposition to the project will be heard by the Supreme Court of Canada. Their earlier, albeit temporary, win in August 2018 at the Federal Court of Appeal delayed the project. As Tsleil-Waututh Chief Leah George-Wilson has explained, "Tsleil-Waututh participated in the consultation in good faith, again. But it was clear that Canada had already made up their mind as the owners of the project . . . The federal government is in a conflict of interest as the owner, the regulator and enforcer."[11]

But the assertion of title in opposition to TMX hasn't just been in the courts. The protests on Burnaby Mountain, which in 2018 saw hundreds arrested, have been led by Tsleil-Waututh people whose territories encompass the terminus of the proposed pipeline,[12] with strong support from the leadership of the Union of B.C. Indian Chiefs and other First Nations. As Rueben George of the Tsleil-Waututh Nation's Sacred Trust initiative and a long-time leader of TMX opposition has made clear, they view this project as a fundamental threat to the inlet they are duty-bound to protect: "We are going to do what it takes to stop this."[13]

Members of the Secwepemc Nation, whose unceded territories in the interior of B.C. cover about a third of the TMX pipeline's proposed route, have sought to physically block construction by setting up tiny houses in the path of the project. The Tiny House Warriors initiative, led by Kanahus Manuel (the daughter of the great Indigenous leader and intellectual Art Manuel), has issued a warning to pipeline investors that "the Trans Mountain pipeline project and any other corporate colonial project that seeks to go through and destroy our 180,000 square km of unceded territory will be refused passage through our territory. We stand resolutely together against any and all threats to our lands, the wildlife and the waterways."[14] The ultimate outcome of the TMX fight, whether in the legal domain or at civil disobedience blockades, remains to be seen at the time of writing.

Before the battle over TMX, British Columbia was embroiled in the fight over Enbridge's Northern Gateway Pipeline. That battle too was led by First Nations all along the proposed route and those whose territories would be impacted by oil tanker traffic along the north coast — the Haisla, the Wet'suwet'en, the Yinka Dene Alliance, the Carrier Sekani Tribal Council, the Gitxsan, the Gitga'at, the Gitxaala, the Heiltsuk and others. In the course of that struggle, 44 Indigenous nations signed the Save the Fraser Declaration. These nations, all with territorial connections to the Fraser River watershed, collectively declared, "Our inherent Title and Rights and legal authority over these lands and waters have never been relinquished by treaty or war . . . We will not allow the proposed Enbridge Northern Gateway Pipelines, or similar Tar Sands projects, to cross our lands, territories and watersheds, or the ocean migration routes of the Fraser River salmon." Ultimately, this strong opposition led to the project's cancellation by the Trudeau government in 2016.

The Save the Fraser Declaration became a model for an even larger Indigenous treaty — the 2016 Treaty Alliance Against Tar Sands Expansion, "an expression of Indigenous Law prohibiting the pipelines/trains/tankers that will feed the expansion of the Alberta Tar Sands." This treaty, with 50 initial signatory nations from across Canada and the U.S., and jointly launched from Musqueam territory in Vancouver and Mohawk territory in Montreal, took aim at all proposed new oil sands pipelines in all directions — Northern Gateway (then still in play), TMX, Keystone XL, Line 3 and Energy East. The Treaty states: "Our Nations hereby join together under the present treaty to officially prohibit and to agree to collectively challenge and resist the use of our respective territories and coasts in connection with the expansion of the production of the Alberta Tar Sands, including for the transport of such expanded production, whether by pipeline, rail or tanker."[15] Notably, the Treaty Alliance opposition to oil sands infrastructure isn't only based upon protecting Indigenous land and water from harm, but also grounds its resistance in the need to protect our shared climate. Indigenous opposition may well have played a role in convincing TransCanada Corporation (now known as TC Energy) to abandon its plans for the Energy East pipeline.

In and around the oil sands themselves, numerous Cree and Chipewyan activists have played a leadership role seeking to slow the pace of development and drawing international attention to the environmental and health impacts of the oil sands. The campaigning work of Indigenous climate activists like Melina Laboucan-Massimo of the Lubicon Cree, Crystal Lameman of the Beaver Lake Cree, and Eriel Tchekwie Deranger of the Athabasca Chipewyan First Nation and executive director of Indigenous Climate Action has helped to galvanize pipeline protestors on both sides of the Canada-U.S. border.

The battle over fracking in New Brunswick drew international attention in 2013 after an Indigenous-led blockade and camp was established by Mi'kmaq and Maliseet activists and community leaders opposed to the exploration and extraction of fracked gas in the territories of the Elsipogtog First Nation. The protest culminated in a dramatic and violent showdown with police, and dozens were arrested. It produced an iconic image of Mi'kmaq activist Amanda Polchies bravely kneeling before the police and holding an eagle feather above her head. The next year, the New Brunswick government instituted a moratorium on fracking developments in the province, although New Brunswick's new Conservative minority government is, as I write, seeking to lift the freeze.

As our politics continues to languish in the land of climate denialism, these Indigenous-led (and notably, frequently Indigenous women–led) campaigns — the grassroots politics of what Naomi Klein has dubbed "Blockadia" — are, with substantial success, standing in the way of new fossil fuel infrastructure. These efforts are managing to slow down political and industry efforts to keep us on the path to climate ruin, as we wait for our politics to catch up and align with the reality of the climate science. The last few years of these blockadia experiences have powerfully demonstrated that

Indigenous rights are a game changer and represent a fundamental threat to the continued power of the fossil fuel industry.

Take liquified natural gas (LNG). A handful of organizations, including the one I worked for, the CCPA, have been working on this file since 2011, warning of the dangers to the climate of developing these new carbon bombs. We did our best to debunk the economic claims of LNG proponents and to highlight the harms and risks. For a number of years, the industry's efforts did indeed get stalled, though largely thanks to the market — declining global gas prices were making this new industry uneconomic.

But then, in spring 2018, the B.C. and federal governments finally landed the big LNG fish they had been craving. LNG Canada (a consortium of Shell, Petronas, PetroChina, Mitsubishi and the Korea Gas Corporation) announced that their project was a go. Their final investment decision was signed with great fanfare at a ceremony in October that year, with representatives from all levels of government, including numerous local First Nations leaders.

The fact that many elected First Nations band councils had chosen to support the LNG project should not be surprising. These communities wrestle with high unemployment and limited economic development opportunities, and so their leaders often feel pressure to say yes to oil and gas projects. As is the case with the TMX project, numerous First Nations along the path of the proposed pipeline have signed "mutual benefit agreements." Importantly though, just because a First Nation has signed a benefit agreement does not necessarily mean it supports the project. Sometimes it merely means that, in the face of governments telling them that the project is a done deal, they are seeking to ensure their community receives as much financial and employment benefits as the leadership can extract. As lawyers Kate Gunn and Bruce McIvor of First Peoples Law explain, "Across the country, Indian Act band councils are forced to make difficult choices about how to provide for their members — a situation which exists in large part due to the process of colonization, chronic underfunding for reserve infrastructure and refusal on the part of the Crown to meaningfully recognize Indigenous rights and jurisdiction."[16] In the case of LNG Canada, the company and government lauded that they had secured signed agreements with all 20 First Nations along the route of the project, from the wellhead, along the needed new pipeline, to the proposed new plant at tidewater in Kitimat.

I felt resigned. It seemed this project — a huge one with a massive carbon footprint (in the realm of four megatonnes annually from phase one alone) — was going to happen.

But the project had a problem.

The planned Coastal GasLink (CGL) pipeline, owned by TC Energy, is to take fracked gas from Dawson Creek in B.C.'s northeast gas fields to the LNG Canada plant in Kitimat on B.C.'s north coast (just south of Prince Rupert). It is supposed to pass through the territories of the Wet'suwet'en people who have resided from time immemorial around the Bulkley River in the central interior of what is now B.C. Like most B.C.-based First Nations, the Wet'suwet'en never signed a treaty with the "Crown" nor have they ever ceded their territory. The Wet'suwet'en also have a long and proud history of blocking industrial developments on their land, and were among

MAP OF WET'SUWET'EN TERRITORY

Source: CBC News, (https://www.cbc.ca/news/canada/british-columbia/wetsuweten-whos-who-guide-1.5471898)

the lead claimants in the historic Delgamuukw Supreme Court of Canada win of 1997 that established groundbreaking case law for Indigenous title. And the Wet'suwet'en hereditary leadership have refused their consent.

Back in 2010, one house of the Wet'suwet'en people — the Unist'ot'en — set up a camp along the proposed pipeline route, in opposition to any and all fossil fuel pipelines through their territory.[17] Five years later, hereditary chiefs from all five Wet'suwet'en clans declared their support for Unist'ot'en and affirmed their opposition to all pipelines through their lands.

In November 2018, CGL sought a court injunction and served notice on the camp that it must cease blocking construction of the fracked gas pipeline. In response, other Wet'suwet'en hereditary chiefs came to Unist'ot'en's defence. In particular, the Gidimt'en, one of the five clans of the Wet'suwet'en Nation, established a blockade camp of their own — the Gidimt'en Yintah Access camp and checkpoint — preventing access to the road leading to the Unist'ot'en Camp.

As explained on the Gidimt'en Yintah website:[18]

> The Wet'suwet'en Hereditary Chiefs represent a governance system that predates colonization and the Indian Act which was created in an attempt to outlaw Indigenous peoples from their lands. The Wet'suwet'en have continued to exercise their unbroken, unextinguished, and unceded right to govern and occupy their lands by continuing and empowering the clan-based governance system to this day. Under Wet'suwet'en law, clans have a responsibility and right to control access to their territories. The validity of the Wet'suwet'en house and clan system was verified in the Delgamuukw and Red Top Decisions that uphold the authority of the hereditary system on Wet'suwet'en traditional territories.
>
> At this very moment a standoff is unfolding, the outcome of which will determine the future of Northern B.C. for generations to come. Will the entire region be overtaken by the fracking industry, or will Indigenous

people asserting their sovereignty be successful in repelling the assault on their homelands?

It was a standoff. Then, on Monday, January 7, 2019, a large RCMP force, court injunction in hand and, we later learned, authorized to use lethal force,[19] moved on the Gidimt'en Yintah Access blockade. Many will remember seeing the images of that confrontation on social media and on the national news that night. Canada has seen many such moments before. But this one seemed to cause instant alarm across the country — this was not what we understood Canada's newfound commitment to reconciliation or honouring the UN Declaration to look like. The jarring contradiction galvanized solidarity protests in cities across Canada over the following days. On Friday, January 11, one solidarity protest led by Mohawks from Akwesasne blocked the Trans-Canada Highway in Ontario, causing miles of backups.

In the wake of that confrontation, the country received a crash course in Indigenous title and governance. Hadn't LNG Canada assured us they had the support of the Wet'suwet'en? Well, it turned out they had secured agreement from the elected band leadership in a deal worth $13 million. But as we quickly learned, many Wet'suwet'en assert that while elected band councils have authority over reserves and community services, it is the age-old system of hereditary chiefs that has authority over the wider Indigenous territory. The question as to which authority holds the collective rights of an Indigenous nation is a complex and often controversial matter within nations, and it varies from nation to nation.

A year later, in February 2020, the RCMP again raided and took down the Wet'suwet'en camps that were blocking CGL access to their desired construction sites. This time, the national reaction was unprecedented. Solidarity actions in defence of the Wet'suwet'en land defenders instantly sprang up, from Victoria to Halifax. Young Indigenous activists and their allies blocked access to the B.C. legislature the day of the Throne Speech, while others occupied the offices of numerous provincial and federal politicians. Activists in Vancouver blockaded the city's main port for days, followed by rotating street, rail and bridge closures. Tyendinaga Mohawk activists blocked the CN rail tracks near Belleville, Ontario, along the busiest train corridor in Canada, forcing the suspension of both freight and Via Rail passenger service for two weeks. A Gitxsan hereditary chief, Spookw, blockaded a rail line in northern B.C. for a few days. In the face of mounting economic costs, both the federal and provincial governments were forced to re-engage with the Indigenous opponents. The final outcome is unknown at the time of writing. But what is clear is that popular understanding and support of Indigenous title has reached a new level. The terrain has shifted, thanks again to the assertion of Indigenous title.

I have not made the trip to Wet'suwet'en territory myself. But in June 2019, I had the pleasure of meeting the remarkable Gidimt'en leader Molly Wickham (whose Indigenous name is Sleydo).

Wickham has become one of the spokespeople for the Gidimt'en and the Yintah Access checkpoint. Wickham, currently the governance director with the Office of the Wet'suwet'en (the organization representing the hereditary chiefs), is among a new generation of Indigenous leaders. When she speaks, she is calm but powerfully compelling. She left her community in her youth and ended up studying Indigenous governance at the University of Victoria. But after the birth of her first child, she felt compelled to return to her territory, where she now lives with her husband (who is Haida) and two children.

When it comes to the defence of her people's territory from the fracking companies, as anyone who saw her on the news as she confronted and sought to talk down dozens of police on January 7, 2019, can attest, Wickham has been extraordinarily courageous. She was among the 14 people arrested that day.

Wickham gives the clearest explanation of Indigenous governance I have ever heard. And she does not mince words: "It's a war. We are literally fighting for our lives." When the Gidimt'en Yintah checkpoint was breeched by the armed force of the Canadian state that cold January day, as Wickham recounts, "We were invaded by heavily weaponized RCMP." There were attack dogs and busloads of armed police. And while things quietened down somewhat between

January 2019 and February 2020, when I met Wickham in summer 2019, she said, "It still looks like a war zone in our territory," with police helicopters flying overhead and an uneasy tension in the area.

As Wickham explains, the Wet'suwet'en are powerfully motivated by what they have seen happen to the territories of First Nations people in B.C.'s northeast, where the proposed pipeline will originate, and where fracking has ravaged the landscape. They are resolved not to see the same happen in the northwestern territories where they have lived sustainably on the land for thousands of years. They are acutely aware that if a gas pipeline eventually crosses their territory and a right-of-way is established, it may well open the floodgates to other gas developments, including fracking, and the Wet'suwet'en do not want to see those cumulative impacts. As noted earlier, that's the thing about fossil fuel infrastructure — it tends to lock in place a certain economic development path and produces its own logic of "if you build it, they will come." That's why the five clans of the Wet'suwet'en have decided there will be no pipelines in their territory.

That decision has been affirmed at a feast — a traditional potlatch — and is now Wet'suwet'en law. As Wickham explains, "That is how our governance works." There are 13 houses of five clans of the Wet'suwet'en people, and together they claim sovereign governance of 22,000 square kilometers of land in north-central B.C. Those houses have long-established systems and practices designed to sustain the land and the plant and animal species that rely upon it. When the hereditary chiefs of those houses gather at a feast and make a decision, it is the law. Yet their territory was included in the "colonial court's" injunction granted to the CGL project against the Wet'suwet'en checkpoints. The Wet'suwet'en have no intention of abiding this.

Wickham is also clear on another point — their defence of their territory isn't only a fight to protect their land and title. "This is a fight for *all* our lives." Wickham tells people she has "a secret" to share with all those who are worried about climate change and seek to protect our futures and those of our children: "Get behind Indigenous governance in a real way. If we lose, it is the end of life. This is really a war."

A good war.

Yet the wartime and emergency frames for climate mobilization are a source of apprehension for some Indigenous people. There is, in truth, a tension — these frames frequently imply a stronger role for the Canadian state and a centralization of power to expedite needed changes. But such moves risk being at odds with the assertion of Indigenous sovereignty and self-government. And there is an inherent tension with respect to time — on the one hand, undoing the racism in our society and our history of colonization will take time, but on the other, as we face down the climate emergency, time is the one thing in very short supply.

Matthew Norris, a member of the Lac La Ronge First Nation in Saskatchewan who now lives in Vancouver and is a policy analyst currently pursuing a Ph.D. in the areas of Indigenous governance and rights, is among those for whom the climate emergency/wartime frame is fraught. "My concern with a wartime framework is that it was very top-down," Norris told me. "The war empowered the government to impose regulations and policies. My worry today is that if you were imposing a structure without questioning what isn't working within that framework, you will only be re-creating the marginalizations and biases that existing systems uphold. If the government today tried to impose emergency measures for climate change, I don't think Indigenous communities would be supportive, especially if it was a top-down imposition without proper consultation and involvement and without making space for Indigenous communities to have their own plans — you would have resistance to that." And rightly so.

As we seek to move quickly on the climate emergency, a basic principle, Norris says, must be to ensure "that Indigenous communities, many of whom are heavily reliant on fossil fuels, aren't worse off than they are now." That would be especially true for many First Nations in Treaty 8, whose territories encompass what is now the oil sands. As we negotiate new and stronger global treaties to tackle the climate emergency, Indigenous nations need a seat at the table if they are being asked to abide by the outcomes of these treaties. And at a more fundamental level, Norris notes that part of why Indigenous people are leading so much of the climate fight is that "they are operating under a different worldview, one that doesn't see the natural world as a resource for exploitation." He suggests that maybe this alternate framework is what we really need now. He may be right, but I still worry that may take longer than we have.

Others think there may be a way to sort through these conundrums of sovereignty and time. Indeed, they are already showing the path.

Khelsilem is a young Indigenous leader from the Squamish — or Skwxwú7mesh — Nation on Canada's west coast. "Easiest way I think about it is that four of my great-grandparents are Skwxwú7mesh, two are Kwakwaka'wakw, one is Shishá7lh, and one is English," he told me. He grew up in North Vancouver (mainly on Squamish reserve land) and now lives in Vancouver. I first met him when he was an activist organizer during Idle No More. Then, in 2017, at the age of 28, he was elected to the band council of the Squamish Nation and is now one of the nation's two appointed lead spokespeople. He has been one of the main Indigenous leaders opposing the Trans Mountain pipeline expansion project. For years, he's been at the forefront of maintaining and reviving the Skwxwú7mesh language, including founding a Skwxwú7mesh language program in partnership with Simon Fraser University. Smart and charismatic, with a savvy and engaging social media presence, he has become an influential voice in B.C. policy circles.

He comes from a political family. Khelsilem's mother is also a long-time elected council member. His late uncle was on council. His grandfather was on council. And his great-grandfather on his father's side, Andy Paull, helped found the Skwxwú7mesh council and was an Indigenous rights leader provincially and nationally. Indeed, Paull was the founder of the North American Indian Brotherhood (a precursor of today's Assembly of First Nations), which he established in 1945 as a national lobby group to press for Indigenous people to have full and equal citizenship rights and benefits without losing their Treaty Rights (arguably another legacy of the Second World War). "He was trained as a lawyer in the 1910s," recounts Khelsilem. "But under Canadian law, in order to become a lawyer, to join the bar, you had to enfranchise and give up your status. He would not be able to live on the reserve or be part of the community. He would not do that. So he got trained as a lawyer, but he did not actually join the bar."

But not all Khelsilem's relatives share a deep connection to his culture. His grandfather's brother enlisted and fought in the Second World War. This led to his great-uncle being enfranchised, thereby losing his "Indian" status. He had to move off the reserve, and his children and grandchildren have lost much of their Sḵwx̱wú7mesh connection. "He got married and had kids and grandkids, but they never grew up in the community. They never grew up as Sḵwx̱wú7mesh people. I only met them briefly once."

Khelsilem's opposition to the pipeline expansion isn't merely about defending the waters and land of the Sḵwx̱wú7mesh people, although it is that. As with leaders like Grand Chief Stewart Phillip of the Union of B.C. Indian Chiefs and others mentioned here, for Khelsilem it's also about climate and the need to get real about the climate emergency. He's been pushing that point at every level of government, including within First Nations political organizations. He helped draft a climate motion that passed at a 2019 general assembly of the Union of B.C. Indian Chiefs, inserting climate emergency language into the policy. He also helped author another climate emergency resolution for the national Assembly of First Nations adopted in July 2019 (a few months later, in their 2019 federal election priorities document, the AFN identified climate change as the organization's top priority for the new government).[20] He's called out the *Clean BC* climate plan of the B.C. NDP government, both for the fact that its targets are not sufficiently aligned with the 2018 IPCC report and for inadequate engagement with Indigenous groups about the plan.

And he's not alone: "I am meeting a lot of young [Indigenous] leaders who are being elected now, people in their twenties and early thirties, who talk about climate in the sense of it being an emergency and the need to mobilize to fight it," he recounts. "Young leaders are feeling like this is a challenge of our time and we need to take it on. But that's leadership, right? Leadership is knowing where you are at, knowing where you need to get to, and developing a plan to bring people along to that."

I heard Khelsilem speak in 2019 on a panel about reconciliation, at which other Indigenous speakers urged settler politicians in the audience to "go slow" and do reconciliation right. Later, I asked Khelsilem how we navigate the tension between that counsel to do reconciliation carefully and to take the time to listen, on the one hand, and the practical reality that we are facing an urgent crisis on the other. "It has to be okay to make mistakes," he replied. "If you are saying, 'Let us do this work together, let us connect to a set of principles that we are all really going to hold true,' then let us be okay with making mistakes, and let us go forward. To practise mastery, you have to master practising. You just chart a course and go. Whether it is addressing the climate emergency or whether it is reconciliation or something else, you just go. You practise something and then you evaluate — are we going in the right direction? Are we doing the right thing?"

Khelsilem sees the support some First Nations have given to fossil fuel projects as "forced consent" given limited economic options. "First Nations are caught up in the same exploitation that resource workers are caught up in. This is about capital trying to increase their profits. First Nations now represent a new hurdle in that process, just like any other regulatory process. And that is how industry sees First Nations. They do not see it as partnership-building or nation-building. They see us as a regulatory hurdle that they have to get over, and sometimes they do."

But he has not been prepared to offer such consent himself. Support from the Sḵwx̱wú7mesh Nation was required for the Woodfibre LNG plant on their territory (a relatively small facility, but if it proceeds, nevertheless a harmful source of new GHG emissions). His nation's council approved the project, although it did not "endorse" it. But it was a split decision — Khelsilem was on the record along with a minority of other councillors opposed to the LNG facility, notwithstanding the economic benefits on offer to the Sḵwx̱wú7mesh people.

Rather than making Faustian bargains with fossil fuel companies, Khelsilem prefers to pursue a future based on post-secondary training. Not content with the limited funding for post-secondary education provided by the federal government for each First Nation, the

Sḵwx̱wú7mesh Nation is supplementing these transfers with their own revenue sources, to offer universal post-secondary tuition and living allowances for all Sḵwx̱wú7mesh students.

Khelsilem likes the climate emergency frame. He sees it as helpful, even if it's fraught. It's just about who invokes the emergency. In summer 2017, devastating wildfires ravaged the interior of B.C. The province invoked a state of emergency and ordered people to leave their homes. But the Tsilhqot'in Nation near Williams Lake defied the evacuation order. Having not been properly consulted, they refused to surrender their homes. It was a messy situation, but also a learning experience. The following spring, in anticipation of more such summers to come, the Tsilhqot'in signed a "first-of-a-kind" agreement with the provincial and federal governments recognizing the First Nation as a full partner in wildfire response. The nation has trained their own firefighters, they now have extensive experience with wildfires, and in future, they will decide if and when to declare an emergency on their territory and how to respond. The summer of 2018 was again an awful wildfire season, but this time, the local response went much more smoothly.

This new joint emergency protocol is entirely different from what Inuit experienced when facing a health emergency in the 1950s and '60s, when they were removed from their communities without consent aboard *The C.D. Howe*. Perhaps herein lies a model for how to reconcile climate emergency response plans with Indigenous title and rights.

Like the Iroquois Confederacy in the Second World War, numerous First Nations and their political organizations are independently passing their own climate emergency motions.

Like Khelsilem, Chief Dana Tizya-Tramm is part of a new generation of elected Indigenous leaders who are keen not only to defend their people's land and water, but to tackle climate change. His enthusiasm for the change he seeks to drive is infectious, and his curious mind sees constant connections between what this moment demands of us and the traditions and history of his people.

Not long ago, Tizya-Tramm's path looked like it might be quite different. The son of a Gwitchin mother and German-born father, Tizya-Tramm grew up in southern Yukon, but with strong connections to his mother's culture and people in northern Yukon — the Vuntut Gwitchin First Nation based in Old Crow, a fly-in community of fewer than 300 people.

At age 13, Tizya-Tramm left home and soon was messed up with drugs and alcohol. He left Whitehorse and moved to Vancouver. His first few years there were mainly lost to addiction, but eventually he straightened out and found work and stable housing.

After a time, he landed a job at the Bella Gelateria ice cream parlour in Vancouver's upscale Coal Harbour area. And because life is full of surprises, he ended up helping to develop a new ice cream flavour — "a salted caramelized pecan with maple syrup gelato, with different reflections on the palate that warmed taste buds in a gently shifting manner; great stuff," he effuses about what sounds to me like, quite possibly, the ultimate Canadian ice cream flavour. In 2013, his new flavour broke sales records, won two gold medals, and was voted #1 by 250,000 Italians! Tizya-Tramm's life was looking good. His boss offered him a $90k job to open a store in Los Angeles. And then, he had a revelation that he wasn't happy. Like Molly Wickham, Tizya-Tramm wanted to reconnect with his people. As he explains it, he wanted to strengthen the link between his grandparents and future grandchildren. And so, Tizya-Tramm walked away from it all. In 2013 he left Vancouver, and returned to Old Crow. Two years later he was elected to the council of the Vuntut Gwitchin self-government, and then in 2018, at the age of 31, was elected chief.

"While I was in the city, I did not have any roots," he told me. "I was a tumbleweed that was subject to the winds of economy, pop culture, all of these different things. But once I aligned myself with my people, I had a deep root system. Because my life is not my life. My life is the culmination of sacrifice and hard work in one of the hardest territories in the world. I actually owe my people, my territory and future generations my life."

In 1995, the Gwitchin signed an historic self-government agreement with the Yukon and federal governments that set a new standard for modern treaties. The Gwitchin take pride in all the affairs and services now under their self-governed jurisdiction. But as Tizya-Tramm quickly realized, he was coming into leadership just as his people faced the most daunting challenge since colonization — "I came into office during the sixth extinction age; during anthropogenic climate change. The Earth is speaking to us now. All the science is backing up what our elders are saying."

Tizya-Tramm sees parallels between climate change and other challenges he has experienced and witnessed in his life: "I have dealt with drug addiction my entire life, I have dealt with alcoholics my entire life. And I can tell you, without reservation, that people are exhibiting the exact same steps with deep addiction as they are with fossil fuels and our way of life. Not being able to see the truth, or even accept it — they cannot look it in the eye."

In ways that many in Canada's south have yet to comprehend, Tizya-Tramm sees climate change as a clear existential threat to his people. "This ability to deny what is actually taking place is a luxury. Because in my community, we do not have that opportunity, when our territory is warming at almost three times the global rate. I read a lot of studies, but I am also out on the land, and I am in contact with a lot of our hunters and our gatherers. And they always talk about it being different. For instance, it used to be −60 for over a week or two weeks when they were young. And now it may touch −60 once during the winter. What really hit it for me was that elders were telling me there are birds coming to our territory we have never seen before. Then, when I read the IPCC's sixth report, that is what galvanized everything."

That IPCC special 2018 report landed the month before Tizya-Tramm was elected chief. And so, one of his first acts was to develop a climate emergency declaration, passed unanimously by the Vuntut Gwitchin council in May 2019. But this climate emergency motion has a twist. "What I love about it is we are making this an Indigenous declaration." The declaration's preamble speaks to how the climate crisis and the heightened temperature rise in the north have "profound implications for our people, community, lands, waters, animals, and way of life." It highlights the agreement between what the IPCC finds and what the community's elders report. And the declaration expresses the concern that "Indigenous peoples' voices and lands are not being heard."

It then declares that "climate change constitutes a state of emergency for our lands, waters, animals and people, and that we will accordingly utilize our local, national and international forums and partnerships to achieve meaningful progress towards the Paris Accord and the inception of an Indigenous Climate Accord that shall call for coordinated efforts with our relatives around the world." And it commits the community to do what it can to keep global temperature rise below 1.5 degrees.

The Vuntut Gwitchin climate declaration has become a model for all 11 Indigenous self-governing nations of the Yukon, who in February 2020, collectively signed a climate emergency declaration. The Vuntut Gwitchin declaration has also been shared with other Indigenous nations nationally and internationally across the Arctic. Tizya-Tramm's hope is that this will lead to a pan-Arctic international Indigenous climate accord. "This is an opportunity for the Indigenous people to be leaders in the world."

And the Gwitchin of Old Crow aren't just talking the talk. They are changing their own community's energy use and driving down their GHG emissions. For decades, and with no road to the community, Old Crow has been flying diesel in, and burning it in a generator for electricity. Now they are building (with financial support from the federal and Yukon governments) what they believe will be the largest solar farm in the Arctic, which they expect will produce enough electricity to meet a quarter of the community's needs. The solar array will displace about 189,000 litres of diesel fuel annually and reduce CO_2 emissions by an estimated 267 tonnes a year. "On sunny days, we will be able to turn our diesel generators off from early March to late September. With one energy project. But this is just the beginning." They are also investigating biomass and are exploring the feasibility of a wind project. Tizya-Tramm wants to see his community beat the Paris commitment on a per capita basis, and ultimately, "We want to shut those diesel generators right off as soon as possible."

The Gwitchin are now among many Indigenous communities seeking to break free of the Faustian choices foisted by fossil fuel companies, and are instead developing major renewable energy projects on their territories. According to Catherine Abreu of Climate Action Network Canada, about 20% of renewable energy projects in Canada are Indigenous. The T'Sou-ke First Nation on southern Vancouver Island was among the first, developing a three-pronged initiative in 2009 that includes a 75-kilowatt solar system (feeding excess power into the BC Hydro grid when it's not needed locally), 40 solar hot-water systems in member residences and an energy conservation program.[21] At the time of its launch, the T'Sou-ke initiative was the largest solar project in B.C. But now, the Lower Nicola First Nation near Merritt, B.C., has a larger one. And not to be outdone, in October 2019, the Tsilhqot'in Nation cut the ribbon on the largest solar farm in B.C., a 1.25-megawatt solar farm that is wholly developed, built, owned and operated by the Tsilhqot'in Nation.[22] Now that's the kind of arms race we need to see!

An Indigenous social enterprise in Manitoba, Aki Energy, has been engaged in geothermal installations in northern First Nations communities since 2012. Melina Laboucan-Massimo, the climate campaigner mentioned earlier, driven by a desire to model positive alternatives as she also fights against tar sands expansion, led a project in her home community of Little Buffalo, Alberta, that built a 20.8-kilowatt solar installation to power the local health centre. Laboucan-Massimo has since taken on hosting a 13-part documentary series called *Power to the People*, airing on the Aboriginal Peoples Television Network, which profiles more such initiatives.

The Gwitchin, like others, are also tackling the supply side. In 2016, in the face of oil and gas companies pressing for drilling access, the community passed a resolution banning all fracking in their traditional territory, part of a larger and thus far successful effort to ban fracking throughout the Yukon.

When asked if the climate emergency declaration was a tough sell with his people, Tizya-Tramm replied that it wasn't because "everyone is already talking about this." People see the impact of changing weather around them. "What we have done in the declaration is we acknowledged the science and the IPCC to help support what our elders are seeing, what our people know, and our traditional knowledge. The science is giving credence to our observations. The declaration acknowledges our responsibility from our ancestors to our grandchildren. This declaration has already changed the political landscape beyond our community. It has empowered my people, instead of scaring them and making them feel powerless."

The title of the Gwitchin declaration — Yeendoo Diinehdoo Ji'heezrit Nits'oo Ts'o' Nan He'aa — translates as "After our time, how will the world be?"

Toronto Public Library

CHAPTER 10

Civil Society Leadership

"You've convinced me. Now go out and make me do it."

— President Franklin D. Roosevelt, when meeting with a group of labour and civil rights activists in the 1930s

"We've had decades of environmental campaigning, and (yet) things are getting worse. Emissions are still going up. It's come to a point where we have no choice but to force governments to pay attention and ask people to join us."[1]

— Amani Khalfan, Extinction Rebellion Ottawa

"We are in an emergency. You need to address it with compassion and with love for your children: remember, cutting emissions literally means saving lives. Stop treating climate action as a high school assignment to put off and squirm your way out of, because the atmosphere doesn't give due date extensions. Twenty years from now, will you be able to look your grandchildren in the eye and honestly say you did everything you could to secure them a safe future?"[2]

— Rebecca Hamilton, age 16, Vancouver climate strike organizer with Sustainabiliteens, sharing her message to political leaders

When newly minted British Prime Minister Winston Churchill faced "the darkest hour" of the war in May 1940, with France on the verge of surrender and a sizeable chunk of the British army pinned down on the beaches of Dunkirk, he was under tremendous pressure from some senior members of the Conservative Party to negotiate peace terms with Hitler. But instead, Churchill took solace and comfort in the resolve and determination he shared with the "ordinary" British people themselves who, despite the odds, encouraged him to fight rather than surrender.[3] The moral of the story being, in the eye of the crisis, don't trust the elites who are ready to appease as they seek only to protect what they have. Instead, be forthright with the public about the situation, and then trust the people.

This book focuses primarily on political leaders. That's because, ultimately, what needs to happen has to translate into political and policy action, and the scale of what is needed and the speed at which it must occur can be achieved only at the state level (in close partnership with Indigenous nations and respecting their title, as we have just established). This is not to discount, however, the critical role of civil society and social movements. It will always fall to social movements to push political leaders to make needed changes. As the great African American abolitionist Frederick Douglass famously stated, "Power concedes nothing without a demand. It never did and it never will." Moreover, as outlined earlier, the public is ahead of our politics when it comes to the climate crisis.

Other contemporary realities also speak to the need for civil society leadership. People today are less deferential to authority than they were in the Second World War era. The public then was, arguably, more prepared to heed the call of political leaders. So today, leadership needs to be multifaceted.

On September 27, 2019, as part of a global climate day of action called by Greta Thunberg, the youth climate strikers, and 350.org, cities across Canada took up the call. In my hometown of Vancouver, an estimated 100,000 people participated. After years of climate protests I have attended with crowds of 3,000–10,000 people, on this day the numbers exploded into the largest protest in B.C. since the early 1980s (the peace marches during a peak period in Cold War rhetoric). The largest Canadian protest that day was in Montreal where an estimated 500,000 people turned out, and where Greta Thunberg, then visiting North America for the first time since launching her youth strikes, joined the protestors. Across the country, somewhere between 800,000 and one million people took part in the climate strike, likely making this the largest day of protest in Canadian history.

That mobilization was a very welcome shift to be sure. If indeed one million Canadians have joined the climate protests, that represents about 2.6% of the Canadian population. Academics Erica Chenoweth and Maria Stephan, authors of *Why Civil Resistance Works: The Strategic Logic of Nonviolent Conflict*,[4] a global historical review of successful mass nonviolent campaigns, posit that successful social revolutions are virtually guaranteed once 3.5% of the population become engaged.[5] In the Second World War, over one million Canadians enlisted from a population less than one-third what it is today, and the overall level of participation in the war effort, as we've seen, was staggering. So the current emergency is registering, but we aren't quite there yet.

YOUTH LEADERSHIP: THE LEAST SAY, YET HIGHEST PRICE TO PAY

Of the hundreds of thousands of Canadians who went to fight overseas in the Second World War, from all corners of the country and many ethnic backgrounds, the overriding characteristic most had in common was their youth.

According to Veterans Affairs Canada, of the 1.1 million Canadians who enlisted, about 700,000 were under the age of 21 (about 64% of the total).[6] Canada did not keep age data of the war dead. However, approximately 44,000 Canadians died in the war, and if one assumes those killed had the same demographic proportions, then about 28,000 would have been below the age of 21.

These hundreds of thousands of young people signed up at our collective request and were willing to risk death. What they could not do, however, was vote. The federal voting age until 1970 was 21. Just as the young people today who will be most impacted by climate change don't get a say in choosing our elected leaders, so it was then that generations of young people sacrificed themselves in wars decided upon by people they were disenfranchised to select.

Let Youth Vote

Many of the teenage climate strikers, in Canada and abroad, have revived the call to lower the voting age. A new initiative to lower the voting age to 16 has emerged in B.C., led mainly by teenage climate strikers.[7] In the U.K., the young climate strikers include lowering the voting age to 16 in their manifesto.[8] As someone whose political activism started as a teenager in the nuclear age, and who mobilized in the 1980s with hundreds of others who could not vote, I have long believed that lowering the voting age represents the next logical step in the evolution of the franchise. Today, if there was ever a compelling reason to do so, the climate emergency is indisputably it.

We have all watched as these amazing, passionate and well-informed young people have sounded the emergency alarm, roused us into the streets and pushed our political leaders back on their heels. (And conversely, we have all met plenty of older people whose lack of political knowledge and understanding is mind-blowing, yet no one suggests this should deny them the right to vote.) How preposterous then that our society deems someone as articulate and wise as Greta Thunberg or Rebecca Hamilton (cited above), both age 16 as I write, unfit to cast a ballot. Surely, the case is now clear that denying these people the right to vote represents a grave injustice. What's more, we need these voters lickety-split — we need their clear-headed priorities in the political debate, and we need those who seek to be our elected representatives to feel the practical pressure of winning the favour of these young people. Extending the franchise would be politically transformative, and doing so deserves a place of priority in any climate emergency agenda.

Their Time to Lead

The good news, of course, is that young climate leaders are not waiting for the franchise. They are exercising considerable political muscle without it. As in the war, youth are once again mobilizing to secure our collective future.

The demographics are on our side. Millennials are the largest demographic cohort in Canada. This past federal election was the first in which millennials represented the largest voting bloc, and that reality is only going to increase over time (as the baby boomers leave us over the coming decades). And more than any other generational cohort, as we see in polling results, millennials understand the climate crisis and want to see real action. Help is on the way.

Many youth-led initiatives have been vital to climate mobilization thus far: the student climate strikers, 350.org's Our Time initiative and the Sunrise Movement in the U.S. (which has rallied support for the amazing Alexandria Ocasio-Cortez and the Green New Deal, making the old-school fossil-fuel-funded Democrats run for cover). There is a new generation of millennials who are now in leadership roles with the Canadian non-governmental organizations pushing for a made-in-Canada Green New Deal, such as Leadnow, The Leap and Dogwood, and who have emerged as elected leaders within Indigenous communities and in mainstream politics. A transformation is underway.

There are the international superstars such as Greta Thunberg, who in her unassuming way launched an international movement in 2018 that spread to over 120 countries and has now seen millions take to the streets. But closer to home we have young people like 15-year-old Indigenous environmental activist Autumn Peltier from the Anishinabek Nation also gaining global attention.

For many young people, the Second World War is an old story from high school history textbooks; it struggles to find resonance. Yet many of them do get the climate emergency framework and have started to latch on to the wartime frame, answering its call in inspiring ways. Who would have guessed that so many millennials would be responding so enthusiastically to the idea of a Green New Deal, with its roots in the 1930s? In early 2019, as Atiya Jaffer and 350.org Canada were ramping up their pre-election plans, their national call-out for volunteers stated, "A Green New Deal for Canada means mobilizing a WWII scale response to climate change that addresses not just the environmental crisis, but enshrines Indigenous rights into law. It means climate policies that transition us off fossil fuels in line with what the science demands by making massive investments in renewable energy and public transportation and guaranteeing green jobs for workers all across Canada." Similarly, as one of the Vancouver climate strike leaders, Rebecca Hamilton, told the CBC in early 2019, "During World War II, everybody knew there was a war on. It was the lens through which every individual and government decision was made, and I want the climate crisis to be seen with the same level of urgency and addressed with the same level of scale."[9]

In the lead-up to 2019 federal election, young Canadian activists called for ambitious climate action, but they have also seamlessly integrated demands for social and Indigenous justice into their agenda. Climate Strike Canada set seven conditions for any candidates who hoped to earn the youth vote,[10] including:

- legislated emission reduction targets of 65% by 2030 and 100% by 2040;
- "separation of oil and state" by rejecting new fossil projects and ending fossil subsidies;
- a just transition for fossil fuel workers;
- commitment to an environmental bill of rights;
- full implementation of the UN Declaration on the Rights of Indigenous Peoples;
- conservation of biodiversity; and
- protection of vulnerable groups — including recognition of climate displacement as a basis for refugee status.

On October 28, 2019, one week after the federal election, 27 young people were arrested during a sit-down protest in the foyer of the House of Commons. They were all from 350.org's Our Time initiative, and they were calling on all of the newly elected members of Parliament to work across party lines and govern for a Green New Deal.

These young people are giving expression to the emergency. They are rightly shaking up what is *normal*, because we must. And in the face of those who object to their school-time strikes, they are proudly honouring Mark Twain's wise counsel that, sometimes, you can't let your schooling interfere with your education.

In January 2019, Vancouver City Council became the first municipal government in English Canada to pass a climate emergency motion. It passed unanimously, and it was not merely symbolic. The Vancouver motion gave city staff 90 days to come up with a new climate plan that would dramatically ramp up the city's actions and timelines to meet the targets set out in the 2018 IPCC report and asked staff to apply an equity lens to ensure those most vulnerable are central to the solutions. The resulting climate emergency plan, again unanimously passed by council (with one abstention) three months later, contains measures to dramatically reduce the use of natural gas in homes and buildings and to fundamentally shift people's modes of transportation. The plan is among the most ambitious in North America. (Full disclosure: the Vancouver motion was crafted and introduced by Councillor Christine Boyle, who is also my wife.)

But the Vancouver motions also had a fascinating political dimension. When Councillor Boyle brought her climate emergency motion forward, she expected it to pass, but she did not know by how much. Vancouver city council, for the first time in decades, currently has a very mixed political makeup, and no party has a majority. But the motions ended up passing unanimously. According to Boyle and some of her colleagues (from two different parties), that only happened because dozens of high school students skipped school to rally outside council. Many of them signed up to speak before council and gave passionate and compelling speeches. They met with the councillors. And as the votes took place, the gallery was full of young eyes watching decisions made by those they are precluded from electing. In that charged moment of intergenerational reckoning, the votes were unanimous. These young climate strikers had made it politically impossible to say no.[11]

The unanimous outcomes these young activists helped to secure matter. They signal to other municipalities that ambitious climate action can be taken with limited political risk. Moreover, opposing these actions was not opportunistically "weaponized" by some politicians to advance their electoral interests, as we have seen

Conservatives do federally and in numerous provinces. Because the Vancouver climate motions passed with support across the political spectrum, climate action in Vancouver is not a political wedge issue.

Let Youth Serve

Young people crave a sense of purpose — a desire to believe in something greater than themselves. We all do, of course, but this too often gets squashed out of the older among us, as we deal with the daily demands of work and family. That yearning for a sense of purpose is what led many to enlist in the Second World War, and it is what philosopher William James sought to find in a non-destructive cause in his 1910 essay "The Moral Equivalent of War."

Kai Nagata is a former journalist, a very creative and strategic thinker, and a long-time climate activist. He's currently the communications director with the B.C.-based grassroots organization Dogwood. But he's always been an outdoors guy and enjoys hunting, and at age 20, in 2007, Nagata decided to join the Canadian Grenadier Guards infantry reserve regiment. "That seems incongruous," I put to Nagata, given his politics. But there was no contradiction to him. He says he was bored with going to school and liked the idea of a "gap year" with plenty of physical activity. "It was a way to prove my Canadian-ness," he said. "Everyone in my platoon was either a recent immigrant or someone who needed a second chance in life."

His brief time in the military gave Nagata insight into what many young people are looking for — the sense of purpose and discipline and comradery. And like William James, Nagata wishes there was "a way to channel that without killing people." Nagata's take-away, from both his experience and his grandfather's during the war, is that Canada needs a new Youth Corps, a modern climate reboot of the U.S. Peace Corps. Not a compulsory form of youth service, as still exists in some countries, but an attractive one. My CCPA colleague Ben Parfitt has long believed in something similar. (The federally created youth program Katimavik gets at this. Operating off-and-on since the late 1970s, it has placed more than 35,000 young people into community service volunteer work. But it is only a half-year program with just a handful of spots available each year. More recently, the federal government launched the Canada Service Corps to offer volunteer opportunities to young people, but it remains in development at the time of writing. Similarly, the United Nations Association in Canada has a "Canada Green Corps" program for young people, but it is small, short-term and for youth who have already completed post-secondary studies.)

Climate mobilization today would be well-served by the establishment of a large-scale Youth Climate Corps. It could be aimed at young people who have finished their high school education and invite them to spend two years volunteering on projects that seek to expedite our climate transition, with all expenses paid, just like in the military. The program should be open to any youth who wish, although special efforts could be made to encourage at-risk youth to enlist. And like the military, when they complete their service, they could be offered free post-secondary education or training.[12] It would be a win-win — youth would find meaning and gain a leg-up on their future plans and careers, while society gains an eager army of climate warriors.

Climate change will be the defining issue facing today's youth, with profound consequences for their futures, families, communities and careers. How we confront the largest collective action puzzle of human existence will be the story of their lives. We would do well to fully engage them in the exercise as soon as possible.

SOCIAL MOVEMENTS: "MAKING THEM DO IT," THEN AND NOW

Governing from the centre is always the safe political space. That is why such a strong status quo bias prevails in our politics. It is also why big changes usually start outside institutional politics. It is an active and engaged civil society that creates the demand and room for boldness, and in the end, as the saying goes, we get the governments we deserve.

In earlier chapters, we explored how the war years marked the start of a three-decade period of reduced inequality and the establishment of core social programs that have become defining elements of our national identity and well-being. But it wasn't just the war.

It was also the advance work done — the protests led by the labour movement and progressive political and social movements in the 1930s — that laid the groundwork. Those movements demanded a New Deal before the war, and ensured that after the war, a new social contract — a "peace dividend" — would await returning soldiers. These popular mobilizations before, during and after the war were all intertwined and formed a continuum of social movement struggles and demands.

So too, if the climate mobilization is to lead not only to a post-carbon economy but also a more just society, then it will fall to social movements to "make them do it." Indeed, it is among the core contentions of this book that if our social movements are unsuccessful in forcing our political leaders to embed equality and social justice goals into the transformation before us, the necessary climate actions will likely fail. Instead, climate action in splendid isolation will be seen as an elite exercise and will foster a right-wing populist backlash that will damn us all to a fiery hell.

Climate Emergency Mobilization and Motions

The January 2019 Vancouver climate emergency motion set off a domino of similar motions across English-speaking Canada. A couple weeks after Vancouver passed its motion, the city of Halifax followed with an almost identical one. Over the next two months, they were joined by the cities of Kingston and Hamilton in Ontario, and the Victoria Capital Regional District in B.C. So far, 70 communities in Canada outside Quebec have passed climate emergency motions. Some seem mostly symbolic, but others, notably Vancouver's, clearly have consequential teeth.

These motions represent another form of pressure from below. These are local governments responding to the IPCC call, but their moves in turn come in response to pressures from a renewed climate movement — grassroots groups that have pushed their city councils to recognize the emergency. And those municipal voices have recently been joined by professional associations — health professionals and scientists — that have also issued climate emergency declarations.

Quebec is another story entirely. Indeed, in the face of the climate crisis, that province is indisputably leading the country when it comes to popular mobilization. The size of the climate protests in Quebec have dramatically surpassed those in English Canada, and it is in Quebec where opinion polling finds the highest level of public support for bold climate action. And with respect to municipal governments passing climate emergency motions, well, Quebec leaves the rest of Canada in its dust — since summer 2018, over 400 Quebec local governments have passed such motions.[13]

This remarkable effort in Quebec was led by a non-governmental organization called GroupMobilisation. The spread of the emergency motions was partly a product of timing — summer 2018 saw a deadly heat wave in the province, and Quebec has also seen flooding along the Ottawa River. But it also grew out of a continuum of fruitful mobilizations. Many of the same groups had successfully mobilized against the Energy East pipeline, and prior to that had pressured the Quebec government to institute a ban on fracking along the St. Lawrence River (an effort that also pursued a strategy of starting with municipal level motions). One win leads to another, as each successive victory emboldens the climate movement for the next fight.

Indeed, sometimes the positive infectiousness of these wins leaps between movements. Université du Québec à Montréal sociology professor Éric Pineault has studied the Quebec mobilizations. When I asked Pineault how it is that Quebec's climate movement is so large and effective, he replied, "There is a tradition of mobilization, of organization, and of victories. And there is a structured student movement in Quebec. It is very unique in the sense that it is a real union-type movement. So that means that mobilization is always easier. And once they decide that something is a priority, they are good at it."

Quebec has a long history of student activism and strikes, and the legacy of that vibrant student movement is a province with the lowest post-secondary tuition in Canada. Many of today's Quebec climate activists are alumni of the Maple Spring — the extraordinary student mobilization against tuition hikes that shook the province in 2012. Pineault notes that one of the largest Maple Spring demonstrations was on Earth Day, after which people from the environmental and student movements began working together more closely. "And there was a link made between this idea of Révolution Tranquille and the idea that the new quiet revolution had to be an environmental, green transition," he told me. "You could actually make the hypothesis that a lot of what was being said in that time was very close to the Green New Deal that is being proposed today." After the Maple Spring, one of the young leaders of those student strikes, Gabriel Nadeau-Dubois, became a leader in the fight against the Energy East pipeline and is now a member of Quebec's National Assembly and co-spokesperson for Québec Solidaire, arguably the political party with the strongest climate plan in Canada.

The popular grassroots climate mobilization efforts in Quebec have produced a political climate unlike anywhere else in Canada. Like the Vancouver story above, but at a much larger scale, climate is not a political wedge issue in Quebec — climate action has not been weaponized by the political right, for the simple reason that any such efforts would fall completely flat. Instead, all political parties in Quebec now compete for the claim of climate leadership. The right-wing Coalition Avenir Québec currently in power, unlike its Conservative counterparts in English Canada, supports carbon pricing, opposes pipelines and is continuing to slowly advance the climate action agenda in that province. The Quebec climate plan certainly isn't yet aligned with the emergency, but at least divisive politics isn't moving things backwards as it is in other locales. That's what we need to see in the rest of Canada, at the provincial level and of course federally.

The Role of Civil Disobedience

No great social movement that I can think of has ever succeeded without including some form of civil disobedience. This moment is no different. These actions must always be peaceful, and they should be strategically and thoughtfully conceived to ensure maximum effect and minimize a backlash. But let there be no doubt — the climate mobilization has had and will surely continue to need civil disobedience among its arsenal of tactics.

We've seen the Indigenous-led blockades of various fossil fuel extraction and export efforts. The year 2018 also saw the arrival of a new organization, Extinction Rebellion (XR), which started in the U.K. and quickly spread to other countries, now with numerous chapters across Canada.[14] The organization is very grassroots and decentralized, unlike the more professional environmental groups that have long operated in the climate space. XR seeks to fundamentally shift the political debate, using peaceful civil disobedience protests to communicate a sense of emergency. Extinction Rebellion's U.K. protests clearly moved the dial in that country. Shortly after their spring 2019 mass protests shut down traffic at key London sites, the British Parliament became the first national legislature to pass a climate emergency motion.

XR protests have now come to Canada. We should expect more such civil disobedience, and indeed join and support them as we are able. While unlawful, these protests generally have a carnival atmosphere — go have some fun, while disrupting the status quo.

Citizen Assemblies: Giving the Public Voice

Interestingly, while XR calls for climate to be recognized as an emergency, they have been reticent to advocate for an approach that would increase state power. Instead, they have called for the widespread use of "citizen assemblies" to map out a path to climate survival and a more just society. This is a wise call. We are, as a society, less deferential to authority than we were. Moreover, the public deserves a say in these matters. People are reasonable in asking who gets to decide how we will achieve our climate goals. The path forward will find more widespread support if it is seen to be the product of an inclusive and deliberative public engagement. That's the strength of citizen assemblies — while much of the public may be disinclined to trust politicians, people are generally prepared to trust their fellow citizens. There is strong evidence that when we see our fellow civil society members and neighbours volunteering their time and deliberating respectfully, we are more inclined to honour their recommendations.[15] Leadership from below is an easier sell.

The first steps in this direction are now underway in the U.K. In the wake of Britain's national climate emergency motion, Parliament created Climate Assembly UK, at which randomly selected members of the public are invited to deliberate on how the U.K. should respond to the climate crisis and which policies to implement. But while the deliberation is being used to consider different policy options, the core goal — becoming carbon-zero by 2050 — is not up for debate. That target is set in law.

Divestment

A final important social movement tool in the emergency toolbox — fossil fuel divestment campaigns. Like Blockadia, divestment achieves two purposes. First, it buys us time, siphoning the "money pipeline" of investment for oil and gas extraction while we wait for our politics to align with the emergency. And second, it directly confronts the fossil fuel companies, publicly stigmatizing them and making them hurt where it counts — their financial returns. Divestment has another strategic value, which is that it offers people something concrete they can take on in their personal lives. Many in civil society find it hard to engage in high-level political organizing and may not be ready to participate in civil disobedience. They are certainly excluded from international climate treaty-making. So divestment campaigns offer a specific action one can take beyond attending a protest march. A majority of us either invest directly ourselves, or indirectly through our pension plans, faith institutions, unions or financial institutions. We almost all belong to municipalities that invest. Many of us are attendees or alumni of post-secondary institutions that invest. And we are all customers of companies that invest. And so we can all push for these organizations and institutions to pull their money out of fossil fuel companies and the financial institutions — the banks and insurance companies — that serve as their enablers.

There is a long and honourable tradition of social movements employing the tactics of boycott and divestment. Most famously — and successfully — this was done to force an end to the apartheid regime of racial segregation in South Africa. South African Archbishop Desmond Tutu, who was awarded a Nobel Peace Prize for his leadership in South Africa's freedom struggle, has called for an anti–apartheid style global boycott and divestment of the fossil fuel industry, urging all organizations to pull their investments from fossil fuel companies. "People of conscience need to break their ties with corporations financing the injustice of climate change," Tutu has said. "We can, for instance, boycott events, sports teams and media programming sponsored by fossil-fuel energy companies."[16]

The global fossil fuel divestment campaign has been led by 350.org. The call to divest has been taken up by students and faculty at hundreds of post-secondary institutions, many of which have started to experience success in getting their universities to divest. Dozens of faith institutions (including the United Church of Canada) and foundations have already divested, and a handful of municipalities have followed suit. 350.org's Bill McKibben sees divestment as "an additional lever to pull, one that could work both quickly and globally." The fossil fuel divestment campaign has become the largest such effort in history. "Funds worth more than eleven trillion dollars [globally] have divested some or all of their fossil-fuel holdings." And various fossil fuel companies are starting to feel its pinch.[17]

As important as these civil society efforts are, it will ultimately fall to our political leaders to act — to quickly scale up what needs to happen to make this transformation society-wide and, when needed, mandatory. State-led mobilization remains essential. We all have some understandable nervousness about the implications of enhancing state power. So, we will next explore some cautionary tales from the Second World War, and how to ensure civil liberties and human rights are respected as we mobilize in a hurry.

Photo of Japanese Canadian fishing boats impounded in 1942, courtesy of VR 991.192.1, CFB Esquimalt Naval and Military Museum

CHAPTER 11

Cautionary Tales: What Not to Do

"As a visible minority that has experienced legalized repression under the War Measures Act, we urge the Government of Canada to take such steps as are necessary to ensure that Canadians are never again subjected to such injustices. In particular, we urge that the fundamental human rights and freedoms set forth in the Canadian Charter of Rights and Freedoms be considered sacrosanct, non-negotiable and beyond the reach of any arbitrary legislation such as the War Measures Act."

— National Association of Japanese Canadians, in their 1984 submission to the federal government entitled *Democracy Betrayed: The Case for Redress*

"The world was not unaware. The Nazis had early on signalled their intent if not their methods. Yet no nation interceded on behalf of those doomed — not for lack of opportunity but for lack of will . . . Like other western liberal democracies, Canada cared little and did less. When confronted with the Jewish problem, the response of government, the civil service and, indeed, much of the public wavered somewhere between indifference and hostility. In the prewar years, as the government cemented barriers to immigration, especially of Jews,

Immigration authorities barely concealed their contempt for those pleading for rescue. There was no groundswell of opposition, no humanitarian appeal for a more open policy. Even the outbreak of war and the mounting evidence of an ongoing Nazi program for the total annihilation of European Jewry did not move Canada. Its response remained legalistic and cold."

— historians Irving Abella and Harold Troper,
from the conclusion to their book *None Is Too Many: Canada and the Jews of Europe, 1933–1948*[1]

Learning from history isn't merely about remembering what we did well. It is also about recalling what not to do. There are some deep and disturbing ironies in Canada's Second World War history. Even as the country rallied to fight tyranny and to defend democracy and human rights abroad, our government engaged in — and society at large abided or even encouraged — some terrible violations of democratic and human rights. In the name of war, ugly and racist things were done that remain a lasting source of national shame — the forced relocation of Japanese Canadians and internment of political dissidents, the suspension of civil liberties and the turning away of refugees fleeing Nazi persecution.

And so, even as this book urges our leaders today to model themselves on our wartime leaders and to take wartime-scale action, it is vital to recall the cautionary lessons, and to resolve that we will not repeat the sins of the past. We need bold leaders like we had then, but we also need them to be different.

While crisis and emergency frames carry risk, it makes sense to classify climate change as an emergency because, well, it is. The fact that such a framework also comes with hazards doesn't mean it should be rejected, merely that it should be employed with caution, with eyes wide open to how it could be misused by those in authority, so that the illegitimate exploitation of this crisis can be resisted and democratic principles protected.

CIVIL LIBERTIES SACRIFICED TO WARTIME MEASURES

As soon as Canada declared war, the government invoked the War Measures Act, allowing it to censor information and to detain citizens without charge. Even the mayor of Montreal was imprisoned for four years for speaking out against the war and conscription. About 800 German Canadians and 600 Italian Canadians were interned during the war on suspicion of being fascist sympathizers. The Communist Party of Canada was banned.

More than a hundred left-wing leaders and organizers, mostly from the Communist Party, were interned for about two years, most without charges or trials and without any press coverage at the time. Despite all being well known anti-fascists (some had even fought against the fascists in the Spanish Civil War), they were imprisoned in harsh conditions and required to join work gangs at internment camps in Kananaskis, Alberta; Petawawa, Ontario; and Hull, Quebec. They were even forced to live and work alongside Nazis and Italian fascist prisoners of war.[2] Only once Hitler and Stalin's non-aggression pact collapsed with the Nazis' invasion of the Soviet Union, making the Soviets our ally, were those political dissidents eventually freed.[3]

One of those interned was Ben Swankey. He had been a key organizer of various protests on the Prairies during the Depression, including the famous Edmonton Hunger March of 1932. When the Soviet Union joined the war on the Allies' side, members of the Communist Party in Canada like Swankey were forced to rethink their position on the war, even as they waited behind barbed-wire fences. Swankey recounts wrestling with the matter:

> For some, like myself, the need for a change in approach was not easy to accept. How could we support a government that had interned us and banned our party? How could we trust this government to really join with the Soviet Union to defeat Hitler? We debated this question back and forth. I remember one argument put forward by Bill Repka that helped to change my mind. "Look," Bill said, "when your house starts on fire, you run to ask all the

> neighbours to help put it out. You don't ask them what their politics are, as long as they are pouring water on the fire." That made sense to me.[4]

And what do you suppose Swankey did when he was finally released from the internment camp? He enlisted, of course, along with many of his fellow internees. They signed up to fight for the very country that had just imprisoned them, because the battle against fascism wasn't done. Swankey ended up moving to Vancouver in the 1950s. I met him there in the 1990s and we became friends. He was a terrific writer, ended up becoming a seniors' advocate, was a mentor to many younger activists, and he remained politically active himself well into his 90s. He never stopped fighting for a better world.

Others too were caught up in outrageous internments.

When I was hired to open the CCPA's B.C. office in the late 1990s, one of my early mentors was the late UBC economics professor Gideon Rosenbluth, a highly respected economist who served as a president of the Canadian Economics Association. In his retirement, and with an early appreciation of ecological limits, Rosenbluth was one of the first economists to work on modelling a no-growth economy with full employment (with his former Ph.D. student, York University economics professor Peter Victor).

But he came to Canada as an internee during the war. Rosenbluth was a secular Jew, born in Germany in the early 1920s. In the early 1930s, as the Nazi assaults on the Jewish community were beginning, his family escaped to London, where Rosenbluth resumed his schooling. He began his university studies in economics and statistics at the London School of Economics (LSE). Then, when the German bombing raids began at the outset of the war, for safety the entire LSE was relocated to Cambridge, where some of the giants of 20th century economics were teaching, including John Maynard Keynes.

What should have been an unparalleled experience for a young economist, however, proved to be short-lived. One day, two police officers arrived at Cambridge and asked Rosenbluth, then still in his late teens, to come down to the police station to answer some questions. That was to be his last day on the Cambridge campus. For the

British had decided to round up and intern all German citizens. The fact that some of them, such as the Rosenbluths, were Jewish and clearly not Nazi sympathizers mattered not at all — civil liberties and common sense were swept aside in wartime.

After some time interned on the Isle of Man, the British then decided to relocate many of their internees — including actual Nazi prisoners of war — to some of their "colonies." Rosenbluth was shipped off to Canada, while his brother was sent to Australia. His parents were eventually released but spent months unable to locate their sons. Rosenbluth landed initially in an internment camp on Quebec City's Plains of Abraham, before being relocated again to a camp in Sherbrooke.

After about a year and a half in captivity, Rosenbluth's family finally managed to gain his release, thanks to the sponsorship of a Jewish family acquaintance in Montreal. He quickly registered at the University of Toronto to finish his degree in economics. His first job out of school in 1943 was with the Wartime Prices and Trade Board in Ottawa (the agency tasked with wage and price controls during the war). Notwithstanding the injustice he had experienced at the hands of the Allies, like Swankey, Rosenbluth tried to enlist a couple times. But he was turned down, advised that his work with the WPTB was more important to the war effort.

All these wartime experiences were formative for Rosenbluth, not least that his time with the WPTB showed him the value and possibilities of good economic planning. But his story, like those of many others, also reminds us of what can so quickly and far too easily be sacrificed in the name of emergency.

Of all the shameful violations of human rights and liberties that marked the war years, the most sweeping and profound was surely the decision in early 1942 to forcibly relocate Canadians of Japanese descent. The Japanese bombed Pearl Harbor on December 7, 1941. One day later, Canada joined the U.S. in declaring war on Japan. Japanese Canadians were declared "Enemy Aliens," and in the months that followed, over 21,000 members of that community from coastal

British Columbia — thousands of whom had been born in Canada — were removed from their homes and deported to camps in the interior of the province (and some further east) for the duration of the war. Their homes, businesses and property were confiscated and auctioned off at fire-sale prices. Hundreds of Japanese Canadian fishermen had their boats impounded and then sold off at pennies on the dollar. Conditions in the camps were harsh and some families wrestled with hunger and malnutrition. The families were barred from returning to the B.C. coast until 1949. All of this happened without any evidence ever being produced that the community represented a security threat. As the National Association of Japanese Canadians recounts, after the expiration of the 30-year Official Secrets Act, "it was exposed that the forced deportation of Japanese Canadians from the British Columbian coast was not for military security purposes; rather, it was based on racism."

Environmental icon David Suzuki very much likes the wartime framework as a model for how we should respond to the climate crisis, and frequently invokes it. Yet Suzuki himself, who is a third-generation Japanese Canadian (his grandparents immigrated, and his parents were both born and raised in Vancouver), was deported as a young boy with his family. They spent the war in the Slocan Internment Camp in B.C.'s interior. "The war only exacerbated the bigotry that existed within the country, especially on the west coast," Suzuki told me. "It was an opportunity for a lot of people on the west coast to get rid of these guys. The RCMP knew damn well that Japanese Canadians were not a threat, as did Mackenzie King. That is very clear. I think, in the end, it was an important lesson."

David Suzuki wasn't the only internee to leave a lasting environmental legacy. One young internee was a fisherman named Tatsuro "Buck" Suzuki (no relation to David). Just before the war, at age 22, Buck Suzuki played a leadership role uniting B.C.'s Japanese fishermen with the mainly white United Fishermen and Allied Workers' Union (UFAWU). Interned for a time, he still sought to enlist in the war but was rejected by the Canadian Army. Instead, he ended up volunteering with British intelligence. After the war, Buck Suzuki continued as a union organizer with the UFAWU and became an environmentalist before the term was even coined, campaigning to preserve the Fraser River. He was one of the early activists to bridge the labour and environmental movements. The union long ago established an environmental foundation in his name, dedicated to preserving fish and fish habitats. T. Buck Suzuki died in 1977 at age 62, but the foundation continues to operate to this day.[5]

Kai Nagata brings a unique historic perspective to this matter. Nagata has deep if disparate roots in British Columbia. His maternal grandparents were white, and his maternal grandfather spent the war years with a teenage group maintaining nighttime curfews in Vancouver, before joining the air force when he turned 18. His paternal grandparents are Japanese Canadians, one side had a farm on Mayne Island and the other a small corner store in Vancouver. All of it was expropriated in the war when their families were forcibly relocated to the interior.

All of which makes Nagata attuned to the cautionary lessons from war. "The Liberal government in WWII used the War Measures Act to suspend the civil liberties of people because of their race. The historical context today has changed. I don't worry so much about my own community being targeted. But the impulse in a crisis is to narrowly define the 'we.' The rhetoric around migrants and refugees today directly echoes the newspaper editorials and political speeches about Japanese Canadians in the 1940s." And yet, Nagata too believes the emergency frame is appropriate and necessary, and he personally finds the wartime comparison inspiring.

Harold Steves, the Richmond city councillor we met earlier, has vivid childhood memories of the Japanese deportations. His hometown of Steveston, in Richmond, where his family farmed, was home to a large Japanese Canadian community before the war, mostly fishermen and their families. Steves figures about 80% of the children in the local school were of Japanese ancestry before the war. After Pearl Harbour, all their fishing boats were confiscated and the families were all uprooted to internment camps.

Steves recalls that a little girl named Fumiko Kojiro, the daughter of the Japanese school principal, was one of his first childhood friends. He can remember going with his mother to walk the Kojiro family to the train or tram that took them away when he was four years old. "The last thing Fumiko did as she got on, she handed me her teddy bear, which I kept till I wore it out as a little kid. And after that, I remember we had my birthday party, and the whole entire neighbourhood came, white kids, and I refused to play with any of them because my Japanese friends were not there."

The Steves family opposed what happened to their Japanese friends and neighbours. But they were unusual among local white families in holding such views. Indeed, the Steves family had been on the outs with most of their white neighbours for generations, due to their support for local Japanese, Chinese and Indigenous neighbours. Steves has a memory from the war years of his father (Harold Steves Senior) coming home from a community meeting about the evictions and property expropriations, "and I remember him saying to my mother, 'They called me a white Jap.' And I guess he got turfed out of the meeting."

There were a handful of white community leaders who spoke out against the Japanese internment, but they were few and far between.

These historic examples understandably make some anxious about invoking emergencies.

Matt Hern and Am Johal, in their book *Global Warming and the Sweetness of Life: A Tar Sands Tale*, articulate this fear: "In some ways we *are* facing a climate crisis that *does* demand immediate collective responses. But *crisis* invokes an emergency where debate is suspended, reflection limited, and objections marginalized . . . Declaring a crisis creates a state of exception that allows authorities to impose measures or rationalities that would not be possible during a conventional or 'normal' period.[6]

They argue further: "Does it help or hurt to call it an emergency? Does that mobilize or paralyze? . . . A planetary emergency is just that, a time when the law can be suspended, *should be* suspended, when every newsflash begs for a sovereign authority to take charge and save us from ourselves."[7]

Hern and Johal's warnings are well-advised, but I believe overstated. The reality is that climate change is an emergency. The IPCC is giving us until 2030 to at least halve our GHG emissions. Time is of the essence. An emergency framework is needed to galvanize collective and political action. The key is to stay attuned to how the crisis can be misused.

Thankfully, we have slowly learned from past times of state overreach and repression. There remains work to do, and it will always be necessary to stay vigilant in the defence of civil liberties. But Canada has been pushed to put some safeguards in place since the Second World War.

The last time the War Measures Act was invoked in Canada was during the "October Crisis" of 1970. The government of Pierre Trudeau used it to imprison hundreds of people on spurious grounds without charge or trial, wildly violating civil liberties as it sought to crack down on a radical Quebec separatist group — the Front de Libération du Québec — that had kidnapped British diplomat James Cross and killed Quebec deputy premier Pierre Laporte.

While a majority of the public supported the sweeping police powers that were "legalized" during the October Crisis, the government's actions did have its prominent critics (more so than when Japanese Canadians were imprisoned in the Second World War). And in the months and years that followed, recognition grew that the civil liberties violations of that crisis were deeply problematic. So much so that, in 1988, the War Measures Act was repealed and replaced by a new federal Emergencies Act. "Every crisis, every instance of executive overstep, and every use of emergency power thus serves as a test of public commitment to core principles and offers an opportunity to recommit," writes Nomi Claire Lazar, the author of *States of Emergency in Liberal Democracies*. "When Pierre Trudeau invoked the War Measures Act in the 1970 October Crisis, for instance, Canadians responded by replacing it with our sober, careful Emergencies Act."[8]

The new Emergencies Act contains more specific and limited state powers. In particular Cabinet orders must be reviewed by Parliament, so governments can no longer use orders-in-council to act with impunity. And the new act makes clear that government actions cannot violate the Canadian Charter of Rights and Freedoms (just as the

National Association of Japanese Canadians called for four years before the new act).

There is an inherent tension between emergency response and civil liberties, but also an evolution. A more recent example was the one noted earlier, of how, in between the British Columbia wildfire seasons of 2017 and 2018, the Tsilhqot'in Nation in B.C.'s interior developed a new protocol with the provincial and federal governments recognizing the First Nation as a full partner in emergency response.

Both the federal and provincial governments have emergency management acts, and municipal governments have equivalent bylaws, which govern who is to do what in an emergency, the temporary powers the state may need to exercise (such as ordering evacuations or taking possession of property needed to respond to an emergency), and the circumstances under which the military may be asked to help. There will surely be times in the coming years when we need to invoke these provisions in the face of climate disasters and extreme weather events. But while this book calls for us to mobilize to confront the climate emergency, it is certainly not suggesting that we must spend the next few decades in a permanent and formal state of emergency.

Fortunately, while we are confronting an emergency, its manifestations are predictable. We know we will face more frequent forest fires, floods and extreme heat. We can plan for it and develop collaborative protocols for how we want to respond when emergency flashpoints occur. But as Greta Thunberg says, "I want you to act as your house is on fire. Because it is."

Ideally though, we can and should do this together. As noted in the earlier discussion of citizen assemblies, as we mobilize in the face of the climate crisis and plan for a carbon-zero society, we would be well-served by more democracy, not less.

NEVER AGAIN: RESPONDING TO REFUGEES

Historians Irving Abella and Harold Troper published their groundbreaking book *None Is Too Many: Canada and the Jews of Europe, 1933–1948* in the early 1980s. It told a shameful chapter of our history, revealing the deeply racist views of our leaders and the profound xenophobia of much of society. The title of the book derives from a comment made by a senior Canadian immigration official: when asked how many Jewish Holocaust survivors Canada should accept, he quipped, "None is too many." As Abella and Troper document in an exhaustive exploration of government records, articles and correspondence of the time, it was a statement that captured the essence of Canada's immigration and refugee policy in the lead-up to, during and after the Holocaust, notwithstanding the fact that Canada had joined the global war against fascism.

In the face of one of the world's most infamous humanitarian crises, Canada's deeply anti-Semitic response was to slam the door shut. No countries had good track records from this time but Canada's was particularly awful. As Abella and Troper quote one leader of the Jewish community in Canada at the time, "The world is divided between that part where the Jews could not stay, and that part where the Jews could not enter." Troper says of the leadership of the day: "The government regarded compassion as a liability, with no votes to be gained and enormous votes to be lost if the doors were opened to persecuted Jews."[9] Jewish families already in Canada made frantic efforts to sponsor relatives for immigration into Canada from Europe. Virtually all these requests were denied. Even desperate pleas from Jews in Europe to allow children special entry were rejected.

The origins of Abella and Troper's book were documents uncovered about the MS *St. Louis*. The *St. Louis* was a German passenger ship that had been allowed by the Nazis to leave Germany in late 1939 on the eve of the outbreak of war. The passengers were some 900 Jews who had been issued Cuban immigration visas. But when the ship arrived in Havana harbour, the Cuban government had second thoughts and turned it away. A humanitarian appeal was issued to other countries, but all those in the Americas, including Canada, refused. A few dozen prominent community leaders petitioned Mackenzie King to accept the refugees, but their appeal was rejected. The director of the Immigration Branch, Frederick Blair, argued strongly that these Jewish refugees did not qualify for admission under Canada's immigration laws (laws he had written), stating,

"No country could open its doors wide enough to take in the hundreds of thousands of Jewish people who want to leave Europe: the line must be drawn somewhere."

The *St. Louis* was forced to return to Europe. At the eleventh hour, the passengers were divided between Britain, France, Belgium and the Netherlands — notably, the ship's German captain, Gustav Schröder, bravely refused to return the passengers directly to Germany. Those who went to Britain survived. The others, once those countries surrendered to the Nazis, largely ended up in the death camps, where over 250 were killed.

On November 7, 2018, the Trudeau government issued a formal apology for Canada's turning away of the MS *St. Louis*. In a statement read in the House of Commons, the prime minister concluded with this: "Too many people — of all faiths, from all countries — face persecution. Their lives are threatened simply because of how they pray, what they wear or the last name they bear. They are forced to flee their homes and embark upon perilous journeys in search of safety and a future. This is the world we all live in and this is therefore our collective responsibility. It is my sincere hope that by issuing this long overdue apology, we can shine a light on this painful chapter of our history and ensure that its lessons are never forgotten."

A few years ago, I heard a speech by Cindy Blackstock, the amazing Indigenous child welfare advocate, in which she offered a very simple definition of reconciliation: "Reconciliation means not having to say you're sorry — twice." Think about that with respect to these cautionary war tales. Canada has now issued a formal apology for turning away Jewish refugees aboard the MS *St. Louis*. With that in mind, what will our response be to climate migrants in the coming years, and how can we avoid a similar shame in the face of what will surely be a defining issue of the next half-century?

Earlier we discussed the intersection of climate and inequality, but primarily at a national level. At a global level, these dynamics are far more acute, and they lay bare a deep injustice — climate change's most cruel paradox — which is that those who contributed least to causing this emergency are the first to experience the harshest consequences of climate breakdown and disaster. And countries such as Canada, which have historically produced far more than our fair share of emissions and disproportionately benefitted economically from the fossil fuel economy, have, as a matter of justice and reparations, a duty to address the global consequences of the climate crisis.

The ever more devastating impacts of the climate crisis — rising sea level, lost water supplies, crop failures, hurricanes, typhoons, forest fires, etc. — will mean a growing tide of people, mostly the poor, will be on the move over the coming years, their homes no longer tenable. Exact figures do not exist, but a 2018 World Bank report focused on Sub-Saharan Africa, South Asia and Latin America — that combined represent 55% of the developing world's population — found that "climate change will push tens of millions of people to migrate within their countries by 2050. It projects that without concrete climate and development action, just over 143 million people — or around 2.8% of the population of these three regions — could be forced to move within their own countries to escape the slow-onset impacts of climate change."[10]

That last point is key. Most of those who are forced to move won't make it to wealthier places such as Canada. Rather, they will move within their country or regionally, placing additional burdens on the poorest countries least able to cope.

This is not some estimate about the future. It is already happening. According to the Geneva-based Internal Displacement Monitoring Centre, in 2018, 17.2 million people were displaced due to weather disasters.[11]

And even for those forced to move due to more "conventional" factors such as war and economic hardship, there is often a climate connection. This crisis has its fingerprints on so much of what ails our planet. Numerous analysts have noted that at the root of the Syrian civil war and the refugee crisis it has produced were years of drought that forced thousands off their farms and into cities, fuelling societal disruptions and unrest. Similarly, underlying the thousands of migrants who have moved up through the Americas to the U.S.

southern border, there are again climate-induced weather disruptions that have forced hundreds of thousands off their lands.[12]

We are looking at a future in which the climate emergency will mean thousands more people seeking to come to Canada, and many times more regionally displaced people within the poorest parts of the world needing help from wealthy countries like ours. But our track record in response to such realities has not been good. And as the CCPA's B.C. office found in a 2014 study on preparing for climate migration, the sad reality is that few if any of our public institutions (in health care, housing, education, social services) have begun to seriously turn their attention to how we will respond to a large influx of climate migrants.[13]

Kai Nagata, whose family history informs his thinking about the future, urges that we must stop speaking about Canada as "full." "Canada is too big and underpopulated to think that way," he says. "We are going to have to become a sanctuary for a lot of people whose homelands are no longer habitable or safe."

The MS *St. Louis* was not the first or last time that Canada has responded callously to the arrival of ships filled with people seeking sanctuary. Whether it was the turning away of the *Komagata Maru* in 1914, or the reaction to the arrival of the MV *Sun Sea* carrying Sri Lankan Tamil people seeking asylum in 2010, and other ships in between, when boats of people arrive on our shores, we seem to handle it particularly badly. The popular response brings to the fore our ugliest tendencies, too often animated by politicians content to capitalize on xenophobia.

Jenny Kwan, MP for Vancouver East, has served for a number of years as the federal NDP's critic for Multiculturalism, Immigration, Refugees and Citizenship, and she has been a strong advocate for immigrants and refugees. When I asked her if we are ready to deal with what will be a defining test of our values in the coming years — the movement of people displaced by climate change — she replied, "Have we learned the lessons of the past? I fear that we haven't. I look at the asylum seekers coming from the United

States under Trump and his racist policies, but we refuse to suspend the 'safe third-party agreement' with the U.S., and we have been turning people away. Are we, in 50 years, going to have to say, 'We're sorry for turning people away?' I fear we are. Our policies have not matched the nice rhetoric."

And yet, there have been historic exceptions when as a country we showed our better selves.

In the late 1970s, for example, in the wake of the Vietnam War, hundreds of thousands of Vietnamese refugees fled that country. Known at the time as the "boat people," they escaped on ships of all sizes. Many died during the exodus, as overcrowded boats contended with storms and other hazards, but most ended up in UN-administered refugee camps in nearby countries. In the face of this humanitarian crisis, a call went out to the wealthier countries of the world, asking them to open their doors.

The federal government of Pierre Trudeau initially responded in 1979 by saying that Canada would settle 4,000 Vietnamese refugees. But he had misjudged the goodwill of a public moved by the desperate images of the boat people. Marion Dewar, a deeply compassionate person who was then the mayor of Ottawa, was appalled by what she saw as a very parsimonious offer on the part of the feds. She countered with an offer of her own — the city of Ottawa would take 4,000 Vietnamese refugees, and then she challenged other cities to match them. People rallied to that appeal. Within a few months, Canada had decided to welcome 60,000 Vietnamese refugees — 15 times more than the federal government had initially offered and more than double the number of Syrian refugees we agreed to settle in 2015. And we are a better country for it.

Canada's response to Vietnamese refugees stands out for its exceptionality. In this case, the public played a very different role, demanding and demonstrating more generosity. The media caught the wave and amplified it. Faith institutions stepped up. Municipalities "competed" in a manner we need to see again. And political leadership — most notably that shown by then Ottawa mayor Marion Dewar — played a pivotal role, animating our best selves.

That's the kind of political leadership we need today.

The shameful historic examples in this chapter may feel archaic to some. Surely, Canada is not that country anymore, many of us like to believe. Yet as I write, the Canada Border Services Agency (CBSA) reports that it currently has over 2,000 people in detention, 50–60 of whom are minors detained for an average of 10–20 days. Over 100 of these people have been detained for more than 99 days.[14] Over the past year, more than 8,000 people have been detained. Between 2008 and 2018, CBSA deported about 140,000 people, and the agency has a target of removing 10,000 people a year, mostly "failed refugee claimants."[15] But "failed" merely means they do not qualify under existing rules that do not recognize those forced to flee their homes due to climate disasters. Moreover, under the so-called Safe Third Country agreement between Canada and the U.S., anyone claiming refugee status who has entered Canada via the U.S. can be rejected simply for that reason (because the U.S. is considered "safe" — a Kafkaesque barrier that allows Canada to turn away thousands without consideration of their actual circumstances). Within the mainstream discourse, remarkably little time is spent talking about these thousands of people. That is who Canada is today.

Harsha Walia, a long-time activist with No One Is Illegal, author of *Undoing Border Imperialism*[16] and the executive director of the BC Civil Liberties Association, encourages us to shift our thinking: it's not a migration crisis, "it's a displacement crisis,"[17] a distinction that highlights that people are moving because they must. And she urges that we reject a false distinction between "refugees" (who we consider deserving of our help) and "migrants" (who we see as undeserving and simply trying to skirt "the rules").

Currently, the terms "climate refugee," "climate displaced persons" and "environmental migrant" do not exist in international refugee conventions or domestic law. These are not officially recognized categories, and so we have no allowances for people in these predicaments. Surely, we have now arrived at a place where that needs to change.

Whether during the Second World War, or the years before and since, Canada's immigration rules differentiate between who is considered "desirable" and who is not, and those lines are most often racialized. And even among those "desired" for work, in recent years,

Canada has chosen to classify hundreds of thousands of migrant workers at any given time as "temporary foreign workers," people whose labour Canada is happy to have, but unlike in the past, our country is reticent to extend all the rights and benefits to which Canadians are entitled, nor do our laws offer an easy path to citizenship. These rules are unjust, and need to change. That is why migrant rights activists have called for "status upon arrival." As outlined earlier, the task of rising to the climate emergency is going to be a great job-generator, and we are going to need many more workers. And they all need equal rights.

So, what principles and policies should guide our way forward? What should we be doing to ensure that our climate emergency mobilization is also a just response with respect to those globally displaced by climate breakdown? We need to update our refugee and immigration laws and practices, and we need to substantially up our game when it comes to global financial transfers and support.

Updating Our Refugee and Immigration Laws and Practices

The 2014 CCPA-B.C. report mentioned earlier made a number of recommendations, including:

- That Canada create a new immigration category of "climate displaced people,"[18] and establish targets and programs to ensure we can settle our fair share of those forced to move;
- That all levels of government ensure that key services — legal, settlement, housing, health and education — are preparing for climate migrants and available to meet the needs of a larger group of displaced people, and funding should be increased to reduce the strain on these already-overloaded systems; and
- Canada must open itself to more migrants. A new immigration category would help, and we should ensure that we are opening our doors to the most needy, not merely those with wealth and professional qualifications for whom access

> is generally easier. Those who come to Canada as climate displaced people should be welcomed *in addition* to our existing immigration and refugee numbers.

Opening our doors to more people displaced by the climate crisis is not only the right thing to do, it will also make us better people, and a stronger country. As emergency planner Anthonia Ogundele put to me, "There are things to be learned about resilience from those who were displaced, from the people who left their world to come here." And greater resilience is a skill that will be much needed in the climate emergency years to come.

Those who have experienced climate breakdown also bring values and attitudes needed to expedite emergency action. In the Abacus poll I commissioned in summer 2019, people were asked to identify if they were first- or second-generation immigrants to Canada, and if so, what geographic region they or their parents had come from. Across almost all the questions, the results for these respondents were either the same as for the overall population, or indeed, these first- and second-generation immigrants were found to be more supportive of bold climate action, particularly those from Southeast Asia, where countries like the Philippines have experienced some of the harshest climate-induced weather events.

Increasing Aid to Other Parts of the World Dealing with Climate Displacement

Recognizing that most climate migrants will remain in the Global South, it is also vital that Canada increase its support to developing countries shouldering the burden of climate displacement. Now is the time to quickly get Canada's aid budget to the long-established UN goal of 0.7% of GDP, minimally.

The UN climate negotiations have begun the process of wealthier nations making financial and technology transfers to poorer nations such that they too can meet the climate challenge, but the amounts in play thus far are not nearly adequate. Some, however, have started to talk up the need for a *Global* Green New Deal (or a new Marshall Plan for the Planet). Bernie Sanders in particular has proposed that the U.S. contribute $200 billion to the UN's Green Climate Fund, to help poorer nations transition.[19] This is a welcome development, but not one that any leading politicians in Canada have taken up. Rather, at the G7 meeting in summer 2019, the Trudeau government committed a mere $300 million to the UN's Green Climate Fund,[20] an amount that is, frankly, embarrassing given our wealth and history. Moreover, we should ensure that all of Canada's Official Development Assistance and other overseas initiatives are fully aligned with global climate goals.

During the Second World War, Canada managed to be remarkably financially generous, despite the demands we faced financing our own war effort. Canada provided Mutual Aid to our allies during the war of $3 billion — a massive sum at the time, equivalent to about $45 billion in today's dollars — which went mostly to Britain, but also to other allied countries in need. As the war progressed, the Allied powers, anticipating the scale of rebuilding that would be needed after the war, established the United Nations Relief and Rehabilitation Administration (UNRRA), an international disaster relief organization specifically tasked with aiding victims of war and millions of displaced people. The countries that made up the alliance were all asked to contribute funds, resources and food, which they started to do even before the war was over. In 1943, Canada contributed 1% of its national income (GDP in today's terms) to the UNRRA and did so again in 1945.[21] In contrast, today, Canada gives only 0.26% of our GDP (or about $6 billion a year) to international development assistance.

The point being, once again, that when truly mobilizing in the face of an emergency, in common cause with others, you find the money to do what is needed.

Personally, I am guilty of seeing all these immigration debates through the lens of my parents' experience — the story I recounted in the preface of how my mom and dad arrived as Vietnam War resisters in 1967 and received landed immigrant status at the airport in a mere 20 minutes. As many as 90,000 Americans came to Canada during the Vietnam War.

The early government of Pierre Trudeau welcomed Vietnam "draft dodgers" (as they were called) and conscientious objectors, declaring in one address that "our political approach has been to give them access to Canada. Canada should be a refuge from militarism."[22]

That was our response once before — why not again?

As Abella and Troper write in an epilogue to the third edition of their book, well before anyone was thinking of climate migration, "We do not pretend to possess any magic formula that will solve the deepening refugee crisis or Canada's role therein. But if the story told in *None Is Too Many* has any moral, it is that Canada should never again turn its back on those in such need."

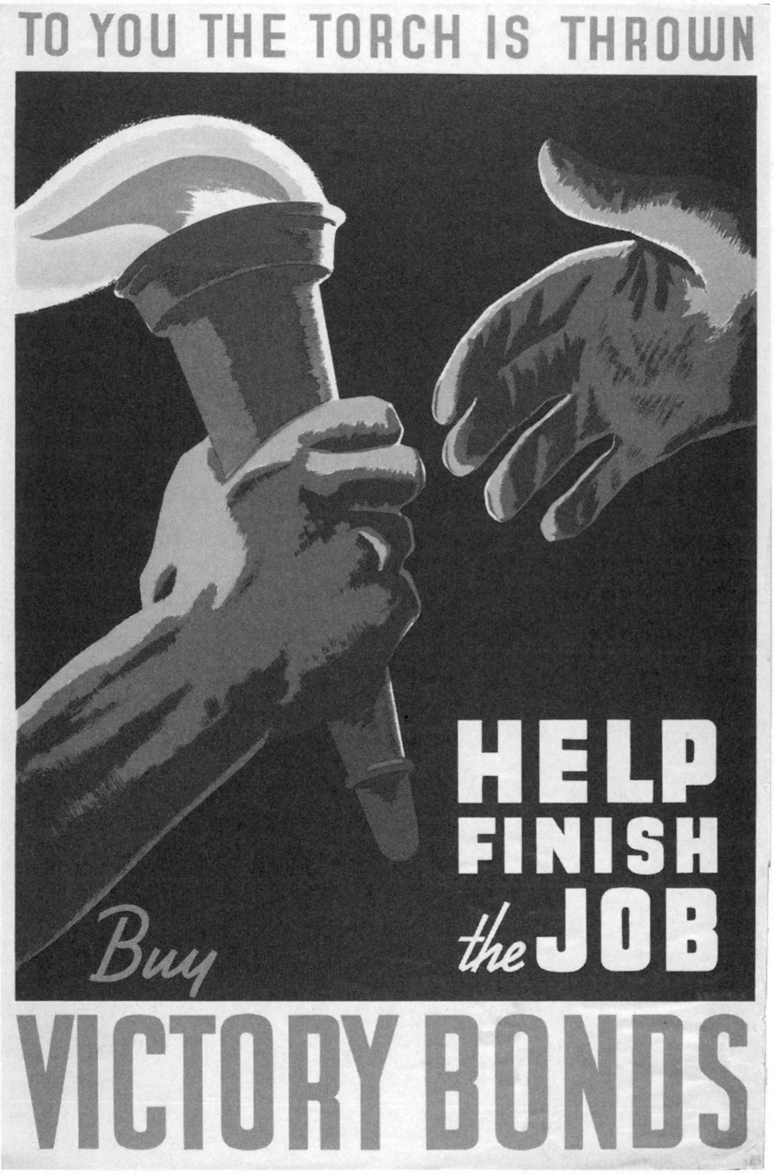
TO YOU THE TORCH IS THROWN
HELP
FINISH
the JOB
Buy
VICTORY BONDS
McGill University Archives

CHAPTER 12

Transforming Our Politics: Bold Leadership, Then, There and Now

"A reluctance to face the full magnitude of our task and overcome it is a coward's part. Yet the nation is not in this mood and only asks to be told what is necessary."

— John Maynard Keynes, from the preface to his short book *How to Pay for the War: A Radical Plan for the Chancellor of the Exchequer*, published in early 1940

"Dear Member of Parliament–elect: We are excited to welcome you to serve in our Parliament. You are now responsible for ensuring Canada meets the climate crisis at the speed and scale that science and justice demands. You ran for a political party that promised to address the climate emergency. Some of you might not have been serious about it. Others pledged that in this parliament, you would be bolder. We expect you to follow through . . . If you are on our side, you can count on us to support you every day. If not, know that you have our entire generation to contend with. Together, we can tackle the climate crisis."

— from the "Mandate Letters" delivered by the 350.org Our Time youth activists to newly elected MPs on October 28, 2019, a week after the federal election

As we know from our wartime story, political leadership matters. In the face of the climate emergency, what does bold leadership today look like? Can we find Canadian political leaders ready to rise to this challenge and truly mobilize us? Will our leaders give our kids and grandkids and those who will live out the balance of this century a fighting chance?

This book has explored three kinds of barriers to the transformative change we need: cultural, economic and political. Solutions have been offered along the way for overcoming each of these barriers. But the obstacles that plague our politics call for a few more solutions — and a new mindset.

THE LEADERSHIP QUALITIES THAT SAW US THROUGH THE WAR

"Study history. Study history. In history lie all the secrets of statecraft," Winston Churchill said in 1953.[1]

Canada in the Second World War was not led by nearly as charismatic a leader as Churchill or FDR. Mackenzie King remains Canada's longest-serving prime minister, which speaks to his political skills. But he was not a great orator, nor was he particularly popular, though he was respected. He was cautious by nature and not always decisive, but skilled at compromise. He was not without faults and renowned for his idiosyncrasies (such as seances and seeking to commune with the dead). Shortcomings aside, Mackenzie King oversaw what endures as the largest mobilization in Canadian history. And he did it while maintaining national unity.

Mackenzie King did not wish to be a wartime leader and resisted being drawn into a battle with fascism. Yet when that could no longer be avoided, he and his team threw themselves into the fight, and rallied the rest of the country along with them. When necessary, they were prepared to take great political and financial risks. They communicated to the public the urgency and severity of the situation and imbued in society a sense of common purpose, a core task of great leaders. Our leaders managed to galvanize and mobilize Canadian society into an all-consuming effort despite the fact that the final outcome was unknown. So much so that over one million Canadians enlisted to join the war effort, prepared to make the ultimate sacrifice in support of that shared imperative. And here we are again as we face the climate crossroads — we know that a society-wide mobilization and wholesale economic transition is necessary, and we are required to act despite not knowing if we will make this leap in time to avoid catastrophic consequences.

When Mackenzie King died in 1950 and his body lay in state in the lobby of Parliament, over 40,000 Canadians lined up to pay their respects.

Of course, government leadership isn't about a single person. There were the dollar-a-year men recruited to help lead the monumental munitions production effort, and the managers and engineers who oversaw that work. And there were remarkable people within the existing civil service. The team of civil servants who oversaw the finances of Canada's war effort (including taxes, war bonds, international exchange matters, financial aid to Britain, price controls and Canada's participation in the negotiations that created the global financial institutions formed at the end of the war) was tiny. Historian Jack Granatstein writes of them, "The extraordinary fact was that all the major financial policy-makers in wartime Ottawa could have crowded around a table in a medium-sized conference room . . . Moreover, what united these men [and they were all men] was their capacity to work hard (for low salaries!), to put in fourteen-hour days seven days a week, to master complex issues quickly, and to argue complicated positions to each other and to explain them to their political masters."[2] Beyond their dedication, these civil service leaders were prepared to take chances and to consider economic and financial ideas that had never been tried, at least not at the scale they undertook. They were prepared to shake up and direct the entire economy. Do we have such people today in the senior ranks of the federal and provincial civil service, ready to meet the current emergency?

Canada's civil service war team reported to a very talented and effective group of ministers in the War Cabinet, notably Howe (in Munitions and Supply), Ilsley (Finance), Ralston (National Defence), Louis St. Laurent (Justice) and of course Mackenzie King himself.

That the years that followed the war were characterized by such economic prosperity speaks volumes to how effectively these people had managed the far-reaching transition to war, and then back again to peacetime. "They had won the economic war at home," writes Granatstein, "and they had properly positioned Canada for the new post-war world."

The takeaway: with political will and courage, a preparedness to embrace ambitious ideas, and the right leadership and talent, undertaking transformative change in the face of an emergency can be done.

BOLD IS CATCHING ON: POLITICAL LEADERS ELSEWHERE WHO ARE THROWING OUT THE RULEBOOK

Outside Canada, a new politics is emerging, one that is attracting huge excitement while practising an unabashedly progressive politics we have yet to witness in Canada. Among our old wartime allies, the Bernie Sanders campaign in the U.S., the U.S. congressional campaign of Alexandria Ocasio-Cortez (AOC) and her subsequent championing of a Green New Deal, and recent U.K. Labour campaigns[3] — all have galvanized enormous intergenerational and multiracial constituencies. Notably, these campaigns speak to the twin crises of inequality and the climate emergency, and have integrated both into a compelling agenda. Their policies are further left than we have seen from these parties in decades.

Beyond our borders a political sea change is underway. In the wake of both economic and climate breakdown, people are increasingly rejecting the centrist elites who uphold neoliberal doctrine. We see hints of this in Canada, in that the two federal mainline parties — the Conservatives and the Liberals — have secured a smaller combined vote share in the last two decades than the historic norm. But these trends are more pronounced outside Canada.

The direction of this political upheaval is unknown. It could go somewhere terrible and hateful (witness the likes of Trump in the U.S., Narendra Modi in India, Jair Bolsonaro in Brazil, Rodrigo Duterte in the Philippines and various incarnations of neo-fascist leaders across Europe), or it could result in something inspirational. Such are the unpredictable and momentous times in which we live. Much like the Depression years of the 1930s, which gave rise to both the terror of fascism and the hope and promise of the New Deal, this volatile era in which we find ourselves could go either way.

One of the refreshing things about these new political campaigns on the left in the U.S. and U.K. is the way in which they approach the relationship between electoral and social movement politics, seeing them not as conflictual but as necessarily complementary. They all see their political projects as deeply rooted in and dependent upon grassroots movements, and this is manifested in how they organize and fundraise, as well as in their platforms. This stands in stark contrast to the traditional U.S. Democratic party establishment, the U.K. Labour party under Tony Blair and his immediate successors, or the insiders who tend to control the Liberal Party and NDP in Canada, all of whom view social movements with a level of suspicion, to be kept at arm's length, tapped when convenient but stifled often, and mostly told to dampen their expectations.

AOC in particular is surely one of the most extraordinary political talents we have witnessed in years — immensely charismatic, a captivating speaker, amazingly fast on her feet and astonishingly clear about her politics and what she is seeking to accomplish. Despite being the youngest-ever person elected to the U.S. Congress, she has a deep knowledge of social movement history, which she weaves into her speeches and her political practice.

Of particular relevance to this book, these campaigns have explicitly run on platforms that name and treat the climate emergency as the existential crisis it is. They have fully embraced the Green New Deal; they are seeking to embed IPCC-compliant targets into law; they have highlighted the role of a GND as a massive job-creation endeavour and an historic opportunity to build a more equitable society; and they are advancing policy solutions that align with what the science demands.[4]

All of which begs the questions: Why aren't these leaders plagued by the cautionary impulse that afflicts progressive parties in Canada? Where is our AOC? These contemporary campaigns are breaking with the long-established consensus within their parties and throwing out the rulebook that warns leaders not to stray too far from the status quo or pique the ire of the mainstream media. Yet these leaders are filling stadiums in a manner we haven't seen in Canada for decades, and their campaigns are catching fire, particularly among millennials. They have clarity about who they are there to fight for, and who they seek to confront. And with respect to the climate emergency, these leaders have managed to break free from the ties that bind and limit the climate debate. They do not suffer from the new climate denialism.

OTHER JURISDICTIONS TAKE ACTION

Our sense of what is possible is contained by what we know. So it is important to know that, beyond our borders, other jurisdictions are taking action on the climate emergency in a manner that is considerably more aggressive than we've seen in Canada. In other industrialized democracies such as ours — Sweden, Norway, Denmark, Finland, the U.K., Ireland and New Zealand — we've seen leaders willing to take political risks in the face of the emergency. In some of these countries, unlike in Canada, they have adopted "keep it in the ground" policies (fracking bans, offshore drilling prohibitions and wind-down plans), walking away from known fossil fuel reserves.

Norway has a long history of doing a much better job than has Canada of ensuring that the public captures the bulk of the returns from fossil fuel development. Consequently, Norway now claims the world's largest sovereign wealth fund with which to fund public services in the future and pay for the transition off fossil fuels. The fund is investing heavily in renewables, while divesting from oil and gas (including here in Canada).[5] While the country has long been western Europe's largest petroleum producer and, distressingly, its offshore production is still slated to climb through 2030, Norway is now walking away from billions of barrels of oil, having decided in 2019, with both Conservative and Labour party agreement, not to pursue offshore oil exploration in the Lofoten Islands area of the Arctic.[6] Norway has an ambitious plan to electrify its ferry fleet (using battery technology from Richmond, B.C., no less), and in 2018 also announced that all of its short-haul plane trips (flights under 1.5 hours) would be 100% electric by 2040.[7] Whereas in Canada about 3.5% of new vehicle sales are electric, in Norway, about 50% of new cars and 10% of all cars on the road are electric — the highest share in the world — and its parliament has legislated that all new cars must be electric by 2025, 15 years sooner than British Columbia, the only province in Canada with a legislated date of this kind. This is supported by an extensive system of charging infrastructure. Various policies (the removal of parking spots, banning cars from many streets, heavy investment in public transit, and new bike infrastructure) have also removed much of the vehicle traffic from downtown Oslo. Interestingly, this is occurring under a Conservative prime minister leading a minority government.

Unlike Canada, Sweden, home of Greta Thunberg, has actually exceeded its Kyoto Protocol commitments. Their GHG emissions have dropped by 26% since 1990. Sweden has banned all new fossil fuel vehicle sales as of 2030. Like Norway, Sweden is swapping out its ferry fleet to electric power by 2045, with the conversion of the first ships already underway.[8] The country seeks to be entirely off fossil fuels by 2050.

In November 2019, the New Zealand parliament voted to enshrine in law a commitment to achieve net-zero GHG emissions by 2050. Unlike the political divisions that have marked the climate debate in Canada, the bill was introduced by Prime Minister Jacinda Ardern's Labour government (a coalition that governs in cooperation with the Green Party) but also received cross-party support from the conservative National Party. In 2018, Ardern's government banned new offshore oil and gas development. About 80% of New Zealand's electricity is already derived from renewables (mostly geothermal and hydro), and the country is aiming to increase that to 90% by 2025, with a new focus on wind power.

Climate action in the U.K. has similarly managed to secure support from across the political spectrum. In spring 2019, the U.K. Parliament became the first national government to pass a climate emergency declaration. In November that year, the country announced a fracking moratorium. GHG emissions in the country have fallen 41% since 1990 (while Canada's have climbed 19% over the same period, although that growth all occurred in the 1990s), an impressive feat driven mainly by the phasing out of coal-powered electricity, but also by lower fuel consumption by industry, lower electricity consumption by residential and business users, and lower transportation emissions. Climate researcher Barry Saxifrage notes that the U.K. produced almost 200 megatonnes more GHGs than did Canada in 1990. But by 2017, the U.K. was producing almost 250 megatonnes less than us, despite having a larger economy and nearly double the population.[9] The U.K. has a carbon budgeting law, overseen by an all-party committee and independent audit, helping to ensure the country meets its GHG reduction commitments.

Much more is happening elsewhere at a pace that is leaving Canada behind. The Social Democrat–led minority government in Denmark has upped its GHG reduction target to 70% by 2030, while the Social Democrat–led minority in Finland is now aiming for carbon-zero by 2035. Unlike Canada, both these countries have already achieved steep GHG reductions since the year 2000, and in Denmark's case, its state-owned energy company has dramatically shifted from coal to offshore wind power. California, with a population roughly the same as Canada, has seen its GHG emissions drop over the last ten years, and currently its total emissions are about 60% of Canada's. In fall 2019, California governor Gavin Newsom announced a moratorium on fracking.[10] Cities in California are now passing bylaws disallowing any new buildings to tie into natural gas lines, while New York City has passed a law requiring all existing large buildings to dramatically lower their GHG emissions. Nearly 100 cities and towns around the world are now providing fare-free public transit.[11]

To be clear, none of these national plans are yet sufficient. Nevertheless, they are a good deal more ambitious than what we've seen so far in Canada, and they represent the kind of meaningful progress upon which even greater ambition can be credibly built. So it is worth asking: what has allowed politicians in places such as these to take more decisive action?

Angela Carter is a professor of political science who is studying jurisdictions that are instituting "keep it in the ground" policies and laws, seeking to understand the political conditions that have resulted in some political leaders turning away from the powerful oil and gas industry and consciously choosing to leave known reserves of fossil fuels where nature buried them away millions of years ago. Now a professor at University of Waterloo, Carter hails originally from Newfoundland, where her childhood experience of seeing the economic and social costs of the collapse of the cod fishery sparked her interest in how societies can escape the boom-and-bust roller coaster ride of economic dependence on a single resource sector. Carter's home province hasn't been giving her much hope these days, given the new climate denialism that prevails there. So Carter has shifted her gaze elsewhere, looking for inspiration, and has found it in the places where, in fits and starts, "keep it in the ground" policies are taking hold.

Take, for example, the case of Ireland.

Despite its known and potential fossil fuel reserves, Ireland is turning away from extracting them. In 2017, the Irish Parliament passed an anti-fracking bill. The next year, it passed a fossil fuel divestment bill, the first national government to do so.[12] Also that year, the legislature began debating a Climate Emergency Measures Bill — a one-page piece of legislation that is "beautiful in its simplicity," according to Carter — which states that, so long as global carbon in the atmosphere remains above 350 parts per million, Ireland will ban new offshore oil and gas drilling, effectively closing the door on the development of major new oil or gas projects. Irish Taoiseach (the term used for prime minister) Leo Varadkar and his Fine Gael party initially attempted to block the bill's advancement, but then, under pressure from rising mobilization for climate action, in late September 2019, Varadkar announced at the UN Climate Summit that his government would be ending offshore oil exploration. Each new act — from anti-fracking to divestment, and now a "keep it in the ground" bill — builds upon the last. "These things are scaffolded on each other," describes Carter.

In 2016, the Irish government struck a citizens' assembly tasked with exploring a number of contentious issues, including what policies the country should adopt to drive down GHG emissions and make Ireland a climate leader. In 2019, the citizens' assembly issued its climate report with 13 strong policy recommendations.[13] The assembly's willingness to make such recommendations may give political cover to the politicians as they consider their next moves.

There is a small oil and gas industry in Ireland — the Corrib offshore gas field will be producing for another 10 to 15 years — and the industry did indeed attempt to defend offshore drilling and block passage of the Climate Emergency Measures Bill. "Oil industry lobbying of parliamentarians increased seven-fold" in reaction to the bill, notes Carter. But the industry is not powerful (as it is in Canada).

Ireland's minority coalition government was another piece of the equation, giving the left parties and Green Party more influence. But conservative pro-business parties still dominate Ireland's legislature. The key, in Carter's view, was pressure from below, from local residents and social movements.

The first notable law to pass, the anti-fracking bill, "originated from pressure by local communities, which happened to be in more conservative ridings, that did not want to risk their water being contaminated by fracking. They were concerned about property values and quality of life, really local issues," recounts Carter. "There were some very politically savvy local people who had close political connections to the leading Fine Gael and Fianna Fáil conservative parties, which are by no means tree-hugger parties. So local citizens were able to effectively press their politicians to make sure fracking never happened in their region, and they scaled that up to a ban on fracking across the country." According to Carter, there was a minor but noteworthy Canadian angle to this story too, as Irish community activists opposed to fracking connected with Albertan anti-fracking activist Jessica Ernst — the industry insider whose story is told in Andrew Nikiforuk's 2015 book *Slick Water*. Ernst was invited to go on an Irish speaking tour, and her talks warning of the devastation caused by fracking in rural Alberta left people aghast — and mobilized to fight it.

After they scored this win, recounts Carter, "some of the activists and organizers set their sights higher." The momentum from this political win went into the push for the divestment bill, and later the Climate Emergency Measures Bill banning offshore development. This evolution of wins is similar to what has happened in Quebec.

The role of students and youth has been fundamental. The divestment act grew out of student-led divestment campaigns at the university level. And in the more recent debates over the bill, the politicians, according to Carter, "really feel the pressure of the students, the school children at the gates of the legislature. They are really hearing that call. In fact, one of the Fianna Fáil [one of Ireland's dominant conservative parties] members I spoke to in Dublin said his daughter is one of the school strikers. Mobilization for bold action to confront the climate crisis is now at the breakfast tables of political leaders who would otherwise ignore it." Again, this echoes Vancouver's recent experience with its climate emergency motions.

This widespread social mobilization has forced Irish politicians to act. Carter said that, "while the industry lobbying has certainly intensified, at the same time there has been an intensification of the Extinction Rebellion and climate strike movements, which has caught the attention of parliamentarians. Irish people generally have not been ones to protest in the streets, but suddenly school children and grandmas are coming out in weekly strikes all across the country. Politicians now realize they need to demonstrate real climate leadership."

Also key to the climate movement's success in Ireland has been the role of Catholic faith leaders, and specifically Trócaire, "a Catholic organization that is well-regarded in Ireland. They do missionary work globally and they have made a commitment to acting on the climate crisis. In the countries in which they work in Africa, they see first-hand the impact of drought caused by climate change. They've connected those issues to Ireland's role in the crisis, and are now pressing for change at home. This is the organization that has been behind the scenes, and sometimes very much in front, bringing together coalitions to get these bills passed," said Carter. Pope Francis's 2015 encyclical *Laudato si'*, imploring action on climate change, may have helped Trócaire move this issue forward in a country where the Catholic Church continues to be prominent. And it raises the question of why faith leaders in Canada have not been more vocal in the fight for climate action. Carter's advice: "Find your Trócaire."

CURBING THE POLITICAL POWER OF FOSSIL FUEL CORPORATIONS

Earlier we explored the work of the Corporate Mapping Project (CMP), which has systematically exposed the power and influence of the fossil fuel industry in Canada. Recall that the CMP highlights four "modes" of industry power — political power, economic power, cultural power and colonial power. Having revealed this "regime of obstruction" to climate action exercised by the oil and gas companies, the CMP is also developing concrete ideas and policy proposals for how this power and influence can be reined in.

With respect to the industry's *political power*, most jurisdictions in Canada have already banned corporate and union donations to political parties, and we have public lobbyist registries that, when followed at least, require that industries record their contacts with government officials and the broad topics of their conversations. But more is needed to restrict the more nuanced forms of political influence.

A 2019 CMP report recommended tougher rules for lobbyist registries: "The current federal Registry of Lobbyists does not require lobbyists to provide detailed information about their communications with state officials and there is a lack of detailed description in the current registry of the nature of meetings held. Additionally, the names of the individual lobbyists involved in meetings and the full disclosure of the costs of lobbying should be reported."[14]

We need to curtail the ability of industry lobbyists to consume the agendas of government ministers and senior civil servants. CMP co-director Shannon Daub acknowledges doing so is challenging. In theory, everyone should have access to government, and we wouldn't want government to regulate an industry without first talking to it. But we should seek to *limit* lobbying. "You could put an onus on government for 'proportionality,'" suggests Daub. "So for every meeting that you have with one particular type of interest, you would have to have a meeting with the other key interests. If you are going to meet with an energy corporation, then you would have to meet with a relevant Indigenous group, and with environmental groups."

We also need tougher laws to not only slow but prohibit the revolving door between government and industry; a person employed by government cannot also lobby on behalf of the fossil fuel industry any time within recent years. We must ensure the government agencies or ministries responsible for regulating the fossil fuel sector are not captured by industry; the public interest cannot be corrupted. It is particularly in the shadowy realm where regulations are written, with much less public scrutiny, that industry lobbyists act to protect corporate interests and dilute the effectiveness of new laws. And we need tougher conflict of interest rules, more robust access to information rules and stronger whistleblower protections.

With regards to *economic power*, we need our economy to be less reliant on fossil fuels, thereby supplanting the industry's economic power. There are policy agendas, such as a Green New Deal, that are "transformative" in this regard, says Daub, "that just change the game fundamentally. The Green New Deal, while not ignoring industry and its role, completely changes the conversation. It recaptures the jobs narrative away from the industry."

Divestment campaigns also help to curtail the industry's economic power. And those campaigns should target both the fossil fuel companies and their financial sector enablers. University of Manitoba sociology professor Mark Hudson notes that, while the oil sands are dependent on financing from Canadian financial corporations (as international investors are pulling out), the relationship is not reciprocal — Bay Street is not dependent on the oil sands. Therefore, public divestment campaigns should seek to make it politically and reputationally too risky for financial institutions in Canada to capitalize fossil fuel expansion.

To curtail the industry's *cultural power*, it will take social movements and people power to really change the dominant narrative about climate and the future of fossil fuels. But the news media should also change its reporting of the climate crisis, and new advertising rules would help — banning fossil fuel car and gas station ads from TV, radio and films, and tightening up truth in advertising rules. People have a hard time imagining a zero-carbon future, and so the arts and culture industries can certainly help to paint a picture of what that new life could look like.

And to begin to rein in the industry and government's *colonial power*, "we have to decolonize," says Daub. "We need to return land and resources back to Indigenous people and Nations, and then we have to figure out how to live with the consequences, which will not always be what non-Indigenous people think should happen." Truly implementing the United Nations Declaration on the Rights of Indigenous Peoples would be transformative — fundamentally shifting economic power away from resource corporations.

ALIGNING OUR POLITICAL PARTIES WITH THE EMERGENCY

So far, we are living in our Phony War phase of this fight. Politicians increasingly speak to the urgency of the crisis and are passing climate emergency declarations, but this fight has been rather bloodless, or indeed rather one-sided. The climate-denying right has been taking off the gloves — setting up "war rooms" to defend the oil industry and marshalling an army of trolls to spread misinformation and to harass people on social media. But the fight has not been enjoined. The response has been too polite, technocratic and lacklustre, not brash and visionary. Too much Chamberlain and not enough Churchill.

When this book is published, we will have ten years remaining on the IPCC clock to reduce Canada's GHG emissions by *at least* 50%, but preferably by considerably more if we are to do our fair share and increase the odds of keeping global temperature rise below 1.5°C. We will also be one year into the life of the federal Liberals' latest minority government. It is *vital* that civil society actors and our political leaders make the most of this minority government moment. The good news is that a sizable majority of Canadians voted for strong climate action in the last federal election, even if that desire was fractured across four political parties.[15]

Now is the time for our purportedly progressive political leaders

— from all those parties that claim to understand and accept the reality of the climate emergency — to push past their cautionary impulses, to stop seeking appeasement with the fossil fuel industry and to become the leaders this moment demands. We need them to take up the call of this good war, and to fight.

With this in mind, some reflections on where our main federal political parties are and need to go with respect to the climate emergency.

The Liberals

The federal Liberal party and cabinet are populated by people who get the climate crisis — people who understand the science and claim to feel a sense of urgency. The Pan-Canadian Framework on Clean Growth and Climate Change begins to move us in the right direction. Yet the level of ambition that has marked the Liberals' time in government has been woefully inadequate, and the willingness to invoke a true sense of emergency and rally the Canadian public entirely absent. The Liberals seem fundamentally unwilling to challenge us. They see themselves as acting at the outer edge of the envelope of what is politically possible — of what the public is prepared to hear and of what confederation is prepared to sustain. But they show no willingness to expand that envelope — what is sometimes called "shifting the Overton window" in politics — let alone to ask, "How can we blow open the collective sense of possibility?"

Like so many before them, the Liberal government is practising the politics of appeasement, guided by considerations such as how to keep the fossil fuel industry satisfied, all the provinces on board, the U.S. on friendly terms and the public happy.

In times of emergency, these are the wrong questions. This moment calls for a different kind of leadership.

The Liberals' Trans Mountain pipeline policy was seen as necessary to secure support from the Prairie provinces. Instead, it proved a grave political mistake. It won the Liberals no love in the Prairies. In the 2019 election the Liberals lost all four of their seats in Alberta, their only seat in Saskatchewan, yet also paid the price in other provinces, with seven seats lost in B.C. and six in Quebec, as progressive climate voters turned their backs on the Liberals.

Today's federal Liberals have convinced themselves that the public is not ready for truly bold action. But as UBC political scientist Kathryn Harrison notes, in our divided politics, it is often just a minority of voters in the middle who determine political outcomes. "You do not need everyone. You need enough. You need the marginal voters." Harrison sadly concludes that the Liberals are simply not prepared to be IPCC-consistent. They have continued to peddle the false message that we don't have to choose between protecting the environment and oil sands expansion. That we don't have to fundamentally change our lives. But we do. If we are to have a Liberal prime minster who leads us through the current crisis, as we did the last one, we need to see a different kind of emergency leadership.

The NDP

The NDP is also plagued by that cautionary impulse — an insecure desire to not appear too radical, even when, as occurred in the 2015 federal election, that results in being outflanked on the left by the Liberals. The NDP's 2019 election campaign was certainly an improvement; the party advanced some bold proposals, the overall platform was the most proudly progressive we have seen for some time and leader Jagmeet Singh situated the party's positions in contrast with the wishes of big corporations. That's the direction the party needs to keep pursuing.

But amplifying such an orientation won't be easy. Federally and provincially, core party insiders — those with the ears of party leaders — and some of the leaders themselves simply do not get the climate emergency. They presume that the public is not ready for ambitious action, and that saying what needs saying would be political suicide. They envision a slow process of "bringing people along" that is simply not aligned with what the science requires of us. They correctly understand that workers currently tied to the fossil fuel economy need to see good alternative jobs, and that most of the public needs to see that the transition is fair and affordable, yet they have been reluctant to advance the bold jobs and progressive tax reforms needed to make this so.

For many of these leaders, senior staffers and insiders, the prospect of landing major fossil fuel investments — an LNG plant or oil sands development or pipeline or offshore project — remains too enticing. They accept a core neoliberal presumption that only the for-profit sector creates wealth and jobs — they can't see an exciting role for government itself, as occurred in the Second World War. And so, they latch on to these fossil fuel projects as the only major possibility on offer.

As one senior insider told me in regard to LNG Canada's project, "We just felt we couldn't turn it down," although this belies the reality that the B.C. NDP didn't merely say okay to LNG Canada — it substantially sweetened the pot with new subsidies and tax breaks beyond what the B.C. Liberal government had previously offered. "When an industry comes in, they do get breaks. That's the reality. But politically we couldn't walk away from this." Yet the math and science of that project simply do not work if one is serious about the IPCC goals — the economics of LNG Canada see it producing well beyond 2050, the date at which we must be net carbon-zero. "I guess that's something we'll have to reconcile," my source conceded.

Some NDP leaders allow segments of the private sector labour movement (albeit a shrinking minority of the labour movement) to dictate support for fossil fuel projects. And some remain overly influenced by the fossil fuel lobby, which, as discussed, is extraordinarily organized and effective, notwithstanding the recent bans on corporate donations to political parties. "People open the door for them," that same source made clear.

All these factors — and people — hold the NDP back from doing what this moment demands of us. These folks seem prepared to hold these views into political irrelevance. And given the choice between landing investments in the present and choosing a safe future, they are making the wrong call.

When I spoke to Libby Davies, my MP for 18 years, she insisted the rank-and-file membership of the NDP "get it," and at party conventions express a clear desire for strong climate action. But she sees the party leadership and insiders as considerably more cautious — worried about offending provincial wings of the party, anxious about the reaction from some traditional elements of the labour movement and intent on holding MPs and even the party leader to much more tightly controlled messages than in the past. Given this, she sees a combination of internal and external pressure as necessary to produce bold climate leadership.

There is no disputing the fact that when an NDP government is elected provincially, it faces unique challenges, with competing demands from activists on the left that they be more ambitious, and pressures from the right that they not disrupt the economy. Often, when an NDP government is elected, it quickly starts to map out a pathway for securing a second term in office.

In contrast, there is a story about the provincial NDP government of Dave Barrett that those seeking audacious policy action are fond of recalling. In an unexpected outcome, Barrett won government in British Columbia in 1972. As occurred for Notley years later, the right had been divided, allowing the NDP to surprise almost everyone — including itself — and secure a majority of seats.

As the story goes, at the first meeting of his new cabinet, Barrett, who was a very humorous and charismatic fellow, removed his shoes, climbed onto the cabinet table, took a running slide across the table in his socks and then looked down upon his newly elected team and asked, "Are we here for a good time or a long time?" (five years before those words would be immortalized in the 1977 Trooper hit song). It was a rhetorical question, of course. The implied answer was the former, and the Barrett government, arguably, then proceeded to enact more progressive legislation and establish more progressive institutions that have stood the test of time — the Agricultural Land Reserve, public auto insurance, major new provincial parks, a public ambulance service and much more — than any B.C. NDP government since.

Today, of course, we only have a short time.

The Green Party

Not surprisingly, the Green Party of Canada has the strongest climate policies of any federal party — packaged ahead of the 2019 election in a set of proposals called "Mission Possible" — although even its fiscal commitments to the emergency leave something to be desired (hamstrung as they are by a misplaced commitment to balanced budgets). In their 2019 federal platform, the Greens went further than any other party in raising taxes on corporations and the financial sector, although their taxation policies remain unnecessarily cautious.

Of course, I have a soft spot for the fact that the Greens framed their entire 2019 election platform around the Second World War analogy. Then-leader Elizabeth May's introductory message was all about what wartime can and should teach us about the need to rally in the face of an emergency, even when surrounded by naysayers who insist it cannot be done.

The party declared in 2019:

> Mission Possible is less about the original New Deal, Franklin D. Roosevelt's massive public works program to lift post-Depression America out of poverty, and more about Churchill's courageous World War II campaign to defeat fascism. It places Canada on something equivalent to a war footing to ensure the security of our economy, our children and their children — our future. It is a call for "all hands on deck." It incorporates all the requirements for economic justice, just transition, the guarantee of meaningful work, while also respecting the United Nations Declaration on the Rights of Indigenous Peoples. It recognizes that we cannot achieve climate security in the absence of equity.

This is great language. My analysis of the Green Party's costed platforms in both the 2015 and 2019 federal elections, however, was that they suffered from a gap between the lofty words and the budgetary commitments. And while they speak to the need to integrate climate action with economic security, just transition and Indigenous rights, the policy specifics with respect to the latter social justice dimensions are insufficient.

And therein lies the conundrum for progressives — the federal NDP and Green programs each have their strengths and weaknesses, and both struggle to win seats in our electoral system when the Liberals play the "vote for us to prevent a Conservative win" card. I would be happy to see the Green Party win more seats in Parliament, but it's hard to see how that happens under our first-past-the-post system, and it is mightily discouraging to see the NDP and Greens expend resources and goodwill seeking to win seats at the other's expense. That is why I dearly want to see proportional representation in Canada. We would be well-served by a legislature where the balance of power relies on both the NDP and the Greens, each bringing their own best ideas and covering off the other's shortcomings.

The Conservatives

There is little if any common ground to be found on the climate front with most of today's Conservative political leaders, the likes of Jason Kenney, Scott Moe, Doug Ford or former federal leaders Stephen Harper and Andrew Scheer.

So instead, an appeal to the small-c conservative voter — you who take pride in your country and are committed to protecting your children and grandchildren, who have faith in our country's capacity for innovation, who treasure good governance and care about national security. All deep-seated conservative values.

It is time to abandon the Conservative "leaders" mentioned above. These climate deniers who throw temper tantrums when told we need to take urgent action in the face of the climate emergency. These people don't deserve you.

Today's Conservative leaders say we cannot speedily transition our economy to meet the greatest existential threat of our time. Where is the courage and imagination of their predecessors?

In the economic and societal transition that is now urgently needed to shift our country off fossil fuels, the Conservative Party of Canada has, for now at least, sadly taken itself out of the game. This despite the Abacus poll I commissioned finding that a majority of conservative voters believe climate change to be a serious problem and "a major threat to the future of our children and grandchildren."

There was a time in the not too distant past when the former Progressive Conservative party could legitimately claim to have environmental and climate leaders among its top ranks. Witness the role of the Mulroney government in securing the Montreal Protocol to protect the ozone layer, and the leadership that government displayed at the Rio Earth Summit in 1992. No more. Today's Conservatives have chosen to opportunistically campaign against genuine climate policies, and to conspire with those who would block real action.

Upon the release of the Conservatives so-called "climate plan" in the lead up to the 2019 federal election, even small-c conservative commentator Andrew Coyne, writing in the *National Post*, described it as "a prop" rather than a plan — "a work, essentially, of mischief — an intentionally pointless bit of misdirection."[16] The Conservatives offered no estimates of how much greenhouse gases would actually be reduced as a result of any of the policies promised (few as they were). Perhaps with good reason. Leading environmental economist and emissions modeller Mark Jaccard, a professor at Simon Fraser University's School of Resource and Environmental Management, predicted the Conservative plan would actually result in an *increase* in GHG emissions.[17] When crises, such as a war, call for real plans, we see clear actions and timelines and expected outcomes. Andrew Scheer's climate platform contained no such thing.

As noted previously, at the outset of the Second World War, the U.K. Conservative party determined it needed to replace Chamberlain with Churchill, a leader ready to rise to the challenge. It's that time again.

Our grandparents' Conservative leaders, in the face of an ominous threat, rallied us and declared, "We can do this!" In the face of today's clear and present emergency, the current Conservative leaders whine, "Don't make me do it!" And so my message to small-c conservatives: you're better than them.

Notably, the prosecution of the Second World War mostly saw cross-partisan cooperation. Hopefully, we will see that again. And no matter who is elected, we in civil society always need to "make them do it."

INSTITUTIONAL POLITICAL REFORMS

Beyond partisan politics and the need for our elected leaders to embrace ambitious policy action, there are a number of reforms to our political institutions and practices that would also help to advance the climate mobilization now needed. For example, I have proposed new carbon budgeting practices and structures that would expedite action, provide a mechanism for allocating carbon budget shares, allow for public input and ensure independent oversight. Other new structures are also needed.

A Climate Mobilization Secretariat

At both the federal and provincial levels, we need Climate Mobilization Secretariats. These secretariats would operate with the understanding that climate mobilization is an "all of government" imperative — crossing all ministries and agencies. In recognition of the importance of its work, the secretariats should reside within the Prime Minister's Office (or Privy Council office) at the federal level and the office of the premier of each province. This would send a clear signal that the work of this secretariat is of the highest priority for government. It should not live within the Ministry of Environment, as the secretariat should be seen as guiding the efforts of multiple ministries — Environment, Finance, Natural Resources, Infrastructure, Employment, Industry, Trade, Foreign Affairs, etc. And like the War Committee of the Second World War, the Climate Mobilization Secretariat should report to a Climate Emergency Committee of Cabinet, made up of all the most relevant ministers and perhaps co-chaired by the minister of finance and the minister of environment, if not by the prime minister.

The Climate Mobilization Secretariat would coordinate climate action across government, and in collaboration with Indigenous nations, other levels of government and various public and private sectors (education, health, social services, housing and relevant industrial sectors).

The secretariat's work could be organized into three streams: research and policy development; public engagement, outreach and communications; and GHG mitigation and adaptation policy implementation. At the federal level, an international affairs stream would also be advisable to coordinate climate negotiations, cross-border measures (such as the possible need for carbon tariffs), international climate transfers to poorer countries and climate migration issues.

The secretariat should publicly report annually or perhaps quarterly on Canada's progress towards its legislated GHG reduction targets, so as to ensure accountability. A climate justice lens should also be incorporated into the work of the secretariat, to ensure climate actions are fair and enhance equality, to meet the needs of particularly vulnerable populations, to provide for just transition for impacted workers and to maintain public support for bold action. The secretariat's work should be monitored by an independent Parliamentary Carbon Budget Officer.

Because climate action, energy sources and industrial makeup look different in each region of Canada, the federal government should establish Just Transition agencies in each province, funded by a new federal transfer, and jointly governed with local and Indigenous governments, with representation from labour, business and academic/non-governmental experts.

The tasks of the secretariat would include:

- setting policy and target dates ("war plans," after all, need to be guided by actual plans);
- establishing new crown agencies and companies as needed, and staffing of their leadership;
- carbon budgeting and revenue planning (ensuring we have the right fiscal architecture in place, such as carbon pricing and systems to sell and promote Green Victory Bonds);
- public engagement and information (like the Wartime Information Board), including the use of citizen assemblies;
- coordinating climate infrastructure planning and delivery;
- climate adaptation planning;
- intergovernment planning and coordination with provincial, Indigenous and municipal governments;
- forestry planning;

- electricity planning;
- transportation planning (including air and ship);
- buildings planning;
- industrial planning;
- technology planning;
- labour and just transition planning;
- climate justice planning; and
- creation and oversight of a Youth Climate Corps.

With structures such as this in place, we would be signalling to the population at large that a major mobilization is underway.

Electoral Reform

While not essential to climate mobilization, there is no question that changing our electoral system to one with proportional representation (PR) would be a huge help.[18]

As one surveys the countries listed above that are taking bolder climate action than Canada, with the exception of the U.K., they all have one thing in common — their legislatures are elected based on some form of PR. This has a few benefits: PR almost always produces minority or coalition governments, which then usually rely on progressive and/or Green parties to govern; these minority governments encourage more policy innovation, as parties need to cooperate and adopt the best ideas from each other's platforms; and PR avoids policy backsliding. The curse of our current and archaic first-past-the-post (FPTP) electoral system is that it sees regular swings between progressive and conservative false majority governments. These governments then have full rein to implement their programs and to undo the work of the previous government. Hence we see the Ford government in Ontario undoing the climate policies of the previous Wynne government, the Kenney government undoing the climate policies of the Notley government, etc. These policy pendulum swings are terribly discouraging, and, in the era of climate emergency, we don't have time for it. Now is a period requiring forward momentum.

Our current FPTP system also exaggerates regional differences, making the confederation quagmire needlessly more complicated.

It paints a false map, making us think that all of Alberta and Saskatchewan is opposed to climate action because, in our elections, those provinces are painted Conservative blue. But this is not so. In the 2019 federal election, even in Alberta, where the Conservatives had a sweeping win, 28% of the popular vote went to a mix of the Liberals, NDP and Greens. Likewise, in Saskatchewan, despite the Conservatives wining 100% of the seats, 34% of the popular vote went to the NDP, Liberals or Greens.

Electoral reform means giving true political expression to the fact that a solid progressive majority in Canada wants climate action, but that position is split among multiple parties. This is, therefore, a matter of some urgency. The federal government should strike an independent citizens' assembly to explore what electoral system would best serve Canada and make a recommendation, and we should implement its proposed system before the next election. And after Canadians have had the chance to experience a new system over a couple of elections — road-testing it in practice — a referendum could then be held to decide whether we want to keep it. Notably, no country in the world that has switched from FPTP to some form of PR has ever chosen to switch back.

A NEW GENERATION OF CLIMATE EMERGENCY LEADERS

Avi Lewis speaks of "a new generation of political first responders." Those are the kinds of leaders one needs in an emergency. And they are out there, across Canada, alive to the crisis and the urgent timeline before us.

Keep your eye on emerging political climate champions: new members of Parliament such as Winnipeg Centre NDP MP Leah Gazan, Hamilton Centre NDP MP Matthew Green, Nunavut NDP MP Mumilaaq Qaqqaq, Victoria NDP MP Laurel Collins, Fredericton Green MP Jenica Atwin and Montreal Liberal MP and now cabinet minister Steven Guilbeault, and rising young Indigenous leaders like Khelsilem, Molly Wickham, Melina Laboucan-Massimo, and Dana Tizya-Tramm. Survey the landscape of emerging leaders among provincial legislatures, such as Québec Solidaire MNA Gabriel Nadeau-Dubois, Ontario NDP MPP Joel Harden, and B.C. NDP MLA Bowinn Ma and Green MLAs Sonia Furstenau and Adam Olsen. Watch municipal leaders like Montreal's Valérie Plante and Marie-Josée Parent, Vancouver's Christine Boyle, Regina's Andrew Stevens, Nelson's Rik Logtenberg and the dozens of councillors from across the country who have joined "The Climate Caucus." There are others, and more are coming from the ranks of young climate leaders, a substantial majority of whom, it bears noting, are women. Our AOCs are out there. They are among a new generation of politicians who are in it to truly win a Green New Deal and defiantly confront the climate emergency. We need to see them and support them. Time is not on their side.

Many politicians I interviewed for this book say they get the climate emergency. But they are not leading like this is indeed an emergency. They have not surrounded themselves with people who act and plan like this is an emergency. They are not communicating that this is an emergency.

The leaders we remember from the Second World War led like it was a damn emergency.

It is an emergency.

And so, to our current political leaders, a final message: like the leaders who saw our country through the war, the climate crisis is your generational mission, your defining challenge. It will determine if and how you will be remembered. The climate emergency, like the peril we faced before, is a direct threat to our security and well-being, which you are entrusted to protect. Be the leaders we need you to be.

Meital Smith

CONCLUSION

Forward to Victory!
Or Making Peace with Our Planet

"I am totally confident not that the world will get better, but that we should not give up the game before all the cards have been played . . . There is a tendency to think that what we see in the present moment will continue. We forget how often we have been astonished by the sudden crumbling of institutions, by extraordinary changes in people's thoughts, by unexpected eruptions of rebellion against tyrannies, by the quick collapse of systems of power that seemed invincible. What leaps out from the history of the past hundred years is its utter unpredictability."

— historian Howard Zinn, written just before his death in 2007. Zinn was a U.S. air force bombardier in the Second World War, but became an outspoken peace activist. The U.S. army paid for his university education after the war, and he became a professor of history and author of the acclaimed *A People's History of the United States*

"The history of disasters demonstrates that most of us are social animals, hungry for connection, as well as for purpose and meaning . . . Two things matter most about these ephemeral moments. First, they demonstrate what is possible or, perhaps more accurately, latent: the resilience and generosity

of those around us and their ability to improvise another kind of society. Second, they demonstrate how deeply most of us desire connection, participation, altruism, and purposefulness. Thus the startling joy in disasters."[1]

— writer Rebecca Solnit

"Thou shalt not be a victim. Thou shalt not be a perpetrator. Above all, thou shalt not be a bystander."

— the immortal words written above the United States Holocaust Memorial Museum in Washington, D.C.

This concluding chapter is a final appeal to you, the reader of this book. All war stories have their protagonists, villains, victims, naysayers and delayers. In these next 10 years, we will all need to be climate heroes. Whoever we are and whatever our role, will we be able to look young people in the eyes years from now and say, with assurance and pride, that when fate called, like those before us, we took up the call?

These militaristic appeals aside, I hope you will not draw the wrong conclusion from the analogy at the heart of this book. Let me be clear — there is nothing glamorous in war. War is never something to wish upon ourselves or others. It is horrific.

In Canada's war story, people did not always show their best selves. Our citizens and soldiers sometimes failed to treat each other well or act honourably. With respect to civil liberties and compassion for refugees, Canada did things that remain a deeply shameful legacy. Mistakes were made, both in the economic planning and production on the home front and on the battlefront. Inevitably in war, atrocities are committed on all sides.[2] By the time the war ended, about 44,000 Canadians had been killed and 54,000 wounded.

Globally, approximately 75 million people were killed during the bloodbath of 1939–1945. Seventy-five million. Its final acts were the horrific atomic bombings of the Japanese cities of Hiroshima and

Nagasaki. When the great physicist J. Robert Oppenheimer, the director of the Manhattan Project, watched as the first atomic bomb they had developed was tested in the New Mexico desert in the last year of the war, the words of the Bhagavad Gita came to him: "I am become Death, the destroyer of worlds."

The coming decades look scary. Things could get very grim. If we do not manage to dramatically bend the curve on GHG emissions and swiftly decarbonize our economy and society, our children and grandchildren will be living through a hellscape. We have to figure out — quickly — how to confront this existential crisis, but without the destruction and devastation that marked the last one. Yes, we need a good war. But we also need to make peace with our planet. The motivation in grand struggles like the one before us is always both negative and positive. We are driven in this task by love as well as fear, just as an earlier generation was in the face of fascist conquest.

The structural device used by this book — lessons from the Second World War — has been employed for the helpful examples and inspiration it provides. The analogy, of course, is imperfect.

For one thing, the focus of attention then was always on a clear enemy. In contrast, the issue of "enemy" is more opaque when it comes to the climate emergency. But perhaps that doesn't matter. Far more insidious in the present have been the new denialists, the appeasers and the defeatists. These people have cost us decades of lost time and sap our will and determination.

Paradoxically with respect to the focus of this book, historian Alex Souchen notes that the Second World War was a huge boost to the fossil fuel and chemical industries, and in some respects "accelerated the Anthropocene and climate change." The war amplified the very mess we are now trying to clean up.

Some contend that the climate challenge is greater than what we faced in the war, as it will inevitably be a multi-decade struggle that involves completely remaking our economy, not just in the short-term, but for good. But is the task before us now really more demanding? Canadian climate writer Chris Hatch wrote the following in a lovely Remembrance Day piece in 2019, and it offers helpful perspective:

> The sacrifices of those who last fought to save civilization dwarf the price now asked of us to defend civilized life on an inhabitable Earth. None of us is being asked to die, although many are already dying. None of us is being asked to leave home, though refugees are already moving. None of us is being asked to go hungry, although many are already desperate. The price to defend our children and our homes today is not blood, but money.
>
> Those among us now working in fossil industries will soon have to find new ways to earn a living. The far greater number of us are being asked merely to support them with transition programs while tolerating some annoyances to upgrade our economy. The financial sacrifice our families offered last century was many times the level asked of us to avoid climatic catastrophe . . .
>
> There is one front on which our challenge mirrors that of our forebears: we are being called to be brave at home. To be unflinching — not to avert our eyes from the gathering storm. Like those before us, we must find the strength to resist false hopes and demagogues. To reject leaders bent on appeasing the fossil barons, and to resist the blustering Vichyists quisling their way to power. We are reminded today that civilization is never more than one generation from extinction.[3]

The task won't be over even once we eliminate GHG emissions. We will still need a concerted campaign to extract the overaccumulation of GHGs from the atmosphere until it returns to a safe level of 350 parts per million. But at least the mobilization program described in this book will have given our children and grandchildren a fighting chance.

About a dozen years ago, my mother observed that my daughter, then about four years old, saw an old cigarette ashtray and asked what it was. This struck my mother as notable, because when she was young, and indeed when I was a child, no one ever would have asked such a question. We all just knew. Many of us made ashtrays in school pottery classes! In contrast, students today have never known a time when smoking was permitted in restaurants, or when seatbelts weren't mandatory. And the next generation of students won't know what a gas station is, except what they see in old movies.

We are about to live through an industrial revolution in high speed, once our leaders are finally forced to act. The day is coming soon when we will all sleep in homes heated without fossil fuels, we will live in communities where most of what we require for work and services is within an easy walk or roll, where we move around our cities and towns in affordable if not free electric buses and trains or in electric vehicles, and we travel between cities by electric high-speed rail, where our rail and tanker freight is also transported by electric power, we eat food that is mainly locally and sustainably produced and where new technology allows us to track our diminishing carbon usage with ease. A good new life awaits us.

A CLIMATE MOBILIZATION PLAN FOR CANADA: BRINGING IT ALL TOGETHER

Fully decarbonizing our economy and society is indeed possible, and can be done more quickly than many presume. Minimally, we must halve our GHG emissions by 2030 and reach net-zero emissions by 2050. Is a more aggressive target like net-zero by 2030 or 2040 achievable? Possibly, and we should aim for maximum ambition. When a tipping point comes — our "fall of France" moment — we are going to witness a level of activity and change at a speed that, today, seems hard to contemplate. But we will.

The second core contention of this book is that climate action cannot and should not be separated from the urgent tasks of tackling inequality, decolonization, poverty, homelessness and economic/job insecurity. Indeed, linking these issues into a compelling overall vision is vital to mobilizing the public and keeping the maximum number of people engaged in this struggle. We can and must build an economy that is climate consistent, and meets people's employment and economic security needs, while upholding Indigenous rights.

A key takeaway from this exploration of our past and present is that the transition before us needn't be only about the hard work ahead. Nor is it merely about a transition of our economy from fossil fuels to renewables. Its most precious opportunity is the possibility presented to transform who we are. Like all crises, it is an opening for us to become our best selves and to create a more just society.

To summarize, here are 20 key takeaways to guide us today as we confront the climate emergency:

1. The first step is to fundamentally shift our mindset — to truly acknowledge and treat this threat as the emergency that it is, and to recognize climate attacks on our soil for what they are. Then we need to let that new mindset liberate us from the neoliberal thinking that has so constrained how we approach this crisis. This applies not just to all levels of government, but to civil society organizations as well — environmental organizations, faith organizations, labour unions, businesses, educational institutions, media organizations, community service organizations and legion halls (your work ain't done, folks!). Whatever we've been doing isn't working, so we can't keep doing more of the same. It's time for an all-hands-on-deck response. Adopting a wartime approach applies to our institutions, but also to all of us as individuals. As has often been said, the Second World War was total war — everyone had to do their bit.
2. A wartime-level mobilization means that a host of changes — at the household, community and industrial level — must be *mandated*, not voluntary or merely "incentivized." We should use regulatory fiat as needed to order changes that have to happen.
3. We must legislate clear and ambitious targets, both for overall GHG emission reductions and for various sectors. For example, embed in law that we will ban the sale of new gas-powered vehicles by 2025, and ban the use of fossil fuels in all homes and buildings by 2040.

4. Ensure that all the institutions and machinery of government are focused on this national task. Appoint a Climate Emergency Cabinet Committee. Embed a Climate Emergency Secretariat in the Prime Minister's Office and each premier's office. Adopt a whole-of-government approach in responding to the emergency, and establish agencies proposed in this book as needed to get the job done.
5. Rally the public at every turn. In frequency and tone, this needs to look and sound and feel like an emergency. The news media and educational institutions need to reimagine their approach in this regard, and we must all demand that they do so. Ban the advertising of fossil fuel vehicles and gas stations, and prohibit ads and sponsorships from the Canadian Association of Petroleum Producers (CAPP) and its oil and gas corporate members. Require that all media companies divest from fossil fuels, to ensure they are not compromised. And marshal the cultural and entertainment sectors, supplying public funding for educational and arts initiatives that seek to rally the public.
6. Take on the fossil fuel corporations. We must not seek to appease them or their financial enablers. If CAPP and its members are not feeling deeply anxious about our climate plans — if they do not see our climate policies as their own existential threat — then they are not plans worth having. And let our collective deployment of peaceful civil disobedience, divestment and regulations hasten their flight.
7. Convene citizens' assemblies in each province, to guide how this transformation will proceed.
8. Spend what is necessary to win this good war, because if we lose, nothing else matters. The funding for this should come from a combination of new taxes (particularly on the wealthy and corporations) and the sale of Green Victory Bonds.
9. Undertake massive public investments in needed climate and social infrastructure.

10. Create the economic institutions needed to get the job done, including new crown corporations as needed.
11. Conduct a national inventory of conversion needs (just like C.D. Howe did). Determine how many heat pumps, solar panels, wind turbines, electric buses, etc., we will need and plan for how those items will be produced and deployed.
12. Ensure visionary and creative people are in key leadership positions in the civil service and bring in outside experts, civil society leaders and entrepreneurs as needed to drive change and oversee the necessary scale up.
13. Ensure a rigorous just transition plan is in place. And match workers to needed jobs. In the war that wasn't left to the market, nor should it be now.
14. Establish a new federal Climate Emergency Just Transition Transfer, to fund the work of provincial just transition agencies, with funding going disproportionately to the provinces with the most heavy lifting to do.
15. Respect Indigenous title and rights and embed the UN Declaration on the Rights of Indigenous Peoples into law at all levels of government.
16. Ensure Indigenous communities and nations are full partners in the development of our climate emergency plans.
17. Lower the voting age.
18. Establish a Youth Climate Corps, open to all high school grads who wish to enlist.
19. Substantially increase our financial transfers to the poorest countries and regions that are being hardest hit by the climate crisis. This is not a matter of charity, but of necessity and justice, given our historic emissions.
20. Ensure Canada is willing and prepared to welcome tens of thousands of climate-displaced people annually, so that we do not repeat the shameful legacy of how Canada treated refugees before.

We can do this! And as we take bold climate action in the ways described, we will also be addressing and fixing other core societal challenges — inequality, homelessness, job insecurity and the shameful legacies of our colonial past. A massive collective effort such as this, on the scale needed, will result in full employment — just as occurred in the Second World War. This reality should give all whose employment is currently linked to the fossil fuel sector comfort that they will find secure and decent work as we transition to a new economy and society.

THE TASK OF OUR LIVES

Earlier, I invited you to stay alert to the people who defied orders, rules and norms, and in doing so changed the course of events: C.D. Howe, who threw out the rulebook when it came to economic planning; the Scottish merchant ship captain who disobeyed orders not to undertake rescues in waters prowled by German submarines and instead saved Howe and his fellow travellers when their ship was torpedoed en route to Britain; the German captain of the MS *St. Louis*, who refused to return his ship to Germany; Japanese diplomat Chiune Sugihara, who defied orders and issued thousands of transit visas for Jews to escape the Holocaust; legendary American reporter Edward R. Murrow, who defied his bosses at CBS to cover the Nazi invasion of Poland; the civil servants at Canada's Department of Finance, who exploded what was deemed possible in financing the war effort; the father of Harold Steves, who stood apart from his neighbours in opposing the internment of Japanese Canadians, reminding us that we are not all simply "products of our time"; and Elsie MacGill, Canada's first female aeronautical engineer, who oversaw the mass production of wartime planes and went on to advance women's rights in Canada.

We have such people again in the present: the climate strikers, who are not letting their schooling interfere with what needs to be done; the Extinction Rebellion civil disobedience protestors; the Our Time youth who were arrested in the House of Commons lobby the week after the 2019 federal election; the Wet'suwet'en and other Indigenous land defenders, and all the people engaged in blocking fossil fuel infrastructure projects; and the political leaders prepared to adopt "keep it in the ground" policies with respect to fossil fuels.

There are many others. All heroes in our midst, then and now.

Kai Nagata worries that "people were a lot tougher in the 1930s." Are Canadians today too comfortable and self-absorbed compared to the so-called "greatest generation" that faced down the Depression and the war? Maybe. But these traits are not static. In the end, we become the people that circumstances require us to be.

We are, all of us, a confusing and contradictory mix of values — virtues as well as vices. Human nature is a mixed affair, and our culture and politics is capable of animating either the best in us or the worst. But our values are, in the main, good and caring. We just need a politics and leadership that allows people to give those positive values proper expression.

As many have warned, and as we are already experiencing, climate change–induced disasters and disruptions in growing frequency and severity are coming. We are going to be pressed into service one way or another, and our character tested. The only question is whether we will be mobilized exclusively in moments of disaster not of our choosing, or whether we will assume agency and mobilize pre-emptively on our own terms?

While this task frequently feels daunting, recognize that once we truly see this time as an emergency, and properly throw ourselves into this fight, an all-hands-on-deck experience like this will transform us in many ways, individually and collectively. And much of that will be for the better. Canada's Second World War experience transformed our politics, our social welfare systems, our economy and businesses, the labour movement, our social relations and our sense of possibility. The next mobilization will do the same.

Rebecca Solnit reminds us, "The word *emergency* comes from *emerge*, to rise out of . . . An emergency is a separation from the familiar, a sudden emergence into a new atmosphere, one that often demands we ourselves rise to the occasion."[4]

Pivotal fights or struggles — wars, strikes, historic marches or occupations, or natural disasters — are transformative; they teach us and embed in us a new understanding of solidarity and collective action. That events such as these are transformational is not to be resisted but embraced. Just as Canadians did not emerge from the war the same people as they were in 1939, so too, we will not emerge from the climate mobilization the same people we are today, nor would we want to. The endgame here shouldn't merely be to "solve" the climate emergency. This is an opportunity to emerge a better, more just and more democratic society, to remake a better world, to reset how we approach challenges and to remind ourselves of what we can accomplish together.

This book has sought to underscore that tackling this emergency isn't just an obligation, it is a generational chance to re-find a sense of shared purpose. In a 2019 speech about the Green New Deal, Avi Lewis listed among the top reasons to support this transformative idea that "the Green New Deal will be good for our souls. In fact, we may never have needed a collective project — a higher purpose — more than we do right now."[5] The forthcoming mobilization can liberate us from feelings of isolation and hopelessness, and reweave our social fabric.

When we act in common purpose, it diminishes our mental anxieties and strengthens our sense of self-worth. Indeed, this was an important element of the wartime experience. British writer Dan Hind, in his book *The Return of the Public*, exploring how neoliberalism has undermined social connection, writes:

> We have known for a long time that the individual's opportunity to flourish depends on social factors. During the Second World War the British population became less mentally distressed, contrary to the predictions of many psychiatrists. The sense of shared jeopardy and the introduction of universal services both served to reduce levels of what the social scientist Richard Titmus called "social disparagement." The war brought a sense of common purpose and sharply reduced the levels of economic inequality. People felt dignified and elevated during the six years in which they worked together to save their country

> and their way of life; the war qualified hierarchies by asserting a civic equality. Overwhelmingly committed to the immediate task of national survival and to the creation of a new society after the war, the British became happier as they came to resemble an autonomous public.[6]

Here's a little secret: when I am giving public talks, the question I most dread is "What keeps you so hopeful?" I dread the question because one's impulse is to lie. The person asking the question is seeking hope, and the urge is to oblige. But the truth is, each of us willing to see the world as it is, and with the conviction to remake the world as it should be, walk a razor's edge between hope and despair. We feel and know both. And maybe we should be more open about that.

To delve into the realities of the climate crisis is often to wrestle with despair. The more one learns, the more one fears what this world will look like for future generations. Yet as I dove into the research for this book, I did find hope. It became clear that the technologies and policies needed to tackle these interlocking ecological and social crises are not unknown or awaiting us in the future; they exist now, waiting to be picked off the shelf. And the story of the war woven through this book is a helpful reminder that we can do this, because we have done it.

So while we do indeed live in a time of uncertainty and high anxiety, there is always hope. The new and beautiful world we crave may be closer than we think. I believe it, because, as the quote from Howard Zinn above reminds us, history is full of surprises.

Let us then end where we began — in the ambiguous time in which we live. This book has sought to offer hope. Does that necessarily mean we will succeed in this task of our lives? No. The odds aren't great. But it is also too soon to give up the game. Even if our eventual efforts prove too late, it's good to go out fighting, in solidarity with other good people.

Few of you reading this book were alive during the Second World War. But many of us with an interest in such things look back on history and pivotal times like those recounted here, and at how a previous generation called upon their best selves, and we ask ourselves, "What would I have done, if I had lived then and there?" But the answer to that question is really no mystery at all. The answer to that question, dear reader, is — whatever you are ready to do now.

EPILOGUE

The COVID-19 Pandemic and How a Recognized Emergency Makes the Impossible Possible

"Canada hasn't seen this type of civic mobilization since the Second World War. These are the biggest economic measures in our lifetimes, to defeat a threat to our health... We all need to answer the call."

— Prime Minister Justin Trudeau, April 1, 2020, during one of his daily pandemic briefings outside his home

Talk about awkward timing. Just as this book was going into production, the world was struck by a pandemic crisis that pushed the climate emergency, for now at least, off the front burner. Life is full of curveballs.

I chose to frame this book around Canada's Second World War experience because I sought an historic reminder of how quickly we have transformed society and our economy in the past. As I write, however, we have all been witness and party to such evidence in real time. Suddenly, everyone is drawing comparisons to the Second World War, and our leaders have been immersed in a crash course in wartime economic planning as they seek guidance for confronting the pandemic.

Similarities abound between our wartime experience and the current pandemic response. This is what emergency looks and feels like, particularly when it catches us off guard. The status quo is suspended. Government leaders and public officials hold daily emergency briefings. Emergency Acts are invoked. Federal and provincial cabinets form emergency response committees of key ministers (like the War Committee). Resources and personnel are redeployed. Manufacturing capacity is requisitioned to produce essential products (today, protective medical gear, hand sanitizer and ventilators), and governments assume the power to direct necessary supply chains.[1] Public facilities are repurposed as needed (such as community centres turned into makeshift clinics and service centres). We honour the frontline people making extraordinary sacrifices, some of whom are public service leaders, civil servants and of course healthcare professionals, but most of whom turn out to be lower-wage workers whose labour we so often devalue — hospital cleaners and other sanitation workers, bus and truck drivers, grocery store clerks, child-care workers looking after the children of other essential workers, and numerous others whose work we are reminded we cannot manage without.

The news media and journalists have risen to the occasion, quickly retooling their kitchen tables so that they can continue to provide vital public information and rally our collective morale from their homes. In a welcome shift from what has marked climate reporting, in this crisis the media seem to feel no obligation to give credence or space to those who question the scientific evidence. Artists too have rapidly mobilized to keep us entertained and encouraged, even as their incomes dry up and they are forced to creatively share their work online. And while society's defenders of civil liberties rightly urge caution and vigilance in the allowance of governmental emergency measures, we are on the whole prepared to abide reasonable intrusions in the interests of public health, provided democratic oversight is maintained. Even some Indigenous organizations, in the early days of the pandemic, urged the federal government to declare a state of emergency, knowing the vulnerabilities of their communities.[2]

As in the war, our governments at every level have appropriately dispensed with the fetishization of balanced budgets and are spending what is required — deficits be damned. Not only do we newly value and appreciate the role of government and public services, but we have also come to understand that the vulnerable must be urgently looked after, or we are all more vulnerable to this invisible foe. As with the climate crisis, the most vulnerable are hit hardest. Consequently, we have witnessed a newfound collective willingness to house the homeless (commandeering hotels rooms as required) and to offer income support to those who cannot work, so that people can properly shelter in place for the benefit of all. In numerous provinces, evictions have been banned and rents frozen (albeit temporarily). The federal government is boosting Canada's international aid, despite our heightened domestic demands, given the global nature of the pandemic. The emergency response to protect the most vulnerable has been far from perfect — the desperation of Canada's poorest communities has only heightened, Canada has turned away asylum seekers at the border (allowing fear to trump our human rights obligations) and support for migrant workers within Canada has been weak. The coronavirus pandemic has revealed and exacerbated all the existing inequities in our economy. It has laid bare the interconnections of the crises we face — the inequality crisis, the poverty and homelessness crises, the opioid crisis, the employment precarity crisis and the ecological crises. Yet now we seem prepared to at least recognize and begin to repair these interlocking fissures.

In times of emergency, it turns out, we are all socialists.

Just as social solidarity was vital for wartime mobilization, so it has been in this crisis. And our displays of such solidarity have been beautiful. Yes, just like in the war, the early days were marked by a minority of people who responded in selfish or panicky ways, engaging in antisocial behaviour — hoarding, attempted profiteering or willfully ignoring public health appeals to keep physical distance. But these have been the exception. Just as Rebecca Solnit foretold, the large majority of us have shown our best selves. We have heeded the instructions of public health officials, done our best to protect the vulnerable, organized "mutual aid" networks and looked out for and delivered food to our housebound neighbours.[3] We've stayed calm and entertained our children. We have cooperated.

We have also shown ourselves capable of changing our ways with remarkable speed, figuring out how to work from home and to employ technology to socialize remotely with friends and loved ones. As in the war, this pandemic has forced a change in our daily household and employment routines and practices. After rigidly insisting for years that people must commute into work and meet in-person, employers and employees have quickly discovered the potential (and even the benefits) of telecommuting and videoconferencing. Even parliamentary committees, city council meetings and other governmental decision-making have moved to virtual platforms, and the prime minister himself worked and held meetings and media briefings from home while in self-isolation during the first weeks of the pandemic. Many are quickly moving to online grocery delivery services. And, of course, everything but the most essential domestic air travel has come to an abrupt halt. A welcome side effect of all these COVID crisis adjustments has been that GHG emissions and harmful air pollution in Canada and globally have dropped (but in the absence of other actions, this will only be a temporary effect).

The evolving governmental response to the pandemic in Canada has been, by and large, impressive and bold. Our governments have shown themselves willing and able to pivot to emergency mode with laudable speed and flexibility. We have seen a level of cooperation across confederation and across partisan lines that is unprecedented in modern times. Prime Minister Trudeau, almost all premiers, government ministers and countless municipal and Indigenous community leaders have shown real leadership thus far.

In March 2020, Parliament quickly and unanimously passed a $107 billion aid package. Of that, $55 billion was for deferred taxes and loans for businesses, while $52 billion was direct aid to households, mainly in the form of the Canadian Emergency Response Benefit — $2,000 per month for workers who lost their incomes due to the crisis — but also one-time boosts to the GST credit and the Canada Child Benefit. This was followed a few days later by the announcement of the Canada Emergency Wage Subsidy — a 75% wage subsidy on the first $58,700 of earnings for businesses and non-profits that experience a major drop in revenues, designed to encourage them not to lay off staff, a very smart and audacious move. The CEWS is estimated to cost $71 billion. Combined with other response measures, total direct spending on the crisis by the federal government is currently approaching $200 billion. The design of these programs is imperfect, but the federal government has not allowed such details to derail swift action.

As our governments spend billions beyond what they originally budgeted and as the economy heads into a deep recession, 2020 will see a large spike in Canadian government debt-to-GDP. Yet this jump is perfectly manageable. Our government debt levels, as we have seen, are historically low, leaving Canada well-placed to weather this storm. And with much of this new debt being owed to the Bank of Canada, which throughout the COVID crisis is mass purchasing government bonds on a weekly basis, our public finances can withstand this crisis just fine. Indeed, with the stock market in free fall and investors spooked and looking for a safe harbour, now is the perfect time to launch an ambitious public "Victory" bond drive like the one urged in this book. Let's invite the public to contribute to an economic recovery plan. And with interest rates at an historic low, our governments at all levels should be borrowing like never before to finance the climate and social infrastructure investments now urgently needed. As economist Jim Stanford has written, "The Government of Canada can now issue 30-year bonds for well below 1% annual interest. That is negative in real terms (i.e., lower than inflation). So quite literally, the government will save money by borrowing more (paying back less in real terms, after 30 years, than they borrowed) — to say nothing of the economic and social good that would be done by putting that money to work in emergency public projects and services."[4]

Overall, while the economic situation may feel chaotic and ominous, in real economic terms, the spending now underway merely shows what we could have done in response to the climate emergency, poverty and homelessness all along.

All of this represents proof positive that, given the will, we are indeed capable of rapidly rising to the climate challenge. The curse of the climate crisis, it turns out, is that, in comparison to the pandemic and the war, it moves in slow motion, and has thus failed to galvanize us — so far.

There is a key difference between the wartime and climate crises, on the one hand, and the coronavirus pandemic on the other. The war effort and climate mobilization demand that we get out and build what was/is required for the transition, while the pandemic obliges us to stay home. Consequently, whereas the war and climate action were and can be a boost to the economy and job creation, the COVID crisis represents a massive hit to both, as people are prevented from working. The fact that climate action is a positive on these fronts, however, is welcome news; just as the Second World War ended the Great Depression, as we rebuild from this pandemic, an ambitious climate plan with massive green infrastructure spending — the Green New Deal — can be just what the doctor ordered.

The vital and urgent challenge now is to ensure that, once we emerge from the coronavirus crisis, we use this experience and the opening it creates to catapult our societies into the post-carbon economy. As writer Arundhati Roy urges, this pandemic, like past ones, should be seen as "a portal, a gateway between one world and the next."[5] We must not return to yesterday's normal, with all its inequalities and fossil fuel reliance. This pandemic has the potential to dramatically jump-start our efforts to decarbonize — to accomplish massive emission reductions in a few short years. But this is not assured. We are certain to see a great battle over what the return to post-COVID "normal" looks like. The new denialists in industry and government are already seeking huge public bailouts for the fossil fuel sector, airlines, traditional auto manufacturing and more. Will we seek to quickly restore the main industries of before, or will we embrace this historic moment and the massive government expenditures any rebuilding efforts will inevitably require to permanently remake our economies?

As I write these final words, the global and Canadian pandemic story is still unfolding. The full extent of the crisis to both human health and the economy remains unknown. (For the latest version of this epilogue, visit sethklein.ca, where I intend to keep these lessons updated.) What remains a clear conclusion from all these crises, past and present, is that humans are amazingly resilient and capable of accomplishing great things with remarkable speed when the circumstances require.

The COVID pandemic has reaffirmed the role and value of ambitious government action. It has reminded us of our mutual reliance. Social solidarity and support for public services is likely at a generational high. There is a new spirit of national cooperation in the land upon which we must capitalize. The public will emerge from the COVID crisis with new respect for scientists and scientific evidence. We collectively now understand that the more our economy is localized, the more resilient we are to disruptions. The economy urgently requires public investment. Combine all these new realities, and the time has never been more opportune for true climate emergency mobilization.

ENDNOTES

Preface

1 See poll results from the Australia Institute at https://www.tai.org.au/content/australians-want-gov-mobilise-against-climate-change-world-war.

2 Livia Albeck-Ripka, Isabella Kwai, Thomas Fuller and Jamie Tarabay, "'It's an Atomic Bomb': Australia Deploys Military as Fires Spread," *New York Times*, January 4, 2020.

3 Joanna Santa Barbara, "Psychological Impact of the Arms Race on Children," *Preventive Medicine* 16, no. 3 (May 1987): 354–360. [For reference: https://doi.org/10.1016/0091-7435(87)90034-X].

4 I first developed this concept in a 2016 article with my former CCPA colleague Shannon Daub. See "The New Climate Denialism: Time for an Intervention," Policy Note, September 22, 2016, https://www.policynote.ca/the-new-climate-denialism-time-for-an-intervention.

Chapter 1

1 Interview on CBC Radio's *The Current*, December 13, 2018.

2 The Mackenzie-Papineau Battalion of Canadian volunteers who fought in the Spanish Civil War was named for William Lyon Mackenzie and Louis-Joseph Papineau, the leaders of the Upper Canada and Lower Canada rebellions of 1837–1838.

3 Contemporary historians have given Chamberlain more credit than previously. They note that Hitler wanted war as of 1938, and the Allies weren't ready. So rather than take the bait, Chamberlain stalled for time to kick-start rearmament efforts.

4 Bill McKibben, "A World at War," *The New Republic*, August 15, 2016, https://newrepublic.com/article/135684/declare-war-climate-change-mobilize-wwii.

5 See http://www.climatecodered.org/2019/08/at-4c-of-warming-would-billion-people.html.

6 From Professor Hayhoe's Twitter thread: https://twitter.com/KHayhoe/status/1167851841041981440.

7 Environment and Climate Change Canada, *Canada's Changing Climate Report 2019*, https://changingclimate.ca/CCCR2019. For more on projected Canadian impacts, see Council of Canadian Academies, *Canada's Top Climate Change Risks* (Ottawa: The Expert Panel on Climate Change Risks and Adaptation Potential, Council of Canadian Academies, 2019).

8 Cited in a 2019 interview with Robert Hackett, "Noam Chomsky: 'In a Couple of Generations, Organized Human Society May Not Survive,'" *National Observer*, February 12, 2019.

9 Rebecca Solnit, *A Paradise Built in Hell: The Extraordinary Communities That Arise in Disaster* (New York: Viking, 2009).

10 Solnit, *Paradise Built in Hell*, 2–3.

11 Solnit, *Paradise Built in Hell*, 4.

12 Solnit, *Paradise Built in Hell*, 53.

Chapter 2

1 The classic history text on this subject is C.P. Stacey's *Arms, Men and Governments: The War Policies of Canada, 1939–1945* (Ottawa: Published by Authority of the Minister of National Defence, Queen's Printer, 1970).

2 My thanks to Alex Himelfarb for encouraging me to think about and organize the challenge into these three pillars.

3 Oreskes's interview with Wen Stephenson, "Is the Carbon-Divestment Movement Reaching a Tipping Point?" in *The Nation*, April 22, 2015.

4 For more examples and some theoretical framing of the new climate denialism, see Shannon Daub, Gwendolyn Blue, Zoë Yunker, and Lisa Rajewicz, "Episodes in the New Climate Denialism," in *Regime of Obstruction: How Corporate Power Blocks Energy Democracy*, ed. by William K. Carroll (Edmonton: Athabasca University Press, 2020).

5 Adam Walsh, "Oil and Water: N.L. Tries to Balance Economy and Environment — to Mixed Reviews," CBC News, June 19, 2019. See: https://www.cbc.ca/amp/1.5181134. For more on the contradictions facing Newfoundland, see also Angela Carter, "Oil Is On the Way Out, but NL Is Going All-In," *The Independent*, May 3, 2019, https://theindependent.ca/2019/05/03/oil-is-on-the-way-out-but-nl-is-going-all-in.

6 Trudeau acknowledged that support for Trans Mountain pipeline expansion in exchange for a carbon tax was always part of a grand bargain with the Notley Alberta government in an interview with the *National Observer*. See Elizabeth McSheffrey, "EXCLUSIVE: Trudeau Says Kinder Morgan 'Was Always a Trade Off,'" *National Observer*, February 13, 2018, https://www.nationalobserver.com/2018/02/13/news/exclusive-trudeau-says-kinder-morgan-was-always-trade.

7 Paul Wells, "A Carbon Tax? Just Try Them," *Maclean's*, November 7, 2018, https://www.macleans.ca/politics/ottawa/a-carbon-tax-just-try-them.

8 Environment and Climate Change Canada, *Canada's Greenhouse Gas and Air Pollutant Emissions Projections, 2018* (Ottawa: Environment and Climate Change Canada, 2018).

9 Jeff Lewis, "Canada Not Doing Enough to Fight Climate Change, Federal Environment Commissioner Warns," *Globe and Mail*, April 2, 2019, https://www.theglobeandmail.com/canada/article-canada-isnt-doing-enough-to-fight-climate-change-federal-environment.

10 Office of the Parliamentary Budget Officer, *Closing the Gap: Carbon Pricing for the Paris Target* (Ottawa: PBO, June 2019).

11 See Carl Meyer, "Oil Sands Polluted More Than Entire Economies of B.C. or Quebec," *National Observer*, April 16, 2019, https://www.nationalobserver.com/2019/04/16/news/oilsands-polluted-more-entire-economies-bc-or-quebec; and Barry Saxifrage, "Surprise! Most of Canada Is On Track to Hit Our 2020 Climate Target," *National Observer*, May 27, 2019, https://www.nationalobserver.com/2019/05/27/analysis/surprise-most-canada-track-hit-our-2020-climate-target. The same dynamic held for 2018, the last year for which we have GHG data — the oilsands was responsible for 84 megatonnes that year, still higher than total emissions in both Quebec and B.C., despite the fact that emissions went up in both those provinces that year.

12 Kathryn Harrison, "How 'Serious' Is a Climate Plan That Relies on Pipelines?," *National Observer*, July 4, 2019.

13 J. David Hughes, "Trans Mountain Expansion Project: Partisan Pipeline Politics Versus Canadians' Best Interest," Policy Note, June 26, 2019, https://www.policynote.ca/trans-mountain-expansion-project-partisan-pipeline-politics-versus-canadians-best-interests.

14 See https://cleanbc.gov.bc.ca. For more analysis of *Clean BC*, see Marc Lee's critique at https://www.policynote.ca/clean-bc.

15 Marc Lee, "Goin' Slow: BC Budget Fails to Make Meaningful Investments in Climate Action," Policy Note, February 22, 2019, https://www.policynote.ca/goin-slow-bc-budget-fails-to-make-meaningful-investments-in-climate-action.

16 See https://www.policynote.ca/clean-bc.

17 Marc Lee, "LNG's Big Lie," *Globe and Mail*, June 30, 2019, https://www.theglobeandmail.com/opinion/article-lngs-big-lie. See also https://www.cbc.ca/news/business/lng-climate-investment-1.5192148.

18 For a rundown of provincial subsidies to LNG Canada, see this analysis by CCPA-BC senior economist Marc Lee: "BC's LNG Tax Breaks and Subsidies Offside with the Need for Climate Action," Policy Note, May 9, 2019, https://www.policynote.ca/bcs-lng-tax-breaks-and-subsidies-offside-with-the-need-for-climate-action.

19 Shawn McCarthy, Brent Jang and Justine Hunter, "Ottawa Clears Way for Proposed LNG Terminal on B.C. Coast With Tariff Exemption," *Globe and Mail*, September 25, 2018, https://www.theglobeandmail.com/canada/british-columbia/article-ottawa-clears-way-for-proposed-lng-terminal-on-bc-coast.

20 For more about Chiune Sugihara, see Jennifer Rankin, "My Father, the Quiet Hero: How Japan's Schindler Saved 6,000 Jews," *The Guardian*, January 4, 2020, https://www.theguardian.com/world/2020/jan/04/chiune-sugihara-my-father-japanese-schindler-saved-6000-jews-lithuania. There is also a feature movie about him entitled *Persona Non Grata: The Story of Chiune Sugihara*.

21 At the time of writing, the climate change division within the federal government's Ministry of Environment and Climate Change (the Pan-Canadian Framework Implementation Office and Carbon Pricing Bureau combined) had 148 employees, while the BC government's Climate Action Secretariat within the Ministry of Environment had about 75 staff. (Numbers provided after specific requests to these ministries.) That strikes me as rather small. As we will see in Chapter 6, these numbers don't hold a candle to the staffing-up that occurred in the Second World War.

22 See https://www.facebook.com/notes/tzeporah-berman/turning-the-corner-on-denial/2251395404924053.

23 Shannon Hall, "Exxon Knew about Climate Change Almost 40 Years Ago," *Scientific American*, October 26, 2015, https://www.scientificamerican.com/article/exxon-knew-about-climate-change-almost-40-years-ago.

24 Sandra Laville, "Top Oil Firms Spending Millions Lobbying to Block Climate Change Policies, Says Report," *The Guardian*, March 22, 2019, https://amp.theguardian.com/business/2019/mar/22/top-oil-firms-spending-millions-lobbying-to-block-climate-change-policies-says-report.

25 Donald Gutstein, *The Big Stall: How Big Oil and Think Tanks Are Blocking Action on Climate Change in Canada* (Toronto: James Lorimer and Company, 2018).

26 Kevin Taft, *Oil's Deep State: How the Petroleum Industry Undermines Democracy and Stops Action on Global Warming — in Alberta, and in Ottawa* (Toronto: James Lorimer and Company, 2017).

27 For more on the work of the Corporate Mapping Project, see https://www.corporatemapping.ca.
28 William K. Carroll, ed., *Regime of Obstruction: How Corporate Power Blocks Energy Democracy* (Edmonton: Athabasca University Press, 2020).
29 Shannon Daub and Zoë Yunker, "B.C.'s Last Climate 'Leadership' Plan Was Written in Big Oil's Boardroom (Literally)," CMP Commentary, September 18, 2017, https://www.corporatemapping.ca/bcs-last-climate-leadership-plan-was-written-in-big-oils-boardroom-literally.
30 Nicolas Graham, William K. Carroll and David Chen, *Big Oil's Political Reach: Mapping Fossil Fuel Lobbying from Harper to Trudeau* (Vancouver: Canadian Centre for Policy Alternatives and Corporate Mapping Project, 2019).
31 For more on how we all practise a form of climate denial, see Kari Marie Norgaard, *Living in Denial: Climate Change, Emotions, and Everyday Life* (Cambridge, MA: MIT Press, 2011).
32 For the latest GHG data and global rankings, see the World Resources Institute site at https://www.wri.org/blog/2014/11/6-graphs-explain-world-s-top-10-emitters.
33 SEI, IISD, ODI, Climate Analytics, CICERO and UNEP, *The Production Gap: The Discrepancy between Countries' Planned Fossil Fuel Production and Global Production Levels Consistent with Limiting Warming to 1.5°C or 2°C (2019)*, 36. Available at http://productiongap.org.
34 Marc Lee, *Extracted Carbon: Re-Examining Canada's Contribution to Climate Change through Fossil Fuel Exports* (Vancouver: Canadian Centre for Policy Alternatives, Parkland Institute and Corporate Mapping Project, January 2017).
35 Lee, *Extracted Carbon*, 19.
36 For more on the idea of a fossil fuel non-proliferation treaty, see https://www.fossilfueltreaty.org.

Chapter 3

1 Rebecca Solnit, "Are We Missing the Big Picture on Climate Change?" *New York Times*, December 2, 2014, https://www.nytimes.com/2014/12/07/magazine/are-we-missing-the-big-picture-on-climate-change.html.
2 The National War Monument in Ottawa was unveiled during the Royal Tour in May 1939.
3 J.L. Granatstein, *Canada's War: The Politics of the Mackenzie King Government, 1939–1945* (Toronto: Oxford University Press, 1975), vi.
4 Granatstein, *Canada's War*, 18.
5 Granatstein, *Canada's War*, 22.
6 Mark R. Wilson, *Destructive Creation: American Business and the Winning of World War II* (Philadelphia: University of Pennsylvania Press, 2016), 47.

7 For more on this topic, see Hugues Théorêt, *The Blue Shirts: Adrien Arcand and Fascist Anti-Semitism in Canada* (Ottawa: University of Ottawa Press, 2017).

8 C.P. Stacey, *Canada and the Age of Conflict, Vol 2: 1921–1948* (Toronto: University of Toronto Press, 1981), 372.

9 For more on Canadian news media reporting during the war, see Timothy Balzer, *The Information Front: The Canadian Army and News Management during the Second World War* (Vancouver: UBC Press, 2011).

10 William R. Young, "Mobilizing English Canada for War: The Bureau of Public Information, the Wartime Information Board and a View of the Nation during the Second World War," in *The Second World War as a National Experience*, ed. Sidney Aster (Ottawa: National Defence Canada and The Canadian Committee for the History of the Second World War, 1981), 189.

11 Young, "Mobilizing English Canada," 190.

12 Young, "Mobilizing English Canada," 193.

13 Young, "Mobilizing English Canada," 194.

14 Christopher Adams, "Polling in Uncertain Times: The Canadian Polling Industry in the 1940s" (presented at the Annual Meeting of the Canadian Political Science Association, Regina, Saskatchewan, May 30, 2018).

15 Results of this opinion project were shared during an April 4, 2019, webinar entitled "Talking Climate," featuring George Marshall (of the UK-based Climate Outreach organization) and Louise Comeau (an expert in climate communications and director of the Environment and Sustainable Development Research Centre at the University of New Brunswick). The analysis drew upon approximately 45 recent climate polls conducted by, among others, the Canadian Surveys of Energy and the Environment, EcoAnalytics, Ekos Research, Abacus Data, Iris Communications and the Angus Reid Group.

16 See http://angusreid.org/election-2019-climate-change.

17 See https://globalnews.ca/news/5860959/canadians-society-politics-ipsos-poll.

18 Interview on CBC Radio's *The Current*, December 13, 2018.

19 The survey was led by Dr. Ellen Field of the Faculty of Education at Lakehead University and the Learning for a Sustainable Future project, in collaboration with the Leger research polling firm. For more details on their survey findings, go to http://www.lsf-lst.ca/en/cc-survey.

20 See, for example, Simon Enoch and Emily Eaton's report *Crude Lessons: Fossil Fuel Industry Influence on Environmental Education in Saskatchewan* (Victoria and Regina: Corporate Mapping Project and Canadian Centre for Policy Alternatives, 2019), https://www.corporatemapping.

ca/crude-lessons-fossil-fuel-industry-influence-on-environmental-education-in-saskatchewan; Bob Weber, "Alberta's Energy 'War Room' Singles Out Climate Activist Steven Lee," *Huffington Post*, December 21, 2019, https://www.huffingtonpost.ca/entry/alberta-war-room-activist_ca_5dfe6835e4b0b2520d0bf655; and Dean Bennett, "Alberta Panel Suggests Schools 'Balance' Lessons about Climate Change, Oilsands," *National Post*, January 29, 2020, https://nationalpost.com/pmn/news-pmn/canada-news-pmn/alberta-curriculum-review-urges-focus-on-basic-learning-standardized-tests.

21 Al Gore's Climate Reality Project offers a model for training people to lead community climate talks, as does the David Suzuki Foundation's Blue Dot initiative.

22 *The Leap Manifesto* can be found at https://leapmanifesto.org/en/the-leap-manifesto.

23 See https://www.theglobeandmail.com/amp/opinion/editorials/article-no-you-dont-have-to-choose-canada-should-say-yes-to-both-carbon.

24 Cited in Bryan Labby, "Is Climate Change Actually a 'Climate Crisis'? Some Think So," CBC News, June 10, 2019, https://www.cbc.ca/news/canada/calgary/climate-change-journalism-language-guardian-cbc-1.5166678.

25 Labby, "Is Climate Change Actually."

26 Robert A. Hackett, Susan Forde, Shane Gunster and Kerrie Foxwell-Norton, *Journalism and the Climate Crisis* (New York: Routledge, 2017), 110.

27 Bill Moyers, "What If We Covered the Climate Emergency Like We Did World War II?," Columbia Journalism Review, May 22, 2019, https://www.commondreams.org/views/2019/05/22/what-if-we-covered-climate-emergency-we-did-world-war-ii.

28 Holman's open letter can be found at: https://thetyee.ca/Mediacheck/2019/05/28/Start-Reporting-Climate-Change-Emergency.

29 Charlie Smith, "Extinction Rebellion Demonstration Shuts Down Traffic to Protest New York Times Climate Coverage," *Georgia Straight*, June 22, 2019, https://www.straight.com/news/1258226/extinction-rebellion-demonstration-shuts-down-traffic-protest-new-york-times-climate.

30 See Naomi Klein's *Intercept* article accompanying the video "A Message from the Future" at https://theintercept.com/2019/04/17/green-new-deal-short-film-alexandria-ocasio-cortez.

31 For example, see https://www.instagram.com/climemechange.

32 Marsha Lederman, "In a Climate Crisis, Artists Have a Duty to Speak Up — but What Should They Say?," *Globe and Mail*, August 24, 2019.

33 The video can be found at https://theintercept.com/2019/04/17/green-new-deal-short-film-alexandria-ocasio-cortez.

Chapter 4

1 J.L. Granatstein, *Arming the Nation: Canada's Industrial War Effort, 1939–1945* (Ottawa: Canadian Council of Chief Executives, May 2005), https://thebusinesscouncil.ca/publications/arming-the-nation-canadas-industrial-war-effort-1939-1945-by-j-l-granatstein.

2 See Ruth Roach Pierson and Marjorie Griffin Cohen, "Educating Women for Work: Government Training Programs for Women Before, During, and After World War II," in *Modern Canada 1930–1980's: Readings in Canadian Social History*, eds. Michael S. Cross and Gregory S. Kealey (Toronto: McClelland and Stewart, 1984), 208–243.

3 J.L. Granatstein, *Canada's War: The Politics of the Mackenzie King Government, 1939–1945* (Toronto: Oxford University Press, 1975), vii.

4 There had been an earlier attempt to create unemployment insurance, but it was scuttled by federal-provincial jurisdictional disputes. There was also an Old Age Pension that was first introduced in 1927, with various provinces joining the plan over the following decade. But it was a very modest means-tested program that offered benefits to people over 70 with little or no other income (quite different from Old Age Security today).

5 Granatstein, *Canada's War*, 265.

6 Granatstein, *Canada's War*, 412.

7 Granatstein, *Canada's War*, 275-276.

8 This correlation was explored in Richard Wilkinson and Kate Pickett's *The Spirit Level: Why More Equal Societies Almost Always Do Better* (London and New York: Allen Lane, 2009) and was made strongly in Naomi Klein's *This Changes Everything: Capitalism vs. the Climate* (Toronto: Knopf Canada, 2014).

9 Marc Lee and Amanda Card, *Who Occupies the Sky? The Distribution of GHGs in Canada* (Ottawa: Canadian Centre for Policy Alternatives, 2011).

10 For more on the distributional analysis of B.C.'s carbon tax, and a progressive alternative, see Marc Lee, *Fair and Effective Carbon Pricing: Lessons from B.C.* (Vancouver: Canadian Centre for Policy Alternatives, 2011).

11 The specific description for a U.S. Green New Deal read: "Some members of Congress are proposing a 'Green New Deal' for the U.S. They say that a Green New Deal will produce jobs and strengthen America's economy by accelerating the transition from fossil fuels to clean, renewable energy. The Deal would generate 100% of the nation's electricity from clean, renewable sources within the next 10 years; upgrade the nation's energy grid, buildings, and transportation infrastructure; increase energy efficiency; invest in green technology research and development; and provide training for jobs in the new green economy." For more see https://earther.gizmodo.com/new-poll-shows-basically-everyone-likes-alexandria-ocas-1831158171.

12 See https://www.foxnews.com/politics/most-voters-support-70-percent-tax-hike-on-richest-americans-poll.

13 Daniel Tencer, "Loblaw Co. Gets $12-Million Carbon Reduction Subsidy from Ottawa, Sparking Online Outrage," *Huffington Post*, April 9, 2019, https://www.huffingtonpost.ca/amp/2019/04/09/loblaws-12-million-carbon-fund_a_23708770.

Chapter 5

1 Granatstein, *Canada's War*, 28. The quote translated into English is "The federal government is using its declaration of war as a pretext to foster a campaign of assimilation and centralization."

2 For more on this subject, see https://www.corporatemapping.ca/who-owns.

3 Armina Legaya, "CIBC CEO Calls Energy Canada's 'Family Business,'" *National Observer*, November 4, 2019, https://www.nationalobserver.com/2019/11/04/news/cibc-ceo-calls-energy-canadas-family-business.

4 *Alberta 2019 Provincial Budget*, 142.

5 For more on the Alberta Narratives project and to see its full reports, go to http://albertanarrativesproject.ca.

6 See in particular this report: https://www.corporatemapping.ca/boom-bust-and-consolidation.

7 For a deeper comparison of Alberta's and Norway's approaches to oil wealth, see Bruce Campbell, *The Petro-Path Not Taken: Comparing Norway with Canada and Alberta's Management of Petroleum Wealth* (Ottawa: Canadian Centre for Policy Alternatives, 2013).

8 John Ibbitson, "Trudeau Is to Blame for National Unity Crisis over Pipelines," *Globe and Mail*, June 12, 2019, https://www.theglobeandmail.com/politics/article-trudeau-is-to-blame-for-national-unity-crisis-over-pipelines.

9 Andrew Parkin, "Most Canadians Don't Want a Province-First Approach to Climate Change," *Policy Options*, July 16, 2019, https://policyoptions.irpp.org/magazines/july-2019/most-canadians-dont-want-a-province-first-approach-to-climate-change.

Chapter 6

1 This list is from a fact sheet produced by Veterans Affairs Canada entitled "Canada Remembers: Canadian Production of War Materials."

2 Granatstein, *Arming the Nation*, 15.

3 Granatstein, *Arming the Nation*, 1.

4 For much of the historic background on C.D. Howe, I am indebted to Robert Bothwell and William Kilbourn, *C.D. Howe: A Biography* (Toronto: McClelland and Stewart, 1979).

5 Bothwell and Kilbourn, *C.D. Howe*, 354.

6 Robert Bothwell, "Minister of Everything," *International Journal Canadian Institute of International Affairs* 31, no. 4 (December 1976): 694.
7 Bothwell and Kilbourn, *C.D. Howe*, 51.
8 Bothwell, "Minister of Everything," 695.
9 Bothwell and Kilbourn, *C.D. Howe*, 130.
10 As quoted in Bothwell and Kilbourn, *C.D. Howe*, 133.
11 Bothwell and Kilbourn, *C.D. Howe*, 134–135.
12 Granatstein, *Arming the Nation*, 4. The dollar-a-year men were not paid as federal employees and were appointed by order-in-council. They received per diem allowances and expenses but not a salary. There is some mythologizing of the dollar-a-year notion. Many continued to be paid by their home companies. They did not work for free. But they certainly worked for a bargain.
13 Bothwell, "Minister of Everything," 695.
14 Bothwell and Kilbourn, *C.D. Howe*, 12.
15 Stacey, *Canada and the Age of Conflict*, 359.
16 Bothwell and Kilbourn, *C.D. Howe*, 160.
17 Bothwell, "Minister of Everything," 701.
18 Ironically, given the focus of this book, Howe's political undoing was the creation of the TransCanada natural gas pipeline in the mid 1950s. The private structure of the company and the degree of American corporate involvement became a source of great controversy. Howe's championing of the pipeline and the government's decision to invoke closure in the Parliamentary debate over the pipeline contributed to the Liberals' and Howe's electoral defeat in 1957. See https://thecanadianencyclopedia.ca/en/article/pipeline-debate.
19 Cited in Granatstein, *Canada's War*, 99.
20 Granatstein, *Arming the Nation*, 2.
21 Granatstein, *Arming the Nation*, 6.
22 Jeremy Stuart, "Captains of Industry Crewing the Ship of State: Dollar-a-Year Men and Industrial Mobilization in WWII Canada, 1939–1942" (MA thesis, Centre for Military and Strategic Studies, University of Calgary, January 2013), 39, https://prism.ucalgary.ca/handle/11023/463.
23 Rod Mickleburgh, *On the Line: A History of the British Columbia Labour Movement* (Madeira Park, B.C.: Harbour Publishing, 2018), 107.
24 Howard White, *A Hard Man to Beat, The Story of Bill White: Labour Leader, Historian, Shipyard Worker, Raconteur* (Vancouver: Pulp Press Publishers, 1983).
25 For an extensive history of Canada's shipbuilding program during the war, see James Pritchard, *A Bridge of Ships: Canadian Shipbuilding during the Second World War* (Montreal and Kingston: McGill-Queen's University Press, 2011).

26 From the website of the Bomber Command Museum of Canada.
27 Granatstein, *Arming the Nation*, 12.
28 Mark R. Wilson, *Destructive Creation: American Business and the Winning of World War II* (Philadelphia: University of Pennsylvania Press, 2016), 61.
29 Paul Gilding, *The Great Disruption: Why the Climate Crisis Will Bring On the End of Shopping and the Birth of a New World* (New York: Bloomsbury Press, 2011), 130.
30 White offers numerous colourful examples of this in *A Hard Man to Beat*, 24–26.
31 Sandford F. Borins, "World War Two Crown Corporations: Their Wartime Role and Peacetime Privatization," *Canadian Public Administration* 25, no. 3 (September 1982): 380–404.
32 Borins, "World War Two Crown Corporations," 380–404.
33 Interestingly, the man tasked with overseeing price controls in the U.S. during the war was Canadian-born economist John Kenneth Galbraith.
34 A.F.W. Plumptre, "Organizing Canada's Economy for War," in *Canadian War Economics*, ed. J.F. Parkinson (Toronto: University of Toronto Press, 1941), cited in Dennis Bartels, "Wartime Mobilization to Counter Severe Global Climate Change," *Human Ecology Special* Issue 10 (January 2001): 229–232.
35 Department of Munitions and Supply, *Canadian War Orders and Regulations 1944: Wartime Industries Control Board* (Ottawa: Printer to the King's Most Excellent Majesty, 1944).
36 Bothwell and Kilbourn, *C.D. Howe*, 160.
37 Matthew Evenden, "World War as a Factor in Energy Transitions: The Case of Canadian Hydroelectricity," *Energy Transitions in History: Global Cases of Continuity and Change* 2 (2013), 91.
38 Stacey Jo-Anne Barker, "Feeding the Hungry Allies: Canadian Food and Agriculture during the Second World War" (Ph.D. thesis, Univeristy of Ottawa, 2008), 239.
39 Barker, "Feeding the Hungry Allies," 294.
40 Jeffrey Keshen, "One for All or All for One: Government Controls, Black Marketing and the Limits of Patriotism, 1939-47," *Journal of Canadian Studies* 29, no. 4 Winter 1994–95, 10.
41 Barker, 197.
42 Ian Mosby, *Food Will Win the War: The Politics, Culture, and Science of Food on Canada's Home Front* (Vancouver: UBC Press, 2014), 129. For more on the rationing experience on the home front, also see Jeffrey A. Keshen, *Saints, Sinners, and Soldiers: Canada's Second World War* (Vancouver: UBC Press, 2004); and Graham Broad, *A Small Price to Pay: Consumer Culture on the Canadian Home Front, 1939–45* (Vancouver: UBC Press, 2013).
43 Barker, 362.

44 Granatstein, *Arming the Nation*, 9.
45 Granatstein, *Arming the Nation*, 13.
46 Borins, "World War Two Crown Corporations," 383.
47 For more on Jacobson's work, see The Solutions Project (co-founded by Jacobson) website: https://thesolutionsproject.org.
48 The Solutions Project roadmap for Canada to reach 100% renewable energy by 2050 can be found at https://thesolutionsproject.org/why-clean-energy/#/map/countries/location/CAN.
49 See more at https://davidsuzuki.org/project/climate-action.
50 See https://www.drawdown.org.
51 Conversely, clearcutting and other destructive forestry practices, combined with the impact of climate-induced beetle infestations, can turn our forests from carbon sinks into carbon sources — releasing more GHGs then they sequester. For more on this subject, as it plays out in British Columbia, see Jens Wieting's report *Clearcut Carbon* (Victoria: Sierra Club BC, 2019).
52 For more on the potential for public transit transformation, see Steven Higashide, *Better Buses, Better Cities: How to Plan, Run, and Win the Fight for Effective Transit* (Washington, D.C.: Island Press, 2019).
53 For more on the concept of "complete communities," see this CCPA Climate Justice Project report: https://www.policyalternatives.ca/transportationtransformation.
54 See the CBC story on the initiative: Nick Boisvert, "Autoworkers at GM's Oshawa Plant Ask Feds for More than $1B to Build Electric Vehicles," September 21, 2019, https://www.cbc.ca/news/canada/toronto/oshawa-gm-plant-ev-transformation-1.5291582.
55 For more on this announcement, see https://www.harbourair.com/harbour-air-and-magnix-partner-to-build-worlds-first-all-electric-airline.
56 For more on the issue of air travel, see Barry Saxifrage, "CO_2 from Jet Fuel Is Soaring 4 Times Faster: What Can Save the Day?" *National Observer*, January 7, 2020, https://www.nationalobserver.com/2020/01/07/analysis/co2-jet-fuel-soaring-4-times-faster-what-can-save-day.
57 See https://theenergymix.com/2019/07/14/prefab-passive-solar-offers-simpler-faster-construction-healthier-homes-lower-emissions.
58 For more on the potential for neighbourhood energy utilities, see this paper by the CCPA's Marc Lee: https://www.policyalternatives.ca/publications/reports/innovative-approaches-low-carbon-urban-systems.
59 The story of the British transition from coal gas is told in Malcolm Gladwell's *Talking to Strangers*. Also see https://www.jstor.org/stable/1147403.
60 In Sedalia, Missouri, building is underway as I write for a new $250 million steel plant that will be powered by wind energy and use electric

arc furnaces to transform recycled metals into new steel. See https://www.eenews.net/stories/1061552453.

61 See https://www.theclimatemobilization.org/victory-plan.

62 For more on Delivering Community Power, see http://www.deliveringcommunitypower.ca. And for more on the idea of postal banking, see the CCPA report found here: https://www.policyalternatives.ca/publications/reports/why-canada-needs-postal-banking.

63 Blair Redlin, *Connecting Coastal Communities: Review of Coastal Ferry Service*, prepared for the B.C. Minister of Transportation, June 2018, 98.

64 For more on wartime rationing lessons for the climate emergency, see Eleanor Boyle, "The Climate Crisis Is Like a World War. So Let's Talk about Rationing," *Globe and Mail*, December 14, 2019, https://www.theglobeandmail.com/opinion/article-the-climate-crisis-is-like-a-world-war-so-lets-talk-about-rationing.

65 George Monbiot, *Heat: How to Stop the Planet from Burning* (Toronto: Doubleday Canada, 2006). See Chapter 3, "A Ration of Freedom."

66 John McDonnell and Hilary Wainwright, "The New Economics of Labour," *openDemocracy*, February 25, 2018, https://www.opendemocracy.net/uk/hilary-wainwright/new-economics-of-labour.

67 For an excellent history of crown corporations in Canada, see Linda McQuaig's *The Sport and Prey of Capitalists: How the Rich Are Stealing Canada's Public Wealth* (Toronto: Dundurn Press, 2019).

68 For more on the work of Reclaim Alberta, see http://www.reclaimalberta.ca.

69 For more on the idea of a zero-waste corporation, see this CCPA report: https://www.policyalternatives.ca/publications/reports/closing-loop.

70 For more on this alternative, see https://science.howstuffworks.com/recycled-plastic-waste-creates-roads.htm.

71 Bartels, "Wartime Mobilization."

72 For more, see Lynne Fernandez, *How Government Support for Social Enterprise Can Reduce Poverty and Greenhouse Gases* (Winnipeg: Canadian Centre for Policy Alternatives, 2016), https://www.policyalternatives.ca/publications/reports/how-government-support-social-enterprise-can-reduce-poverty-and-greenhouse.

Chapter 7

1 For a definitive source on Canada's military capacity at the start of the Second World War, see C.P. Stacey, *Official History of the Canadian Army in the Second World War, Volume 1, Six Years of War* (Ottawa: Ministry of National Defence, 1955).

2 Maude Barlow and Bruce Campbell, *Straight through the Heart: How the Liberals Abandoned the Just Society* (Toronto: HarperCollins Publishers, 1995), 19.
3 Dominion Bureau of Statistics, *Canadian Labour Force Estimates, 1931–1945*, produced for The Right Honourable C. D. Howe, Minister of Trade and Commerce, 1957.
4 *Rosies of the North*, directed by Kelly Saxberg (National Film Board, 1999).
5 Ruth Roach Pierson and Marjorie Griffin Cohen, "Educating Women for Work," 213–215.
6 Rod Mickleburgh, *On the Line: A History of the British Columbia Labour Movement* (Madeira Park, B.C.: Harbour Publishing, 2018), 106.
7 Mickleburgh, *On the Line*, 108.
8 James Pritchard, *A Bridge of Ships: Canadian Shipbuilding during the Second World War* (Montreal and Kingston: McGill-Queen's University Press, 2011), 171.
9 Mickleburgh, *On the Line*, 110–111.
10 Mickleburgh, *On the Line*, 121.
11 Peter Neary, *On to Civvy Street: Canada's Rehabilitation Program for Veterans of the Second World War* (Montreal: McGill-Queen's University Press, 2011). An NFB film during the war was also called *The Road to Civvy Street* and sought to let both soldiers and the public know about the benefits in place for returning soldiers.
12 On December 8, 1939, a cabinet committee on demobilization was formed.
13 Notwithstanding the planning that took place, the transition back to a peacetime economy was not easy. For more on those challenges see Joy Parr, *Domestic Goods: The Material, the Moral, and the Economic in the Postwar Years* (Toronto: University of Toronto Press, 1999); Magda Fahrni, *Household Politics: Montreal Families and Postwar Reconstruction* (Toronto: University of Toronto Press, 2005); Peter McInnis, *Harnessing Labour Confrontation: Shaping the Postwar Settlement in Canada, 1943–1950* (Toronto: University of Toronto Press, 2002); and Alex Souchen, *War Junk: Munitions Disposal and Postwar Reconstruction in Canada* (Vancouver: UBC Press, 2020).
14 Neary, *Civvy Street*, 198.
15 Neary, *Civvy Street*, 203.
16 Richard Harris and Tricia Shulist, "Canada's Reluctant Housing Program: The Veterans' Land Act, 1942–75," in *The Canadian Historical Review* 82, no. 2 (June 2001): 253–282.
17 Harris and Shulist, "Canada's Reluctant Housing Program," 253–282.
18 For an example of the "public education" the government produced to popularize these programs, see the short 1944 NFB documentary *Welcome Soldier*, available at https://www.nfb.ca/film/welcome_soldier.

19 Alice Sorby, *A Study on Demobilization and Rehabilitation of the Canadian Armed Forces in the Second World War, 1939–1945*, produced for the Canadian Department of National Defence, 1960.
20 Cathy Crowe, "Canada Needs a New National Housing Program," *Globe and Mail*, October 3, 2019.
21 Naomi Klein, *On Fire: The (Burning) Case for A Green New Deal* (New York: Simon & Schuster, 2019), 26.
22 Clayton Thomas-Müller, "Canada Needs Its Own Green New Deal. Here's What It Could Look Like," *National Observer*, November 29, 2018, https://www.nationalobserver.com/2018/11/29/opinion/canada-needs-its-own-green-new-deal-heres-what-it-could-look.
23 See Clean Energy Canada's report *The Fast Lane*, available at https://cleanenergycanada.org/canadas-clean-energy-sector-set-to-accelerate-amid-fossil-fuel-slowdown.
24 Marjorie G. Cohen, "Using Information about Gender and Climate Change to Inform Green Economic Policies," in *Climate Change and Gender in Rich Countries: Work, Public Policy and Action* (New York: Routledge, 2017).
25 The full 14-page AOC congressional resolution can be found at https://ocasio-cortez.house.gov/sites/ocasio-cortez.house.gov/files/Resolution%20on%20a%20Green%20New%20Deal.pdf.
26 For a history of the Green New Deal idea, see the introduction to Ann Pettifor's *The Case for the Green New Deal* (London: Penguin Random House, 2019).
27 See more here https://www.climatechangenews.com/2019/04/29/spains-socialists-win-election-green-new-deal-platform.
28 The body of GND work produced by the U.K.'s Common Wealth can be found at https://common-wealth.co.uk/Green-new-deal.html.
29 Tony Mazzocchi, "An Answer to the Jobs–Environment Conflict?" September 8, 1993, https://www.greenleft.org.au/content/answer-jobs-environment-conflict.
30 The *Jobs for Tomorrow* report can be found at https://columbiainstitute.eco/research/jobs-for-tomorrow-canadas-building-trades-and-net-zero-emissions.
31 For more analysis of Alberta's coal phase-out just transition experience, see Ian Hussey and Emma Jackson, *Alberta's Coal Phase-Out: A Just Transition?* (Edmonton: Parkland Institute and Corporate Mapping Project, November 2019), https://www.parklandinstitute.ca/albertas_coal_phaseout.
32 According to the Parkland Institute's report, both ATCO and Capital Power have continued to earn profits in the hundreds of millions since the start of the coal phase-out, and recall that most of the companies involved remain active producing gas-powered electricity. See Hussey and Jackson, *Alberta's Coal Phase-Out*, 10.

33 Lunny's just transition ideas come from a presentation he delivered at a Parkland Institute–hosted conference in Edmonton in June 2019.

34 For more Canadian just transition ideas, see the following CCPA reports: https://www.policyalternatives.ca/publications/reports/making-decarbonization-work-workers; https://www.policyalternatives.ca/publications/reports/just-transition; and https://www.policyalternatives.ca/publications/reports/green-industrial-revolution.

35 For a model of how crown corporations in the building of public infrastructure can be used to achieve equity goals in training and hiring, see Marjorie G. Cohen and Kate Braid, *The Road to Equity: Training Women and First Nations on the Vancouver Island Highway — A Model for Large-Scale Construction Projects* (Vancouver: Canadian Centre for Policy Alternatives, 2000).

Chapter 8

1 The full Douglas speech can be found on YouTube at https://youtu.be/MUwRULlgMec.

2 As quoted in Bothwell and Kilbourn, *C.D. Howe*, 133.

3 Cited in Granatstein, *Canada's War*, 97.

4 For more on how the senior public service in Finance and the Bank of Canada managed these considerations, see the history of the period by Robert Bryce, who was a fairly young economist in the Department of Finance at the time (and a former student of Keynes) and went on to be one of Canada's longest-serving and most respected civil servants, ultimately serving as clerk of the Privy Council. See: Robert B. Bryce, *Canada and the Cost of World War II: The International Operations of Canada's Department of Finance, 1939–1947* (Montreal: McGill-Queen's University Press, 2005).

5 Mickleburgh, *On the Line*, 107.

6 The Bank of Canada was initially created as a private central bank under the Bennett government, but was made into a public crown corporation under the Mackenzie King government in 1938.

7 Deborah D. Stine, *The Manhattan Project, the Apollo Program, and Federal Energy Technology R&D Programs: A Comparative Analysis* (Washington, D.C.: Congressional Research Service Report for Congress, 2009), cited in and calculation updated by Laurence L. Delina, *Strategies for Rapid Climate Mitigation: Wartime Mobilisation as a Model for Action?* (New York: Routledge, 2016).

8 Granatstein, *Arming the Nation*, 8.

9 Barlow and Campbell, *Straight through the Heart*, 19.

10 Keshen, *Saints, Sinners, and Soldiers*, 54.

11 Granatstein, *Arming the Nation*. For more on the excess profits tax, see also Keshen, *Saints, Sinners, and Soldiers*, 53.
12 Granatstein, *Canada's War*, 419.
13 My thanks to Ana Sofia Hibon at the McConnell Foundation for her help collecting historical notes on McConnell's wartime efforts.
14 John Gandy, "Environmental Charities in Canada," Charity Intelligence Canada, June 2013.
15 See https://www.policyalternatives.ca/newsroom/news-releases/canadian-dynasties-richer-ever-wealth-gap-continues-widen-study.
16 See Sandy Garossino's calculation of this total in the *National Observer* at https://www.nationalobserver.com/2019/06/21/opinion/serious-70-billion-climate-plan-youve-heard-nothing-about.
17 See https://www.common-wealth.co.uk/public-finance-for-a-green-new-deal.html.
18 The *Alternative Federal Budget* can be found at https://www.policyalternatives.ca/projects/alternative-federal-budget.
19 See Marc Lee, *Fair and Effective Carbon Pricing: Lessons from B.C.* (Vancouver: Canadian Centre for Policy Alternatives, 2011), available at https://www.policyalternatives.ca/publications/reports/fair-and-effective-carbon-pricing.
20 Finance Canada, Historical Series, Fiscal Reference Tables. See https://www150.statcan.gc.ca/n1/pub/11-516-x/sectionh/4057752-eng.htm.
21 John Maynard Keynes, *How to Pay for the War: A Radical Plan for the Chancellor of the Exchequer* (London: Macmillan and Company, 1940).
22 See estimate from economist Robyn Allan here: Robyn Allan, "An Open Letter to Rachel Notley and Jason Kenney," *National Observer*, March 5, 2019, https://www.nationalobserver.com/2019/03/05/opinion/robyn-allan-open-letter-rachel-notley-and-jason-kenney.
23 The International Monetary Fund, in a 2019 working paper that calculated global fossil fuel subsidies using the broadest measure (including estimates of the externalized environmental costs), found annual subsidies in Canada amounted to US$43 billion. See David Coady, Ian Parry, Nghia-Piotr Le, and Baoping Shang, *Global Fossil Fuel Subsidies Remain Large: An Update Based on Country-Level Estimates*, IMF Working Paper (Washington, D.C.: International Monetary Fund, May 2019), 35.
24 For the full report on the ATB, go to https://www.parklandinstitute.ca/albertas_public_bank.
25 Joe Castaldo, "Maple Leaf Foods Lays Out Plan for Taking on Climate Change," *Globe and Mail*, November 7, 2019.
26 See https://www.canada.ca/en/department-finance/news/2019/06/expert-panel-on-sustainable-finance-delivers-final-report-finance-minister-joins-international-climate-coalition.html.

27 For more on the Canadian economic costs of climate disasters, see Appendix I at sethklein.ca. Glen Hodgson, a senior fellow with the Conference Board of Canada, wrote in the *Globe and Mail*, "Canadian insurers are now facing claims on natural catastrophes — floods, forest fires and other extreme weather events — of approximately $1-billion annually, according to the Insurance Bureau of Canada." See https://www.theglobeandmail.com/business/commentary/article-the-costs-of-climate-change-are-rising. But this doesn't capture public expenditures on flood and fire response, and of course these costs will only escalate. Internationally, Nicholas Stern, in his 2006 report, estimated that the costs of inaction on climate change could reach 20% of global GDP. Other economists believe this estimate to be too high. But even if the cost is only half this amount, it would still far outstrip the costs of taking action.

28 Joseph Stiglitz, "The Climate Crisis Is Our Third World War. It Needs a Bold Response," *The Guardian*, June 4, 2019, https://www.theguardian.com/commentisfree/2019/jun/04/climate-change-world-war-iii-green-new-deal.

Chapter 9

1 Melina Laboucan-Massimo, "The Winds of Resistance Are Spreading across America," *National Observer*, March 13, 2017.

2 For more on the Iroquois declaration of war, see http://www.combat.ws/S3/BAKISSUE/CMBT04N4/IROQUOIS.HTM.

3 Sheffield is the co-author of a book chapter "Moving Beyond 'Forgotten': The Historiography on Canadian Native Peoples and the World Wars," in the book *Aboriginal Peoples and the Canadian Military: Historical Perspectives*, eds. P. Whitney Lackenbauer and Craig Leslie Mantle (Kingston: Canadian Defence Academy Press, 2007).

4 Quoted in an interview from CBC Radio's *Unreserved* with guest host Falen Johnson, November 9, 2018, available at https://www.cbc.ca/radio/unreserved/remembering-the-contributions-of-indigenous-veterans-1.4894412/indigenous-veterans-they-fought-for-freedom-democracy-and-an-equality-they-could-never-share-1.4897735.

5 For more on the experience of the Déline Dene, see http://www.firstnationsdrum.com/1998/12/deline-dene-mining-tragedy.

6 For more on the environmental and health impacts of the Sarnia area, see the 2007 report *Exposing Canada's Chemical Valley*, produced by Ecojustice on behalf of the Aamjiwnaang Health and Environment Committee, available at https://www.ecojustice.ca/wp-content/uploads/2015/09/2007-Exposing-Canadas-Chemial-Valley.pdf.

7 Tu Thanh Ha, "Last Surviving Mohawk Code Talker Kept His Secret for Seven Decades," *Globe and Mail*, May 29, 2019, https://www.theglobeandmail.com/canada/article-last-surviving-mohawk-code-talker-kept-his-secret-for-seven-decades.

8 Jessica Deer, "Louis Levi Oakes, Last WWII Mohawk Code Talker, Dies at 94," CBC News, May 29, 2019, https://www.cbc.ca/news/indigenous/louis-levi-oakes-code-talker-obituary-1.5153816.

9 The B.C. Act embedding the UN Declaration into law was unanimously passed in the B.C. legislature in November 2019. It can be found at https://www.leg.bc.ca/parliamentary-business/legislation-debates-proceedings/41st-parliament/4th-session/bills/first-reading/gov41-1.

10 In November 2018, the CCPA and the Union of B.C. Indian Chiefs co-published a report on what UNDRIP implementation should look like in B.C., a project that I co-directed. The final report can be found at https://www.policyalternatives.ca/UNDRIP-BC.

11 Chief Leah George-Wilson cited in this *National Observer* piece: https://www.nationalobserver.com/2019/07/09/news/first-nations-renew-court-battle-stop-trudeau-and-trans-mountain.

12 For more on the Tsleil-Waututh Nation's Sacred Trust initiative, which has coordinated much of the legal opposition to the proposed Trans Mountain expansion, see https://twnsacredtrust.ca/. Much of the civil disobedience in Burnaby was coordinated by Protect the Islet, which is also led by members of the Tsleil-Waututh Nation such as Will George. For more see https://protecttheinlet.ca.

13 Cited from this CBC interview: https://www.cbc.ca/news/politics/powerandpolitics/rueben-george-on-trans-mountain-pipeline-1.3611683.

14 For more on the Tiny House Warriors, see http://tinyhousewarriors.com.

15 For more on the Treaty Alliance, see http://www.treatyalliance.org.

16 Kate Gunn and Bruce McIvor, "The Wet'suwet'en, Aboriginal Title, and the Rule of Law: An Explainer," February 13, 2020, available at the blog of First Peoples Law: https://www.firstpeopleslaw.com/index/articles/438.php.

17 For more on the Unist'ot'en camp, see https://unistoten.camp/no-pipelines/background-of-the-campaign.

18 For more, including how you can support the Gidimt'en checkpoint, see https://www.yintahaccess.com.

19 On the authorized use of lethal force, see this *Guardian* exposé: Jaskiran Dhillon and Will Parrish, "Exclusive: Canada Police Prepared to Shoot Indigenous Activists, Documents Show," *The Guardian*, December 20, 2019, https://www.theguardian.com/world/2019/dec/20/canada-indigenous-land-defenders-police-documents.

20 The full AFN document can be found at https://www.afn.ca/2019-federal-election.

21 For more on the T'Sou-ke solar project, see https://www.policynote.ca/tsou-ke-first-nation-building-a-network-of-clean-energy-systems; and http://www.tsoukenation.com/first-nation-takes-lead-on-solar-power.
22 For more on the Tsilhqot'in solar project, see http://www.tsilhqotin.ca/Portals/0/PDFs/Press.

Chapter 10

1 From an interview with the *National Observer* available at https://www.nationalobserver.com/2019/10/29/news/extinction-rebellion-organizer-how-politics-cant-save-us-climate-breakdown.
2 From Rebecca Hamilton's interview with Greenpeace, available at https://www.greenpeace.org/canada/en/story/7902/12-questions-with-youthclimatestrike-organizer-rebecca-hamilton.
3 For more on this, see the reports on British public morale, as summarized in John Lukacs, *Five Days in London, May 1940* (New Haven: Yale University Press, 1999).
4 Erica Chenoweth and Maria J. Stephan, *Why Civil Resistance Works: The Strategic Logic of Nonviolent Conflict* (New York: Columbia University Press, 2011).
5 See Chenoweth's interview at https://news.harvard.edu/gazette/story/2019/02/why-nonviolent-resistance-beats-violent-force-in-effecting-social-political-change.
6 See https://www.veterans.gc.ca/eng/remembrance/history/historical-sheets/youth.
7 For more on B.C.'s Vote 16 effort, see https://www.facebook.com/vote16bc.
8 For more on the U.K. climate strikers' manifesto, see https://www.theguardian.com/commentisfree/2019/mar/15/uk-student-climate-network-manifesto-declare-emergency.
9 See "Cutting Class to Stop Climate Change: Young Canadians Strike for the Planet," CBC Radio's *Day 6*, February 1, 2019, https://www.cbc.ca/amp/1.5001974.
10 See https://actionnetwork.org/petitions/call-for-emergency-climate-justice-plan.
11 For more on the non-partisan nature of Vancouver's climate motions, see this op-ed by councillors Christine Boyle, Rebecca Bligh and Michael Wiebe, "In Vancouver, Climate Action Isn't Partisan," *Vancouver Sun*, October 17, 2019. To see Vancouver's new climate plan, go to https://council.vancouver.ca/20190424/documents/cfsc1.pdf.
12 The Canadian Armed Forces has numerous paid education programs, and generally will cover one month of tuition for every two months of service.

13 For more on the movement in Quebec to push municipal governments to pass climate emergency motions, an interview with key organizer Normand Beaudet can be found on the website of Climate Action Network Canada: https://climateactionnetwork.ca/wp-content/uploads/2019/02/DUC-2019.02.25-1.pdf.

14 For more on Extinction Rebellion, its origins and demands, see https://rebellion.earth/the-truth/demands.

15 This was a clear finding from British Columbia's experience with the Citizens' Assembly on Electoral Reform in the mid-2000s. A body of academic work found that the public was more supportive of recommendations that flowed from this model. When B.C.'s Citizens' Assembly recommended a particular form of proportional representation and it was put to a referendum in 2005, 58% of British Columbians voted in favour (unfortunately, the threshold for success was placed at 60%). For more on this, see Mark E. Warren and Hilary Pearse, eds., *Designing Deliberative Democracy: The British Columbia Citizens' Assembly* (Cambridge: Cambridge University Press, 2008).

16 Cited in Damian Carrington, "Desmond Tutu calls for anti-apartheid style boycott of fossil fuel industry," *The Guardian*, April 10, 2014.

17 Bill McKibben, "Money is the oxygen on which the fire of global warming burns," *The New Yorker*, September 17, 2019.

Chapter 11

1 Irving Abella and Harold Troper, *None Is Too Many: Canada and the Jews of Europe, 1933–1948* (Toronto: Lester Publishing Ltd., 1983).

2 Canada established numerous prisoner-of-war camps on behalf of the Allies. About 34,000 German POWs spent the war in Canada.

3 The stories of those internees are told in a book called *Dangerous Patriots: Canada's Unknown Prisoners of War*, by William and Kathleen M. Repka (Vancouver: New Star Books, 1982).

4 Cited in Repka and Repka.

5 For more on T. Buck Suzuki, go to http://www.bucksuzuki.org/about-us/t-buck-suzuki.

6 Matt Hern and Am Johal, *Global Warming and the Sweetness of Life: A Tar Sands Tale* (Cambridge, MA: MIT Press, 2018), 23. A similar concern is articulated in Geoff Mann and Joel Wainwright's book *Climate Leviathan: A Political Theory of Our Planetary Future* (New York: Verso, 2018).

7 Hern and Johal, 49.

8 Nomi Claire Lazar, "Trump's Emergency Creates Its Own Crisis," *Toronto Star*, February 22, 2019.

9 From a talk by Harold Troper available at https://www.facinghistory.org/resource-library/video/none-too-many-antisemitism-canadas-past.

10 Kanta Kumari Rigaud, Alex de Sherbinin, Bryan Jones, Jonas Bergmann, Viviane Clement, Kayly Ober, Jacob Schewe, Susana Adamo, Brent McCusker, Silke Heuser and Amelia Midgley, *Groundswell: Preparing for Internal Climate Migration* (Washington, D.C.: World Bank, 2018).

11 See http://www.internal-displacement.org/global-report/grid2019. Similarly, a 2019 report from Oxfam, *Forced from Home: Climate-Fuelled Displacement*, estimates that climate change has, over the last decade, forced approximately 20 million people per year from their homes. See https://www.oxfam.org/en/research/forced-home-climate-fuelled-displacement.

12 Jonathan Blitzer, "How Climate Change Is Fuelling the U.S. Border Crisis," *The New Yorker*, April 3, 2019, https://www.newyorker.com/news/dispatch/how-climate-change-is-fuelling-the-us-border-crisis.

13 Stephanie Dickson, Sophie Webber and Tim T. Takaro, *Preparing BC for Climate Migration* (Vancouver: Canadian Centre for Policy Alternatives, 2014).

14 For CBSA stats, go to https://www.cbsa-asfc.gc.ca/security-securite/detent/stat-2012-2018-eng.html.

15 See https://www.cbc.ca/news/politics/cbsa-deportations-border-removals-1.4873169.

16 Harsha Walia, *Undoing Border Imperialism* (Chico, CA: AK Press, 2013).

17 From Harsha Walia interview on the podcast of the Transnational Institute, *State of Power*, episode 9, https://www.tni.org/en/state-of-power-podcast.

18 Some Democratic congresspeople have introduced a Climate Displaced Persons Act, which would create a new category for entry into the U.S. and set an annual target. It is unlikely to pass anytime soon, but it shows that some lawmakers are beginning to consider such ideas. See https://theenergymix.com/2019/10/29/u-s-house-democrats-to-unveil-climate-displaced-persons-act.

19 Tom Athanasiou, "Only a Global Green New Deal Can Save the Planet," *The Nation*, September 17, 2019, https://www.thenation.com/article/green-new-deal-sanders.

20 See https://theenergymix.com/2019/08/28/canada-uk-boost-their-green-climate-fund-contributions-still-fall-short-of-fair-share.

21 Granatstein, *Canada's War*, 306.

22 John Hagan, *Northern Passage: American Vietnam War Resisters in Canada* (Cambridge, MA: Harvard University Press, 2001).

Chapter 12

1 Cited in Andrew Roberts, *Leadership in War: Essential Lessons from Those Who Made History* (New York: Penguin Random House, 2019).

2 J.L. Granatstein, in the afterword to Robert Bryce's *Canada and the Cost of World War II* (McGill-Queen's University Press, 2005), 324.

3 While 2019 saw Boris Johnson's Conservatives win a majority of seats in the U.K.'s general election and the subsequent resignation of Jeremy Corbyn as Labour leader, this should not be seen as a popular rejection of the bold platform advanced by Labour in the last two U.K. elections. Many analysts believe the messenger (Corbyn) was less appealing than the policy agenda, and that in 2019 U.K. Labour was simply caught in the vortex of Brexit, which presented them with a no-win scenario wherein they were destined to lose core support no matter which side they took on that matter. One should not lose sight of the fact that, in the 2017 U.K. election, the first for Labour under Corbyn, while the Labour party did not prevail, it defied the punditry's predictions and won more of the popular vote (40%) than it had since 2001 — one of the best results since the 1970s — and more than any federal Canadian party has won since the Chrétien win of 2001. U.K. Labour in the last two elections has also won a decisive majority of support among younger voters.

4 The Sanders platform proposed to "directly invest an historic $16.3 trillion public investment toward these [GND] efforts, in line with the mobilization of resources made during the New Deal and WWII, but with an explicit choice to include black, indigenous and other minority communities who were systematically excluded in the past." The detailed Bernie Sanders campaign Green New Deal climate plan can be found at https://berniesanders.com/en/issues/green-new-deal.

5 See https://amp.theguardian.com/environment/2019/apr/05/historic-breakthrough-norways-giant-oil-fund-dives-into-renewables.

6 See https://www.bloomberg.com/news/articles/2019-04-08/norway-is-walking-away-from-billions-of-barrels-of-oil-and-gas.

7 See https://www.theguardian.com/world/2018/jan/18/norway-aims-for-all-short-haul-flights-to-be-100-electric-by-2040.

8 See https://www.reuters.tv/v/P08A/2019/03/14/sweden-s-emissions-free-ferries-lead-the-charge.

9 See https://www.nationalobserver.com/2019/05/21/analysis/canada-vs-uk-lessons-climate-fight.

10 See https://www.latimes.com/california/story/2019-11-19/california-fracking-permits-scientific-review-gavin-newsom.

11 See https://tribunemag.co.uk/2019/08/public-transport-can-be-free.

12 The Irish state government has divested approximately 68 million euros from 38 fossil fuel companies. See https://www.irishtimes.com/business/energy-and-resources/republic-withdraws-public-money-from-fossil-fuel-investments-1.3747740.

13 The recommendations of the Irish Citizens' Assembly can be found at https://www.citizensassembly.ie/en/how-the-state-can-make-ireland-a-leader-in-tackling-climate-change/final-report-on-how-the-state-can-make-ireland-a-leader-in-tackling-climate-change.

14 Nicolas Graham, William K. Carroll and David Chen, *Big Oil's Political Reach: Mapping Fossil Fuel Lobbying from Harper to Trudeau* (Vancouver: Canadian Centre for Policy Alternatives and Corporate Mapping Project, 2019).

15 The Liberals, the NDP, the Greens and the Bloc Québécois all ran on platforms promising strong climate action. In the 2019 federal election, if you add up the popular vote for these four parties, it was 63.5%.

16 Andrew Coyne, "Conservative Climate Change 'Plan' Is Really More of a Prop," *National Post*, June 21, 2019, https://nationalpost.com/opinion/andrew-coyne-conservative-climate-change-plan-is-really-more-of-a-prop.

17 Mark Jaccard, "Emissions Will Rise Under Conservative Climate Plan," *Policy Options*, August 21, 2019, https://policyoptions.irpp.org/magazines/august-2019/emissions-will-rise-under-conservative-climate-plan.

18 There is empirical evidence of a connection between countries with proportional electoral systems and stronger environmental policies. See, for example, Per G. Fredriksson and Daniel L. Millimet, "Electoral Rules and Environmental Policy," *Economics Letters* 84, No. 2 (August 2004): 237–244, https://doi.org/10.1016/j.econlet.2004.02.008.

Conclusion

1 Solnit, *Paradise Built in Hell*, 305–306.

2 For more on the underbelly of Canada's wartime experience on the home front, see Keshen's *Saints, Sinners, and Soldiers*. For more on the errors, miscalculations, needless sacrifices and slaughter stemming from misguided decisions by generals, atrocities and inadequate military equipment on the Canadian side in the Second World War, see the controversial 1992 NFB and CBC co-produced documentary series by Brian and Terence McKenna, *The Valour and the Horror*.

3 Chris Hatch, "Lest We Betray Them," *National Observer*, November 11, 2019, https://www.nationalobserver.com/2019/11/11/opinion/lest-we-betray-them.

4 Solnit, *Paradise Built in Hell*, 10.

5 Avi Lewis's Green New Deal speech can be found at https://theleap.org/top-5-reasons-the-green-new-deal-is-workable-winnable-and-the-idea-we-need-right-now.

6 Dan Hind, *The Return of the Public* (London and New York: Verso, 2010), 142–143. He also draws upon the work of Anna Minton, *Ground Control: Fear and Happiness in the Twenty-First Century City* (London: Penguin, 2009).

Epilogue

1 See https://www.cbc.ca/news/canada/british-columbia/covid-19-bc-update-public-safety-minister-mike-farnworth-live-1.5501769.
2 See https://vancouversun.com/news/local-news/covid-19-first-nations-leaders-call-on-trudeau-to-declare-state-of-emergency.
3 For more on the caring responses and "mutual aid" networks formed during the COVID pandemic in Canada, see David Moscrop, "In Canada, an Inspiring Movement Emerges in Response to the Coronavirus," *Washington Post*, March 24, 2020, https://www.washingtonpost.com/opinions/2020/03/24/canada-an-inspiring-movement-emerges-response-coronavirus.
4 Jim Stanford blog post available at http://www.progressive-economics.ca/2020/03/13/economic-respone-to-pandemic-go-big-go-fast.
5 Arundhati Roy, "The Pandemic Is a Portal," *Financial Times*, April 3, 2020.

ACKNOWLEDGEMENTS

My greatest gratitude goes to my wife, Christine Boyle, who was the first reader and editor of almost every chapter, and whose support and encouragement throughout this process has been magnificent. Chris has been a climate champion and campaigner for years, forging new alliances and breaking new ground in this task of our lives. I could not have done this without her. Despite the intense demands of her own job, Chris gave me time away when needed to complete this work. For years, Chris has taught me how to appeal to both the head and the heart — she does so with tremendous grace, and I've tried to do some of that here.

In many ways, this book consolidates years of what I learned while with the Canadian Centre for Policy Alternatives, and I am deeply thankful to my CCPA colleagues for that journey. It was the privilege of a lifetime to work with such smart, caring and committed people for so many years. In particular, I want to acknowledge the following CCPA folks (past and present) who have contributed to the ideas in these pages: Shannon Daub, Marc Lee, Ben Parfitt, Iglika Ivanova, Alex Hemingway, Bruce Campbell, David Macdonald and Armine Yalnizyan.

The principal funder of this book was the McConnell Foundation of Montreal. And I want to express a special thanks to their president and CEO Stephen Huddart. I spent some time quietly shopping around a book idea in 2018, but it remained only notional until Stephen sent me a letter late that year with the thrilling news that the McConnell Foundation was willing to support my time and research. That leap of faith is what green-lighted the book you now hold, and Stephen has remained very enthusiastic and supportive throughout. My thanks as well to other members of the McConnell team — Ana Sofia Hibon, Nicolina Farella and Niamh Leonard.

I owe an immense debt of gratitude to ECW Press, who took a chance on a first-time book author, and in particular to my wonderful senior editor there, Susan Renouf. Susan was a delight to work with; deeply engaged in and passionate about the content, skilled as an editor and generous with her guidance about the world of publishing. My thanks to the rest of the highly skilled and supportive team at ECW: David Caron, Jen Knoch, Tania Blokhuis, Emily Ferko, Aymen Saidane, Jessica Albert, Susannah Ames, Shannon Parr and Adrineh Der-Boghossian.

The research for this book was also part of and supported by the Corporate Mapping Project (CMP), a research and public engagement initiative investigating the power of the fossil fuel industry in Canada. The CMP is jointly led by the University of Victoria, Canadian Centre for Policy Alternatives and the Parkland Institute, and funded primarily by the Social Science and Humanities Research Council of Canada (SSHRC). My thanks to the CMP steering committee for agreeing to support this project, and in particular to the CMP's co-directors Shannon Daub and William Carroll for their encouragement and to former CMP project manager Thi Vu.

This project also received funding from a number of other organizations to whom I am very grateful: the Metcalf Foundation (and its CEO Sandy Houston); the David Suzuki Foundation (and its CEO Stephen Cornish); and the British Columbia Teachers Federation (and its former president Glen Hansman and executive director Moira Mackenzie).

I owe special thanks to Simon Fraser University's Urban Studies Program, its director Meg Holden, and managers Terri Evans and Karen Sawatzky. After leaving the CCPA in January 2019, Meg and Terri (along with Jane Pulkingham) gave me and this project an institutional home with SFU, and then patiently navigated me through all SFU's research ethics, accounting and various other processes and systems. I am so grateful for their hospitality and support.

SFU Urban Studies also led me to my research assistant on this project, Steve Tornes. I couldn't have pulled this off without Steve. Calm, confident, reliable and a very quick study, Steve gathered a huge amount of the background research for this book, wrote book summaries, assembled all the data and graphs, and transcribed most of the interviews. It was a pleasure to collaborate with him.

Thanks to Kristen Eklow for designing the book's website, and Meital Smith for the original climate emergency poster design.

While I have received encouragement from many people in this project, I want to single out Alex Himelfarb. Alex's early enthusiasm for the book proposal was just the boost I needed. He urged me to approach the McConnell and Metcalf foundations for support, and his wise counsel, and political and policy insights, are reflected throughout these pages. Other friends who provided early encouragement and advice on how to undertake this book project include: David R. Boyd (who directed me to ECW Press), Deena Chochinov, Libby Davies, Charlie Demers, Ian Gill, Matt Hern, Am Johal, Parker Johnson, Geoff Mann, Ross McMillan, Kevin Millsip, Charles Montgomery, Kai Nagata, Matthew Norris, Lyndsay Poaps, Venessa Richards, Jim Stanford and Chris Tenove.

Special thanks as well to Abacus Data and its CEO David Coletto, who shared my curiosity about the public's appetite for bold climate action, and guided me through the process of conducting and analyzing the national survey featured in Chapter 3.

My deep gratitude to the many people who reviewed and edited either some chapters, or in most cases, the entire draft manuscript: Ricardo Acuna, Christine Boyle, Bruce Campbell, William Carroll, Graham Cook, Stephen Cornish, Shannon Daub, Tim Dickson, Alexander Dirksen, Maxime Faille, David Green, Shane Gunster, Ana Sofia Hibon, Alex Himelfarb, Meg Holden, Sandy Houston, Stephen Huddart, Bonnie Klein, Michael Klein, Naomi Klein, Nancy Knickerbocker, Larry Kuehn, Marc Lee, Mark Leier, Avi Lewis, Kevin Millsip, Matthew Norris, Ken Novakowski, Ben Parfitt,

Nadene Rehnby, Steve Tornes, Jon Woodward, and Marcus Youssef. I am an interloper in the field of history, and so my special thanks to historians Peter Neary and Alex Souchen for their careful reviews of the manuscript and detailed feedback. But to be clear, none of these good people are responsible for the opinions, policy recommendations or any errors contained in this book. Those belong to me.

My appreciation to everyone who agreed to be interviewed for the book: Ricardo Acuna, Stephen Buhler, William Carroll, Angela Carter, George Chow, Penelope Comette, Scott Critchon, Shannon Daub, Libby Davies, Liz Hanson, Robin Hanvelt, Kathryn Harrison, George Heyman, Stephen Huddart, Ian Hussey, Emma Jackson, Atiya Jaffer, Jenny Kwan, Lee Loftus, George MacPherson, Gil McGowan, Dale Marshall, Khelsilem, Bill Moore-Kilgannon, Joyce Murray, Kai Nagata, Matthew Norris, Anthonia Ogundele, Anthony Perl, Cara Pike, Éric Pineault, Vanessa Richards, Vera Rosenbluth, Dianne Saxe, Harold Steves, David Suzuki, Dana Tizya-Tramm, and two other anonymous interviewees.

Throughout this endeavour I received tremendous encouragement (and writing hide-a-ways) from my family: my parents Bonnie and Michael Klein, my sister Naomi Klein and brother-in-law Avi Lewis, and my wife's parents Nancy and Jim Boyle. Thanks to Erica Johnson for her collaboration with the child raising and kid juggling.

Final thanks to my children Zoe and Aaron for their patience and inspiration. More than anyone, I hope they really like this book, and feel inspired in return.

INDEX

Page numbers in italics refer to figures and tables.

Seth Klein was the founding British Columbia director of the Canadian Centre for Policy Alternatives for over two decades and has been immersed in climate change and inequality issues for his working life. He is currently an adjunct professor in urban studies at Simon Fraser University and remains a research associate with the CCPA. He lives in Vancouver, B.C.